INVENTING ELIZA

The MIT Press
Massachusetts Institute of Technology
77 Massachusetts Avenue
Cambridge, MA 02139
mitpress.mit.edu

The MIT Press would like to thank the anonymous peer reviewers who provided comments on drafts of this book. The generous work of academic experts is essential for establishing the authority and quality of our publications. We acknowledge with gratitude the contributions of these otherwise uncredited readers.

This book was designed by Stefanie Tam in Brooklyn, New York, and is set in Univers, Iosevka, and Courier. Printed and bound in the United States of America.

Library of Congress Cataloging-in-Publication Data

Names: Ciston, Sarah author | Berry, David M. (David Michael) author | Hay, Anthony C. author | Marino, Mark C. author | Millican, P. J. R. author | Schwarz, Arthur I. author | Shrager, Jeff author | Weil, Peggy author
Title: Inventing ELIZA : how the first chatbot shaped the future of AI / Sarah Ciston, David M. Berry, Anthony C. Hay, Mark C. Marino, Peter Millican, Arthur I. Schwarz, Jeff Shrager and Peggy Weil.
Description: Cambridge, Massachusetts : The MIT Press, [2026] | Series: Software studies | Includes bibliographical references and index.
Identifiers: LCCN 2025028421 (print) | LCCN 2025028422 (ebook) | ISBN 9780262052481 paperback | ISBN 9780262052504 pdf | ISBN 9780262052498 epub
Subjects: LCSH: ELIZA (Computer program) | Weizenbaum, Joseph | Chatbots | Natural language processing (Computer science) | Artificial intelligence--History
Classification: LCC QA76.76.C52 C57 2026 (print) | LCC QA76.76.C52 (ebook)
LC record available at https://lccn.loc.gov/2025028421
LC ebook record available at https://lccn.loc.gov/2025028422

10 9 8 7 6 5 4 3 2 1

EU Authorised Representative: Easy Access System Europe, Mustamäe tee 50, 10621 Tallinn, Estonia
Email: gpsr.requests@easproject.com

Sarah Ciston, David M. Berry, Anthony C. Hay,
Mark C. Marino, Peter Millican,
Arthur I. Schwarz, Jeff Shrager, and Peggy Weil
Foreword by Janet H. Murray

INVENTING ELIZA

HOW THE FIRST CHATBOT SHAPED THE FUTURE OF AI

The MIT Press
Cambridge, Massachusetts
London, England

Software Studies

Expressive Processing: Digital Fictions, Computer Games, and Software Studies, Noah Wardrip-Fruin, 2009
Code/Space: Software and Everyday Life, Rob Kitchin and Martin Dodge, 2011
Programmed Visions: Software and Memory, Wendy Hui Kyong Chun, 2011
Speaking Code: Coding as Aesthetic and Political Expression, Geoff Cox and Alex McLean, 2012
10 PRINT CHR$(205.5+RND(1)); : GOTO 10, Nick Montfort, Patsy Baudoin, John Bell, Ian Bogost, Jeremy Douglass, Mark C. Marino, Michael Mateas, Casey Reas, Mark Sample, and Noah Vawter, 2012
The Imaginary App, Paul D. Miller and Svitlana Matviyenko, 2014
The Stack: On Software and Sovereignty, Benjamin H. Bratton, 2015
Coding Literacy: How Computer Programming Is Changing Writing, Annette Vee, 2017
The Software Arts, Warren Sack, 2019
Critical Code Studies, Mark C. Marino, 2020
How Pac-Man Eats, Noah Wardrip-Fruin, 2020
Live Coding: A User's Manual, Alan F. Blackwell, Emma Cocker, Geoff Cox, Alex McLean, and Thor Magnusson, 2022
Thresholds of Digital Gameplay, Daniel L. Gardner, 2025
100% Utilization: Computation and Labor after Moore's Law, Andrew Lison, 2026
Inventing ELIZA: How the First Chatbot Shaped the Future of AI, Sarah Ciston, David M. Berry, Anthony C. Hay, Mark C. Marino, Peter Millican, Arthur I. Schwarz, Jeff Shrager, and Peggy Weil, 2026

Wendy Hui Kyong Chun
Winne Soon
Noah Wardrip-Fruin
Jichen Zhu

Series Foreword

How do we see, think, consume, and make software? How does software—from algorithmic procedures and machine learning models to free and open-source software programs—shape our everyday lives, cultures, societies, and identities? How can we critically and creatively analyze something that seems so ubiquitous and general yet is also so specific and technical? How do artists, designers, coders, scholars, hackers, and activists create new spaces to engage computational culture, enriching our understanding of software as a cultural form?

These questions drive the Software Studies series. Software is deeply woven into contemporary life—economically, culturally, creatively, politically—in manners both obvious and nearly invisible. Although much has been written about how software is used, and about the activities that it supports and shapes, thinking about software itself is notoriously challenging in no small part due to its unknowability and ubiquity. On the one hand, it seems that we have never been better at making software. From augmented reality to generative AI systems, software today routinely dazzles users with new capabilities or higher performance than what was barely thinkable only years before. On the other hand, we know increasingly less about the software we make and the systems with which we interact. The software stack necessary to run an application grows thick; each layer obfuscates what is actually happening beneath. As a result, even technical experts struggle to fully make sense of these black boxes. Economic interests, rising geopolitical cyber threats, and nationalistic protectionism all exert pressure on software developers to make software ecosystems more closed. Although we can virtually make software do anything, in practice we often lose sight of who decides what gets built, why it's built, and whom it really serves.

To make sense of software, it is crucial that we cultivate multiplicities of critical and creative approaches in unpacking, conceptualizing, examining, speculating, and making software, a cultural object that is not often apparent to our immediate registers but which has significant effects and consequences. As the editorial collective, we are interested not only in the question of what software is about but also in how to study software and its related communities, organizations, cultures, and practices. We are especially committed to publishing work that moves beyond broad statements about software and integrates a wide range of disciplines—from mathematics to critical race theory, from software art to queer theory—to understand the social and cultural implications of software. Our books engage a plurality of practice—from participatory design to critical art—and draw on knowledge from media, visual, game, cultural, and literary studies; history; decolonial theory; new materialism; artistic and critical design practices; and electronic literature and narrative.

Ultimately, this series explores the vast possibilities, histories, relations, and harms that software encompasses. It focuses on archival research and oral histories that capture the "untold stories" of technologies from primary sources—all in pursuit of answering the pressing question: what are our desirable technological presents, pasts, and futures? This is a technical design question as much as a philosophical, ethical, and political one. It requires critical analysis of existing software practices as well as constructive reimaginations of future possibilities, with undisciplined, eclectic approaches that our world needs now.

Wendy Hui Kyong Chun
Winne Soon
Noah Wardrip-Fruin
Jichen Zhu

Foreword

Janet H. Murray

Joseph Weizenbaum cringed when people paid attention to his path-breaking 1966 ELIZA program, but he would have loved this book. As the first artificial intelligence chatbot program ever written, ELIZA caused a sensation similar to the excitement that greeted her great-great-grandchild, ChatGPT, almost 60 years later. In 1966 only a few technically advanced labs had a teletype interface hooked up to a time-sharing mainframe. ELIZA was the first program to take advantage of this interactive environment to set up the kind of conversation that Turing had imagined in his 1950 essay "Computing Machinery and Intelligence," which introduced what is now known as the Turing test for determining if a computer is intelligent. And ELIZA seemed to pass the Turing test. People who conversed with the program thought it was a person, a phenomenon that Weizenbaum devoted much of the rest of his career to lamenting and debunking.

Team ELIZA brings us into new understanding of Weizenbaum's creative achievement, capturing the creative energy, playfulness, and iterative design that made ELIZA so unexpectedly lifelike. They reveal that the famous "DOCTOR" script, which along with the first text adventure game *Zork* was a staple of early mainframe configurations, was meant to be the first of a number of planned scripts, which they have rescued from the archives for us to examine. This proliferation of possible ELIZAs only makes clearer how brilliant the classic DOCTOR scenario was, giving to the computer the role of a rigidly "nondirective" *Rogerian therapist* whose therapeutic method is to parrot back keywords taken from the patient's remarks, turning them into questions that serve as directed prompts to keep the conversation going.[1] An important feature of Weizenbaum's program was the default response—a set of plausible things to say when, as often happens, the computer is at a loss to respond to the last patient utterance. Hence the ELIZA refrain that echoes through this collection: "Please Go On."

Here for the first time we can witness the many design choices behind the making of ELIZA, and trace them back to the program's native implementation in MAD-SLIP (Michigan Algorithm Decoder and Symmetric List Processor) rather than the Lisp code that has long been assumed to be the first implementation, or the BASIC implementation in which it was reborn and spread even more widely on early personal computers. Because of this close attention to the code itself, I feel sure that the Joe Weizenbaum who created this triumph of digital storytelling would have enjoyed seeing this book.

But the post-ELIZA Joe Weizenbaum might have been even more delighted by it. For, despite the general atmosphere of hype surrounding artificial intelligence (or AI, which was in Weizenbaum's research circles as intense in the 1960s as it is in Silicon Valley today), Weizenbaum did not find the success of his creation a cause for celebration. He was appalled by attempts to use it to justify serious research into automated psychotherapy and dismayed by facile claims that he had solved the problem of language understanding. He was famously startled to see how quickly and how very deeply people conversing with DOCTOR became emotionally involved with the computer and how unequivocally they anthropomorphized it.

> Once my secretary, who had watch me working on the program for many months and therefore surely knew it to be merely a computer program, started conversing with it. After only a few exchanges with it, she asked me to leave the room.[2]

What would undoubtedly have been embraced as success by his colleagues seemed to Weizenbaum a dire example of public confusion about computing.

As a result, in his canonical technical article introducing ELIZA, Weizenbaum begins by stating that his purpose in publishing is to debunk it.

> It is said that to explain is to explain away. This maxim is nowhere so well fulfilled as in the area of computer programming, especially in what is called heuristic programming and artificial intelligence. For in those realms machines are made to behave in wondrous ways, often sufficient to dazzle even the most experienced observer. But once a particular program is unmasked, once its inner workings are explained in language sufficiently plain to induce understanding, its magic crumbles away; it stands revealed as a mere collection of procedures, each quite comprehensible. ...
>
> The object of this paper is to cause just such a re-evaluation of the program about to be "explained." Few programs ever needed it more.[3]

Far from claiming to have conquered the Turing test, he notes something that took decades to be widely acknowledged in the AI community—that Turing's test is not a benchmark for machine intelligence, but really a measure of human gullibility. ELIZA's success was evidence that researchers needed to recalibrate the Turing test so that it is not so easily passed:

> Some subjects have been very hard to convince that ELIZA (with its present script) is not human. This is a striking form of Turing's test. What experimental design would make it more nearly rigorous and airtight?[4]

Though the demonstrable success of ELIZA positioned Weizenbaum to be a public major spokesperson for the promise of AI at a time when fledgling laboratories like the one at MIT were engaged in fierce pursuit of sustained sources of corporate and federal funding, he chose instead to establish himself as a leading critic of the social dangers of computation in general and AI in particular. He began publishing short pieces warning of the dehumanizing effect of overvaluing computers and in 1976 published *Computer Power and Human Reason*.

The role was seriously out of step with his colleagues, who had no trouble celebrating ELIZA as a foundational program in artificial intelligence. At some point in the

1980s, I sat in on the legendary introductory lecture by the head of the AI Lab, Patrick Winston, to MIT's intro to AI course. Winston skillfully told the comic anecdote reproduced in this book, about a vice president of BBN, a pioneering private computer science lab, who mistook a telex conversation with ELIZA for a conversation with the human programmer who had installed the program on their system, an early MIT Computer Science PhD, Danny Bobrow.[5] It is a funny anecdote and it played well, but Winston misremembered the event in a way that was common at MIT, transposing it from BBN to the early MIT AI group, and imagining it happening not to Bobrow but to Joe Weizenbaum.[6]

ELIZA was not just the first chatbot, but one of the first times artificial intelligence in the form of an actual functioning interactive computer program, rather than a thought experiment or a sci-fi fantasy, came to public notice. It spawned delusional experiences of animation and also the first coherent arguments against the dangers of AI. Importantly, unlike pioneering AI figures from the machine learning period who later turned to fervently warning the public about the dangers of AI, Joseph Weizenbaum did not share the delusions of his naive users and overoptimistic colleagues. He saw ELIZA with great clarity as a clever human creation but not another form of intelligence. He saw the projection of intelligence upon the computer program as a failure of human cognition. And unlike some of the current generation of AI wizards who describe the processing of their machines as beyond human understanding, Weizenbaum thought it as his responsibility to offer explanations of computational techniques without any accompanying mystification. Team ELIZA is carrying on that tradition of meticulous documentation, by tracing the many versions and iterations of ELIZA and placing it within the cultural context of its critical progeny like Peggy Weil's MrMind and Sarah Ciston's ladymouth.

Weizenbaum decried a certain kind of arrogant but ignorant computer scientist, the culturally obtuse hacker or "compulsive programmer." He was equally skeptical of the proliferation of computers beyond the boundaries of the specialized CS Lab. Such skepticism set him apart from his colleagues, sometimes in surprising ways. In the 1980s MIT embarked on a major educational computing initiative. I was part of that effort, which I joined in part because I was inspired by ELIZA, which I saw as one of the first examples of the narrative potential of computing as a medium of representation.[7] Joe Weizenbaum graciously sent me a floppy disk implementation of ELIZA for use in my teaching, but his public remarks on the general enterprise of flooding the campus with the new computer terminals, and then the first desktop machines, was to compare it to randomly positioning cellos all over campus and expecting great music to result. He believed that computers should be left in the hands of virtuosos, while at the same time holding the hacker community to which he belonged in great suspicion. The paradoxical nature of his position perhaps reflected Weizenbaum's unresolvable ambivalence about the disturbing effects of his own intense virtuosity.

Inventing ELIZA offers a reconciliation of the hacker and the cultural critic. I believe Joseph Weizenbaum would have seen the chapters of this unique volume as deeply consonant with his own values. Team ELIZA combines virtuoso computer expertise with humanistic perspective. They are the cellist antihackers he would have approved of, and they have produced a volume that is both timely and timeless, and a worthy companion to Joseph Weizenbaum's painfully successful masterwork.

COL. PICKERING. We're always talking Eliza.
PROF. HIGGINS. Teaching Eliza.
COL. PICKERING. Dressing Eliza.
MRS. HIGGINS. What!
PROF. HIGGINS. Inventing new Elizas.

– George Bernard Shaw
Pygmalion, act 3 (1912)

ELIZA/DOCTOR

```
DOCTOR  TELL ME MORE ABOUT YOUR FAMILY
user    My mother takes care of me.
        WHO ELSE IN YOUR FAMILY TAKES CARE OF YOU
        My father.
        YOUR FATHER
        You are like my father in some ways.
        WHAT RESEMBLANCE DO YOU SEE
        You are not very aggressive but I think
        you don't want me to notice that.
        WHAT MAKES YOU THINK I AM NOT
        VERY AGGRESSIVE
        You do not argue with me.
```

01.00 Please Go On: An Introduction

USER: MY FATHER.

DOCTOR:
YOUR FATHER

When computers were still room-sized machines, they did not speak or listen. People programmed these giants using paper punch cards and physical typewriter keyboards—there were no screens, no touchpads, not even mice. Computers were used to speed up calculations, manage payroll, automate banking systems, and detect ballistic missiles, among other applications. Although, for centuries,[1] ideas of thinking machines and automata have been, by turns, cautionary tales, fantasies of control, and utopian dreams, the expectations for computation typically far outpaced the reality—from Asimov's three laws of robotics to *The Jetsons*. But at the time the thought of conversing with such a machine would have seemed as awkward as chatting with a toaster.[2]

When it debuted in the mid-1960s, the ELIZA software program transformed the way people thought about interacting with computers. As the first chatbot,[3] ELIZA demonstrated how a calculation machine might engage in conversation, ushering in a host of social and technical questions that still resonate today. Now we don't think twice about interacting with a machine in real time, conversing over text, or even speaking into the air to ask about the weather. In many ways, ELIZA shaped not only the way we think about interacting with computers, but also how we *think about them*. It began to give a reality to the science fiction stories of how we expect computers to work. Although ELIZA was far from a faultless conversation partner, it astonished its users. ELIZA expanded the new technical imaginary, gesturing toward a world to come where we might work *with* computers rather than *on* computers.[4]

ELIZA did not just spark excitement, it also triggered alarm at the trust humans all too easily place in computers that respond to them in a conversational way. This caused the developer of ELIZA, Joseph Weizenbaum, a professor at Massachusetts Institute of Technology (MIT), to think very seriously about how our inclinations towards companionship, social trust, and community could be manipulated by computer interfaces. As he would write repeatedly over the following decades, computers do not understand but only calculate. Computers have no sense of morality, no sense of justice or rights, and they certainly are unable to take on a responsibility toward others. Though they might appear to be intelligent, they are not. Though they might appear to be sympathetic, they are not. Instead, they are instrumental, technical, and calculating, not just about mathematics or computing but about everything, including human life, feelings, and communities. Weizenbaum foresaw correctly that the development of conversational systems would likely result in a severe clash of values between the safeguards of social norms on the one side and, on the other, the priorities of a technocratic class using computers to control and manipulate people.

However, in the sixty years that ELIZA has been influencing computation and culture, popular histories have portrayed it primarily as a software program that could converse as an automated psychologist, modeled on psychotherapy techniques designed by Carl Rogers. ELIZA uses simple pattern matching and script-based responses to simulate a therapist's conversation. In the conventional account, it is the first of what we now call chatbots and is known for "fooling" even the secretary who watched Weizenbaum create it. That's how the story goes. However, in all these accounts, one essential piece has been missing: the ELIZA source code. After a deep dive in the MIT Archives, we have found that source code and it tells a much more complex story. This book represents a major research project into ELIZA as we recover, close read, and discuss ELIZA's source code for the first time.

By examining the program's code, dialogues, and paratexts, *Inventing ELIZA* seeks to challenge the popular story of ELIZA, instead revealing that ELIZA is, in fact, multiple ELIZAs, developed as different versions with differing functionalities, designed to run a wide variety of scripts or personas—not only DOCTOR—and built using a series of technical innovations (like abstraction, API-like software layering, list-processing language extensions, real-time interaction, and scripting functions). We seek to correct and to complicate ELIZA's history and influence by exploring the misconceptions, multiple versions, and missing code of ELIZA. We question a computing history that has lost many of its nuances, along with the artifacts that document these differences. Using the methods of critical code studies, we close read the historic source code as a collaborative media archeology project. We chose this software not just for its technological achievements, which were groundbreaking, but also because ELIZA speaks to this moment when conversational interfaces have returned into public use at a greater intensity and with much more impressive capacities—marked most dramatically by the appearance of tools like ChatGPT. Such systems offer new possibilities for conversational computing and its incorporation into a variety of technological devices and social situations, but their proliferation also calls for the same cautions Weizenbaum foresaw. Our journey into the code and its history has brought us into dialogue with Weizenbaum as an innovator who also saw the impact of his work as a warning. *Inventing ELIZA* is dedicated to understanding how ELIZA works, how and why it was written, what it has meant, and how it continues to influence culture.

ELIZA has set a template for human-computer conversations to come.[5] Its significance lies both in its technical innovation and also in its ability to capture the public imagination and to shape expectations for artificial intelligence. The development of ELIZA's language-processing capabilities and its DOCTOR script's conversational interactions established design patterns that would persist through decades of technological development. ELIZA/DOCTOR adds a critical chapter to the history of computing—including through its entanglements with problematic assumptions of gender, race and ethnicity, and class, as well as philosophical questions about what it means to be human. From Amazon's Alexa to ChatGPT, from HAL in *2001: A Space Odyssey* to Samantha in *Her*, the contemporary vision of conversational AI traces back to those early interactions between people and ELIZA/DOCTOR in Weizenbaum's MIT lab.

01.01 Recovering ELIZA's Code

To appreciate ELIZA's powerful techniques operating in just a few hundred lines of code, imagine the state of computing in the mid-1960s (see chapter 4). These large machines relied upon magnetic tape and punch cards. Weizenbaum's version of DOCTOR was, as Nick Montfort describes, "printed more or less permanently on paper, rather than transiently appearing on the screen, [which might] influence how we understand the system as literary and as psychotherapeutic. A session that leaves a printed record, like a diary, may be experienced differently than a transient on-screen encounter."[6] Today, we are more familiar with ELIZA/DOCTOR on screens, but these printouts are crucial for appreciating the material specificity of what it would have been like to use ELIZA in the 1960s. The computer ELIZA was developed on, an IBM 7094 running Compatible Time-Sharing System (CTSS),[7] did not support mixed upper and lower case letters, unless you were using special libraries.[8] How did these tangible aspects affect users' experience of ELIZA, when they were not immersed in the same kinds of digital environments we are today? Using ELIZA would have been their first experience chatting with a program, with no prior context to draw on. To imagine how different their expectations of chatting with DOCTOR would have been at that early moment, we rely on material traces from the archives.

Our discovery in 2021 of the lost ELIZA source code in the MIT Archives' Distinctive Collections revealed surprising details that transformed our understanding of ELIZA. (For example, in early printouts of the source code, it appears that the program is called SPEAK before it is ever referred to as ELIZA!)[9] This inspired our international, interdisciplinary research team to convene for four years, unravelling the mysteries of the code. It also meant (re)investigating the larger historical context surrounding ELIZA to rewrite the stories, not only of this code, but also of how it has shaped the histories and futures of human-computer interaction. Nevertheless, this work has not progressed along a straightforward path, and the revised stories of ELIZA we present here are not stable, but plural, iterative, and endlessly reinventing themselves.

Despite ELIZA's considerable impact on computer science and culture, with versions of its program widely run and referenced, the original code was never published nor distributed. In his most widely read paper, published in 1966 in the *Communications of the Association of Computing Machinery* (*CACM*),[10] Weizenbaum describes in detail the workings of ELIZA and the DOCTOR script. His description has been used to create the many versions of ELIZA around today—*all made without access to the actual source code*. Even with the many copies of ELIZA, no one knew exactly how the original ELIZA did what Weizenbaum described. The missing code did not suppress appetites for ELIZA-like systems, but it left a large gap in understanding the original program as well as its impact and legacy.

Our investigations in the MIT Archives found one complete version of Weizenbaum's ELIZA code and DOCTOR script. Our comparative analyses reveal that this code is close to but not the exact version that Weizenbaum detailed in his article. Among Weizenbaum's papers, we also found evidence of multiple other versions of the ELIZA program; multiple versions of published dialogues, some of which had been hand-edited; additional unpublished DOCTOR dialogues; as well as previously unknown scripts that were entirely distinct from DOCTOR. With these findings, the notion of a definitive ELIZA and its familiar distinctive DOCTOR persona began to disintegrate into a proliferation of ELIZAs.

The discovery and archaeology of the original ELIZA source code represents a significant intervention in the history of computing. By examining the actual MAD-SLIP[11] implementation of ELIZA rather than relying on later reconstructions and reimplementations, we challenge taken-for-granted assumptions about this key software artifact. For example, importantly, the source code reveals that ELIZA was not merely a simple pattern-matching chatbot but can be better understood as a sophisticated platform designed for multiple "personas," or scripts, with a complex set of capabilities, including script editing and contextual memory. This transforms our understanding of early AI development by demonstrating that Weizenbaum's technical innovations were far more advanced than previously documented.[12] (ELIZA is even Turing complete, see chapter 7.) Moreover, the discrepancies between his published descriptions and the actual implementation help to show the gap between theoretical computational models and their material instantiations in computer source code, a tension that continues to shape digital culture today. By recovering this code, we reposition ELIZA in the genealogy of computational media, not just as a historical curiosity but instead as a complex technical achievement. *Inventing ELIZA* is in fact a complex story of many ELIZAs.[13]

Contrary to another popular misconception, ELIZA and DOCTOR are not one and the same. The name ELIZA refers to the entire system Weizenbaum built, and the DOCTOR script is just one of several scripts written for that system. ELIZA is an interpreter, a program that interprets a particular form of data called a script.[14] The most famous ELIZA script, DOCTOR, inspired numerous unauthorized copies, reimplementations, and confusions. It produced versions that often omitted the behavioral element of the original ELIZA software. This omission was unfortunate—as it ignored both the additional scripts as well as ELIZA's potential for multiple personas and flexibility.

These adaptations implement the DOCTOR script as if it *were* ELIZA, rather than implementing ELIZA as a system for running DOCTOR along with multiple other chatbot personas. For simplicity, and in keeping with their own naming conventions, we will call these adaptations "ELIZA" even though they are technically adaptations of DOCTOR. Versions can be found in early programming languages as well as in most contemporary high-level languages, from Python to JavaScript. Shortly after the publication of Weizenbaum's 1966 paper, Bernie Cosell programmed an ELIZA in Lisp (1966, 1969, 1972), based on Weizenbaum's published algorithm (see chapter 8).[15] Lisp was rapidly becoming the primary language of AI, and Weizenbaum's DOCTOR script is formatted similar to a Lisp S-expression (symbolic expression). This contributed to the persistent misconception that ELIZA was originally written in Lisp, although one need only read the first page of the *CACM* paper to learn that ELIZA was actually written in the language MAD-SLIP. In 1977, Jeff Shrager's BASIC ELIZA was printed in *Creative Computing* magazine, re-creating ELIZA for personal computers.[16] As a result of these coincidences, the ELIZA known in academic communities was the Lisp version, and the one known to the public was the BASIC version.

ELIZA's original code went virtually unseen for almost 60 years. Written in just over 400 lines of code, the ELIZA program is surprisingly sophisticated. When people conflate ELIZA and DOCTOR, they mistake the DOCTOR script to be the sole aim of Weizenbaum's research, when in fact he offered DOCTOR merely as an illustration of what the system could do. By choosing to simulate (or, in his words, "parody"[17]) Rogerian nondirective psychotherapy using computation, Weizenbaum selected a premise that allowed ELIZA to operate with very little context while remaining a plausible conversationalist. This reflexive style of therapy mirrors back the response of the patient to elicit more details. This premise meant that ELIZA could create the illusion of a conversational partner who kept previous comments in mind by storing and recalling certain phrases made by the user as the conversation progressed.

What makes interacting with ELIZA/DOCTOR so enticing is the pattern-matching and substitution algorithm at its core. But what makes this algorithm come to life are Weizenbaum's selection of the psychotherapist persona and his code's elegance and concision in responding to the challenges of producing coherent conversation and seemingly relevant phrase recall. In the *CACM* article, Weizenbaum describes how simple transformations can create the impression of "understanding" complex input: "an input sentence is scanned from left to right. Each word is looked up in a dictionary of keywords. If a word is identified as a keyword, then (apart from the issue of precedence of keywords) only decomposition rules containing that keyword need to be tried."[18] The system responded to any input with a typed response in English, and there is also mention of German and Welsh versions. The algorithm replaces pronouns from first person to second person (and vice versa) and simulates understanding. However, lacking the code itself, misconceptions about how ELIZA produced these simulations sprang up over the years.[19]

In fact, the instructions Weizenbaum gives in the article could not have output the oft-reproduced conversation that the article gives as an example, *but the code we have found can.* Indeed, essential elements of the program that were not described in the article have been located in the recovered code. These reveal new aspects of ELIZA that this book connects to the program's legacy, insights that also allowed us to create a reconstruction of ELIZA that replicates the original 1966 version's dialogue exactly.[20]

In one key insight, our exploration of the ELIZA code revealed the precise workings of "a certain counting mechanism" (see chapter 3.2) that makes this essential memory recall possible. Although Weizenbaum described the mechanism that stores and recalls users' statements in memory, his article's description glossed over two important details about how the memories are selected and recalled, which the study of the code reveals, allowing for the exact recreation of responses the original ELIZA would have made.

In another example, Weizenbaum's article describes two delimiters for the end of clauses—"," and "."—that discard any additional text past the delimiter once a keyword is found in a phrase. A delimiter is a symbol that tells the computer that a break can be made in the input text. However, both the ELIZA code we uncovered and the sample dialogue he published in that article use a third delimiter as well—the word "but." Consider the following exchange from the article:

```
You are not very aggressive but I think you don't want me to
notice that.
WHAT MAKES YOU THINK I AM NOT VERY AGGRESSIVE
```

This response would not be possible unless "but" is a delimiter. If it were not a delimiter, that response would be, "WHAT MAKES YOU THINK I AM NOT VERY AGGRESSIVE BUT YOU THINK I DON'T WANT YOU TO NOTICE THAT." Without the "but" delimiter, the part of the phrase following the word "but" is no longer discarded but is instead included in the transformed output. The discovery of this discrepancy between the code and the article highlights the ways code can stay in flux as programmers revise, extend, and rethink it.

The addition of the "but" delimiter is just one sign that ELIZA was being actively developed over time. In many ways ELIZA should be seen as a process, rather than a product, of software programming—although we can identify specific milestones in its development as major versions of the software (see below). Weizenbaum described how he continued work on ELIZA across multiple versions, as "difficulty with the [original ELIZA] system currently [meant] that it [could] do very little other than generate plausible responses." Weizenbaum explained:

> The [new version of] ELIZA differs from the old one in two main respects. First, it contains an evaluator capable of accepting expressions (programs) of unlimited complexity and evaluating (executing) them. It is, of course, also capable of storing the results of such evaluations for subsequent retrieval and use. Secondly, the idea of the script has been generalized so that now it is possible for the program to contain three different scripts simultaneously and to fetch new scripts from among an unlimited supply stored on a disk storage unit, intercommunication among coexisting scripts is also possible.[21]

These later advances were, sadly, poorly documented and their source code remains unrecovered, as do many of the scripts of the other personas.

However, in chapter 5, we include examples from some of the recovered scripts, showcasing ELIZA's alternative personas. As Weizenbaum notes:

> the most famous script is the one where ELIZA plays the psychiatrist, but there were others. I remember an undergraduate at that time who is now full professor at MIT—Steve Ward by name—who wrote a script that was a parody of the way President Nixon talked. That was not hard to do: Nixon has a very characteristic way of speaking, like "Let me say this about that" and so on.[22]

Although this is often forgotten, Weizenbaum described ELIZA as a family of programs.[23] Most people today assume there was only one version of ELIZA, but our research shows that there were at least five major versions of the ELIZA program (there may be many more). This includes the versions discovered in the archive: a flowchart and outline (1965a), and the complete source code (1965b). ELIZA (1966) is the version described in Weizenbaum's CACM paper. This ELIZA was developed by Weizenbaum during the early 1960s, and it is this version (and its offshoots) that is commonly known as the world's first chatbot. There were also later versions we call ELIZA (1967, 1968+) that had much more sophisticated scripting functions using the language OPL (see table 1.1 below).

As Weizenbaum iterated the system, ELIZA found its way into new fields of research. ELIZA was built under the auspices of MIT's Project MAC, which was funded by the Department of Defense's Advanced Research and Projects Agency (ARPA, now DARPA), with MIT producing annual reports detailing the experiments conducted with the system. According to Project MAC Progress Report II (1964–65), ELIZA was used "for experiments on two-person conversational interaction at the Massachusetts General Hospital," whereas Report III (1965–66) describes ELIZA's use in a "study designed to assess the phenomenon of misunderstanding in man-machine natural-language communication, and establish the conditions and limits under which natural-language communication between man and machine could be made plausible." These project reports emphasize ELIZA's original conception as an educational and research tool rather than merely a conversational chatbot.[24]

ELIZA (1967) added script handling[25] and could run other scripts with names like ARITHM, F29, and FIGURE (see table 1.2). A colleague in the Education Research Center at MIT, Edwin F. Taylor, then built on these versions to develop ELIZA (1968+). This version was greatly extended beyond ELIZA (1966), and was used at MIT and Harvard to build Computer Assisted Instruction (CAI) systems. Each of the versions we have identified are shown in table 1.1. That Weizenbaum called all of these programs ELIZA, in the singular, can be confusing (not least for the authors of this book). Surely, the history of chatbots may have been quite different if the world had seen the code for each of these versions and if the versions had been clearly delineated. As David M. Berry suggested in a Critical Code Studies Working Group on the ELIZA code, it became

> clearer that not only are there multiple ELIZAs but that this has an important bearing on its identity as a research object. ...

> ELIZA is certainly a work in progress, less than a finished software object, and very much like a process of research activity with stop-offs along the way. ... [So] we are looking at both a technical object, a cultural object and a historical object and that these things might not always be the same thing.[26]

As ELIZA merged with DOCTOR to become a cultural object, the programs we group under the name ELIZA have been obscured.

Table 1.1 Comparison of ELIZA versions.

Version	Delimit "."	Delimit ","	Delimit "but"	NEWKEY Function	Script Edit Function	Keyword Stack	Hardcoded Messages	Script Handling	Evidence
ELIZA 1965a	Yes	Yes	—	—	—	—	—	Script	MIT Archive flowcharts
ELIZA 1965b	Yes	Yes	Yes	—	CHANGE Function	—	Yes	Script	MIT Archive Source Code
ELIZA 1966	Yes	Yes	Yes	Yes	EDIT Function	Yes	—	Script	Weizenbaum, "Eliza"
ELIZA 1967	Yes	Yes	Yes	Yes	EDIT Function	Yes	—	OPL[27]	Weizenbaum "Eliza"; Taylor, "Automated Tutoring and Its Discontents."
ELIZA 1968+	Yes	Yes	Yes	Yes	EDIT Function	Yes	—	OPL	Taylor, "Automated Tutoring and Its Discontents."

1 ELIZA 1965a: Delimits sentences with only "." and "," as shown in MIT archive flowcharts.
2 ELIZA 1965b: Version recovered from MIT Archives. Delimits sentences with "." and "," and "but." Lacks NEWKEY function. Includes undocumented CHANGE function and hard-coded messages.
3 ELIZA 1966 CACM: Functionality includes the NEWKEY function and keyword stack, as described in Weizenbaum, "ELIZA," as well as the "but" delimiter, which is omitted from the article but present in the dialogues.
4 ELIZA 1967: Adds sophisticated script handling using OPL, as described in Weizenbaum, "Contextual Understanding by Computers," "OPL-I," and Taylor, "Automated Tutoring and Its Discontents." Also evidenced by the extant ARITHM, F29, FIGURE scripts in the archive.
5 ELIZA 1968+: A description of the planning of this version appears in Taylor, "Automated Tutoring and Its Discontents." The scripts SPACKS, INTRVW, and FVP1 also give evidence of its use in work developed by Walter E. Daniels.

As part of our version mapping, we discovered a section of the code (the CHANGE function) that gives ELIZA extra functionality that Weizenbaum called "teachable" This means it provides a method for a user to revise and extend scripts with new rules for text transformations while the program is running. In subsequent versions of ELIZA, this function is replaced by an EDIT function that calls an operating system editing program named ED (we discuss this in more detail in the ELIZA code annotations, line 10). Weizenbaum alludes to this functionality briefly when describing ELIZA: "Its name was chosen to emphasize that it may be incrementally improved by its users, since its language abilities may be continually improved by a 'teacher'. Like the Eliza of *Pygmalion* fame, it can be made to appear even more civilized, the relation of appearance to reality, however, remaining in the domain of the playwright."[28] Indeed, Weizenbaum saw programming as an experimental practice: "One programs, just as one writes, not because one understands, but in order to come to understand. Programming is an act of design."[29] He valued ELIZA as an evolving work, meant to enable one to understand human-computer interaction through language.[30] That ELIZA is not only a system for engaging conversational agents—it is also a platform for writing and editing scripts—is one of the many aspects of the program lost in the popular ELIZA imaginary.

Even programming languages themselves had, and have, material specificity—the specific hardware mattered particularly in this historical period. In a letter to the editors of the *CACM*, Weizenbaum also said of the scripting language he wrote, "SLIP is entirely host language-dependent. A FORTRAN-based SLIP will look quite different from a MAD-based SLIP, for example. A given host language is in general also dependent on the machine for which it is implemented and even, in some cases, on the monitor system within which that language is exercised."[31] By challenging ideas of ELIZA as a single program, a single chatbot persona, a single legacy, we can expand the many complex narratives and entanglements of ELIZA.

01.02 The "Eliza Effect"

The dialogue that opens this chapter is an excerpt of a published conversation purported to be between a young woman and DOCTOR. That dialogue has been reprinted countless times and has inspired programmers and writers to dream up many of the chatbots that followed. Yet the closer one inspects that dialogue, the more questions arise: Who was this young woman? Was she a real person, or is she Weizenbaum's invention? How exactly did the ELIZA system generate its responses, and how much were they edited? Why did the system work so well to draw people in?

ELIZA/DOCTOR helped catalyze a mode of thought and an anxiety about our relationships with computers. Weizenbaum explored this in his 1976 book *Computer Power and Human Reason*, invoking philosophical, social, and political critiques. The unique machine interaction presented by his program revealed how new forms of human-computer relation would have profound effects that he attempted to explore and to contest. After seeing its public reception, Weizenbaum was startled by the quick and often emotional attachments people would form with ELIZA, which he saw as "clear evidence that people were conversing with the computer as if it were a person who could be appropriately and usefully addressed in intimate terms."[32] The tendency to attribute empathy and invest private feelings into a computer puzzled Weizenbaum. He was concerned by the extent to which people associated rationality with computation, and ascribed understanding and intelligence to computer systems where none existed. This tendency became known as the "ELIZA effect." The term appeared in online forums by 1991 but its use predated that appearance by decades.[33] Sociologist Sherry Turkle defines "the Eliza effect" as "our more general tendency to treat responsive computer programs as more intelligent than they really are. Very small amounts of interactivity cause us to project our own complexity onto the undeserving object."[34] Cognitive and computer scientist Douglas Hofstadter describes it as "the susceptibility of people to read far more understanding than is warranted into strings of symbols—especially words—strung together by computers,"[35] which applies easily to generative AI systems today.

To understand the power and provocation of ELIZA, we can look to the infamous challenge formulated by computer scientist Alan Turing in the essay "Computing Machinery and Intelligence," in which Turing posed the question, "Can Machines Think?" Turing premised his thought experiment on a parlor game—not about technology but about gender: A man and a woman are hidden behind a curtain and an interrogator tries to identify who is which gender by asking a series of questions. The man tries to mislead the interrogator, pretending to be a woman, while the woman tries to convince the interrogator of the "correct" answer. That is, both of them claim they are the "real" woman, a challenge to essentialist notions of gender.

In his revision of this game, Turing replaces the original gender question with what is now called the Turing test.[36] In this challenge a machine pretends to be a man rather than a man pretending to be a woman. More than a mere opening gambit, Turing's choice of the initial gender imitation game has ensured that artificial intelligence would remain intertwined with questions of gender and identity. In this sense imitation, drag, and gender deconstruction laid the groundwork for AI and the performance of intellect. Weizenbaum's ELIZA picks up where Turing left off, not least with the opening line of the dialogue: "Men are all alike."[37]

Nonetheless, even though Weizenbaum references Turing's imitation game in his 1966 paper introducing ELIZA, he explicitly distances his creation from any claims of intelligence:

> The crucial test of understanding ... is not the subject's ability to continue a conversation, but to draw valid conclusions. ... In order for a computer program to be able to do that, it must at least have the capacity to store selected parts of its inputs. ELIZA throws away [most] of its inputs. ... ELIZA in its use so far has had as one of its principal objectives the *concealment* of its lack of understanding.[38]

This assessment shows that ELIZA was never intended to pass the Turing test, but rather to explore the psychological factors that might lead humans to misinterpret its capabilities.[39]

In that tradition, ELIZA continues to play with the notion of *performative identity*, as a vehicle for personas made out of responses that stick to the script, literally and figuratively. By naming the system after Eliza Doolittle—a working class female character who is taught to pass as an upper-class white woman—Weizenbaum continues Turing's provocation about the performance of identity. "I chose the name 'Eliza,'" Weizenbaum says, "because, like G.B. Shaw's Eliza Doolittle of *Pygmalion* fame, the program could be taught to 'speak' increasingly well, although, also like Miss Doolittle, it was never quite clear whether or not it became smarter."[40] As Shaw's Eliza performs race and ethnicity, class, sexuality, and gender through linguistic transformation, the ELIZA system performs an assumed persona through scripted and repetitive linguistic patterns and transformations, but without possessing humanlike understanding. Feminist philosopher Judith Butler's theories of gender performativity offer a possible framework to think about this.[41] Butler argues that gender and sexuality are not innate but are performed through repeated acts. They are iterated.[42] Just as Eliza Doolittle challenges assumptions of class by performing speech that help her pass as among the upper classes,[43] Weizenbaum's system performs personas like DOCTOR through speech acts, or code acts, and sample dialogues that perform gendered, classed, and racialized identities.

In this way, we argue that ELIZA both reinforces and challenges assumptions about gendered and embodied communication. It is interesting to note that in the published dialogues and popular stories about ELIZA, the women who appear alongside the program remain unnamed. They are conversing with a therapist called DOCTOR, a name that, though it does not carry gender markers today, would have sounded like a man's title in the 1960s. Therefore, stories that suggest that these women confessed their secrets to this artificial doctor carry a gendered message, including the fantasy that one can be disembodied.[44] In this way, the ELIZA system becomes a site where questions of identity, performativity, and embodiment play out across historical algorithms in ways that we believe have implications for contemporary AI systems.[45]

01.03 Systems of Promise and Peril

As norms, values, and ideology become embedded within algorithmic forms, exploring details of a software object like ELIZA reveals culturally and historically specific assumptions about what software can and should be, including ideas about what technology is for and how it has developed. While ELIZA might look unsophisticated to a contemporary audience, it was already addressing, in the 1960s, many of the design questions that persist in the systems we use today. Fundamentally, these systems ask how humans and machines should interact, by what means communication can be represented computationally, and to what extent machines can and should be imbued with an ability to influence users.

ELIZA is a common touchpoint not only because it was one of the first chatbots and helped launch a field of computational agents but also because it intersects with so many of the computing innovations that followed. Alongside developments for string processing and text analysis, it influenced research on text synthesis, entity recognition, and sentiment analysis. It emerged alongside research in machine translation, semantic networks, speech recognition, speech synthesis, and other techniques that developed into the group of tasks we now call "natural language processing" (NLP), the area of computation that deals with how computers can parse, interact with, process, and output languages that are used by people as opposed to programming languages used by computers. In practice, these tasks are often combined to create automated agents and many other kinds of systems.

The reappearance of ELIZA-like chatbot interfaces in contemporary large language models shows that early software genealogies can help us understand new technologies. ELIZA is a useful foil for emerging models because—though much has changed, much more remains the same. The history of NLP emerges from and overlaps the era of ELIZA and early AI work at institutions like MIT. In that field, trends have come and gone: Although different styles—*syntactic*,[46] *semantic*,[47] *statistical*,[48] *stochastic*[49]—were popular at different times, all continued to develop in parallel. Even now as the latest large language models (LLMs) have amazed observers with the seeming intelligence of text output, the actual power of contemporary systems like OpenAI's ChatGPT (generative pretrained transformer) are disguised behind their chatbot interfaces, which retain a resemblance to Weizenbaum's original. These enticing facades obfuscate the machinery that often includes a combination of statistical predictions, rule-based procedures, and human labor disguised as machine labor. From the perspective of the user, this leaves limited opportunity to distinguish hype from substance, to understand how each system operates, and to see why it produces certain outputs.

Weizenbaum warned of the many issues this obfuscation could create. It can, for example, lead to exploitation as humans are replaced, harmed, or treated unfairly by these systems. As Weizenbaum says, "an individual is dehumanized whenever he is treated as less than a whole person. The various forms of human and social engineering ... do just that, in that they circumvent human contexts, especially those that gave real meaning to human language."[50] Weizenbaum argued that removing language from its social contexts and treating it as a set of abstract concepts in a computational system can be dehumanizing. It risks ignoring the multiple meanings inherent to language, which are impossible to capture fully in AI systems and which can result in direct harm, rights violations, privacy breaches, exploitation, displacement, and discrimination.[51] That is why it is essential to consider broader ethical and social impacts when designing, deploying, and using automated systems.

Emerging large language models support a new class of knowledge-powered systems.[52] Under the surface, the chatbot interface runs on social labor. Its machinery supported by literally millions of traces of humans' writings and conversations, siphoned into datasets usually without creators' awareness or consent. That labor enables automated cultural production that is managed, controlled, monitored, disaggregated, and reaggregated on demand. This abstracts human cultural production through software, treating it as a *standing reserve* for a computational system (in a similar fashion to electricity or a water supply, yet privatized and monopolized).[53] As Langdon Winner observed in 1978, it is possible "for inanimate instruments to perform their own work 'at the word of a command or by intelligent anticipation,' that is, by a computer program. This development has led to conjecture that the perfection of industrial technology will eventually liberate mankind from toil,"[54] a common story that OpenAI, Anthropic and other AI companies repeat today. Yet computation is always serviced by human labor—even as chatbots currently manage customer queries and problems, help with homework and replace teaching assistants, converse as companions and counselors, and entertain and entangle with notions of identity and human exceptionalism.[55] They also help produce a deluge of AI slop that is cannibalizing its source materials and the planet's natural resources.[56] The idea that ELIZA might inspire computational systems that create cognitive factories and AI slop would likely have horrified Weizenbaum. It is, in a sense, the realization of cybernetic systems that attempt to closely couple machines and humans into tightly linked feedback loops—exactly the threat that Weizenbaum warned about in *Computer Power and Human Reason*.

In that book and other post-ELIZA writings, Weizenbaum speaks out as an early critic of what we now call the tech sector and its mindset of exponential acceleration, regardless of the exploitative relationships it may be encoding and the social and political consequences of abstracting computational systems. His own relationship to technology shifted as he saw the impact of ELIZA and the ways automated systems were being scaled up and leveraged for power, trends that have continued with a wider public fascination and unease with the emergence of "intelligent" machines. This book is a call to revisit Weizenbaum's work, and its warnings, through direct engagement with the code and scripts of ELIZA itself, as well as through Weizenbaum's writing and the works of others it influenced.

01.04 Re-Creating ELIZA

Those who wish to talk with DOCTOR directly can find an accurate reproduction of ELIZA/DOCTOR, created by Anthony (Ant) Hay, on the inventingeliza.com website.[57] This reproduction is based on analysis of the source code and includes the "certain counting mechanism" that allows it to replicate Weizenbaum's ELIZA more accurately.

As part of our software archaeology work, we have also assembled two other notable recreations, one displayed in Paris and another available online on a CTSS simulator. In April 2025 at the Musée Jeu de Paume, museum visitors encountered a realistic simulation of what it might have been like using ELIZA in the mid-1960s. Ant and his son, Max, translated Ant's C++ ELIZA to JavaScript. Jeff Shrager and his son, Leo, created a web-based simulation of Ant and Max's ELIZA running on an ASR-33 teletype, including image, sound, and timing that give a good sense of what it was like to run ELIZA in its original environment.[58]

That recreation led to resuscitating the original ELIZA code on an IBM 7094 emulator. David M. Berry asked Rupert Lane, an aficionado of early operating systems, whether he could run the MAD-SLIP ELIZA code on an emulated version of the Michigan Time-Sharing System (MTS) (where the Michigan Algorithm Decoder (MAD) language that Weizenbaum used to implement ELIZA was developed).[59] It turned out to be easier, and more authentic, to rebuild MIT's CTSS operating system on the 7094 emulator, then reconstruct SLIP, MAD, and finally ELIZA on that stack.[60] Bringing the original ELIZA up on the MAD/CTSS stack required numerous steps of code cleaning, completion, and debugging.[61] Some required functions were missing and had to be rewritten, but 96 percent of the running ELIZA MAD-SLIP code is running exactly as it was found in Wiezenbaum's archive. The entire stack is open so that any user of a Unix-like operating system can run an authentic ELIZA on their version of a pioneering time-sharing system.[62]

We see these reconstructions as part of our ongoing work to help contemporary audiences experience what it was like to talk with the first chatbot. The process of bringing ELIZA back has also helped us to develop a fuller understanding of the software in its original material context, as well as to better grasp its sociotechnical impact on this moment.[63]

01.05 About This Book

Inventing ELIZA has been written by an interdisciplinary team of scholars, artists, philosophers, and programmers, who each inhabit different combinations of these roles. As a collection of investigators with diverse interests, we have constructed this book together, while also highlighting our individual perspectives that made this conversation so rich. Given our individual relationships to ELIZA, its history, and its legacy, certain chapters draw more on one or more authors' experiences, and we have indicated who they are. Of course, we do not always agree or speak with a single voice, so you will find occasional reference to each of us by our names or initials—but we hope this makes reading more interesting.

As part of the Software Studies series, this book uses the methods of critical code studies to explore and interpret the ELIZA code and the DOCTOR script. Critical code studies (CCS) is a set of emerging methods to explore the extra-functional significance of computer source code.[64] *Extra-* here means "growing out from," so first we make clear how the code and scripts operate. However, to understand the code, we need to establish the historical and material context in which it circulated and operated. Once that foundation is laid, we can discuss the implications and further meaning of this code in terms of culture, computer and human history, power dynamics, creative computing, automated therapy, and the further legacy of chatbots including their use in AI. In other words, "a close reading of code can also draw attention to the way in which code may encode particular values and norms."[65] Nonetheless, code here is not treated as a singular, static text but as a snapshot of a broader system at work.Code is both sign and cause of dynamic processes brought into action on hardware and interacting with other software in a field of shifting states. For this reason, our efforts to get the code running and to run other adaptations and approximations are crucial to our reading. The code is the entry point but also the grounding of our investigation. It is the artifact around which we gather to discuss and interpret. However, "the way in which code is actually 'doing' is vitally important, and we can only understand this by actually reading the code itself and watching how it operates."[66]

In this collaborative book, our methods are radically interdisciplinary, developed out of years of conversations among a team with expertise on computer programming, history, literary studies, philosophy, and art. As such, this book adapts its methods from previous collaborative close readings of software including *10 PRINT CHR$(205.5+RND(1)); : GOTO 10* and *Reading Project*.[67] In our analysis of ELIZA and its impacts, we apply theories from across computational media studies (in particular software and platform studies, and media archeology).[68] We also draw on critical AI research (broadly defined, including critical algorithm and data studies, and histories of science and technology).[69] Many of these fields overlap, which makes sense given the rapid, widespread deployment of emerging technologies. There are many entry points for reading more on these topics, including the excellent anthologies such as *Your Computer Is on Fire* and *Uncertain Archives: Critical Keywords for Big Data*.[70]

To illustrate the way code itself engenders conversation, this book is grounded in close readings of the ELIZA program code written in MAD-SLIP (chapter 6) and the DOCTOR script (chapter 7), both of which we have annotated collectively and individually for over three years. We use our understanding of the code as the foundation for the entire project.[71] In studying a sophisticated piece of software such as ELIZA, it is inevitable that we will need to explain some complex technical concepts and ideas in detail throughout the book. We do our best to approach these explanations with enough specificity to satisfy the experienced but enough clarity not to intimidate the newcomer. This reading continues the work begun in several Critical Code Studies Working Groups (particularly 2016 and 2022).[72] The goal of our annotation is not to offer a final, definitive set of interpretations of this code but instead a foundation and model for future discussions. These readings benefit from the diversity of our approaches and backgrounds and will only become richer with an even wider range of readers who become literate in this code. We hope this book serves as an invitation to explore further one of the most widely known, influential, and misunderstood pieces of software.

As such, the book offers many avenues of entry. We have composed all chapters iteratively and conversationally rather than sequentially in isolation. We invite you to read and cross-reference in any order you choose.

Chapter 2 looks at Weizenbaum's life from his role as ELIZA's creator to prominent AI critic. Through historical context and biographical details, we examine how his early experiences shaped his growing concern that computation and artificial intelligence threatened to reduce human judgment to mere calculation. The chapter explores his intellectual evolution and his principled stand against what he saw as technology's encroachment on human reason.

Chapter 3 explains how ELIZA works. We complement Weizenbaum's 1966 *CACM* paper (reproduced in Appendix II) by giving both a high-level overview of ELIZA's operation and some low-level details from the source code that were not present in Weizenbaum's description.

Chapter 4 contextualizes the programming and technological challenges Weizenbaum faced and makes tangible the computing environment at the time ELIZA was written—including physical practicalities from punch cards to magnetic tape storage. We discuss how Weizenbaum's work on SLIP and ELIZA overcome the obstacles presented by the era's material constraints, as well as describe some of the material conditions of interacting with the system via teletype.

Chapter 5 compares a selection of ELIZA's multiple scripts and dialogues, extending beyond DOCTOR and revealing the program's lesser-known applications, from teaching physics to analyzing poetry. Through archival transcripts, we uncover how users engaged with ELIZA across various contexts, including a clinical psychology trial.

In chapters 6 and 7, we reproduce our discovery of the code for ELIZA and the script of DOCTOR (both annotated). These chapters not only offer insight into the working of the code and script but also demonstrate some of the thinking inspired by our close reading through critical code studies.

Chapter 8 reviews many of the offshoots of ELIZA, from its contemporaries to later successors, demonstrating ELIZA's long legacy in software. Rather than a comprehensive catalogue, we contextualize a sampling of bots most in dialogue with ELIZA.

Chapter 9 looks at a program inspired by ELIZA and discusses the continuing impact of ELIZA on our cultural conceptions of self: Peggy Weil's chatbot, MrMind, performed "The Blurring Test," chatting with people about what makes them human.

Chapter 10 takes up ELIZA's influence in education, featuring ELIZA's teaching scripts as well as Mavis Beacon Teaches Typing and Peter Millican's Elizabeth as case studies before exploring later uses of ELIZA in both formal and informal learning contexts.

Chapter 11 looks at the processes behind contemporary large language models, discussing how they build on or diverge from the techniques of ELIZA. We explore the implications of both their technical and their social aspects given what we have learned from ELIZA's code. Though vastly different from ELIZA, conversational models such as ChatGPT still rely upon many of Weizenbaum's approaches, including demonstrating the deceptive "magic" of mechanized conversational exchange.

Finally, the conclusion reflects further on the implications of this research project, particularly at this moment in AI history.

Inventing ELIZA is also interspersed with short dialogues between humans and chatbots, both real and imaginary. These include selections from ELIZA/DOCTOR plus other previously unseen ELIZA scripts. They also include later adaptations of ELIZA, as well as chatbots that respond to the ELIZA form and the current chatbot boom, like Claire James Carroll's "The Imitation Game." These interludes are a glimpse into the dialogic heart of bots, the wordplay, the hits and misfires of these systems. In a book about chatting machines, they illustrate the power of chat.

We invite readers of this book to join us in our exploration of ELIZA, to read the code and the paratexts, and then to continue the conversation about ELIZA with us at our websites inventingeliza.com and elizagen.org.

Programs have a life of their own, and we piece them together with snapshots of code and sample outputs. The program is not the code alone, though any given instance of code can teach us much about a particular moment in its development. The code of any program is always partial. The program is more than the sum of instances of the code and its outputs over time, moving through many ecosystems and across many platforms. Software programs—like any media objects, like any stories—are ongoing, dynamic, versioned, and evolving texts, rather than fixed works.[73] Weizenbaum's work has been often reproduced and taught, read and analyzed, and ported into many other languages and platforms and has inspired countless descendants. We have collected and analyzed not only code but transcripts and scribbles, tangents and offshoots—knitting together a complex story among the numerous voices that make up ELIZA's history and possible futures. In our search for ELIZA, we still have many mysteries to explore, much code to uncover, read, and interpret. Weizenbaum framed our interactions with this dynamic code as conversations. Like the code and the story of ELIZA itself, we consider these conversations to be in process and ongoing, and we invite you to join us in the chat.

In the words of ELIZA, "Please go on."

Table 1.2 Script Genealogy — This is an attempt to document the different scripts that are extant within the archive, mentioned in articles, or referenced elsewhere. We are sure there are many others.

#	FILENAME	Description	Type	Referenced	Archive	Year
1	1965 CONVER-SATION THE FIRST	In which the editor talks to the computer. Author: Andrew T. Weil.	OUTPUT	*Harvard Review* (Spring 1965)[74]	Published	1965
2	1965 CONVER-SATION THE SECOND	In which a distraught young lady consults the machine. Author: Weizenbaum.	OUTPUT	*Harvard Review* (Spring 1965)	Published	1965
3	1965 CONVER-SATION THE THIRD	In which a human psychiatrist meets his mechanical counterpart. Author: Weizenbaum.	OUTPUT	*Harvard Review* (Spring 1965)	Published	1965
4	ANTIPR		NONE	Taylor, "Automated Tutoring and Its Discontents" (WATSNU)		1968
5	ARITHM	Undertakes mathematical calculations within the conversation and is able to evaluate and return the result	SCRIPT, OUTPUT	Archive	Yes	Unknown
6	CANVEC	By this time the student has studied 4-vectors using a textbook of his choice. Notice that the machine is ready to analyze an example provided by the student (although in this conversation the student asks the machine to provide the example). Notice also that toward the end of the interchange the machine misinterprets the student answer "Wwhy not" to mean "No," thus ending the conversation.	OUTPUT	Taylor, "Automated Tutoring and Its Discontents"		1968
7	COLOR	Script to ask about colors of things	SCRIPT, OUTPUT	Archive	Yes	1967
8	COMMNT	"HINTS ON USING, CONTROLLING, AND COMPLAINING ABOUT ELIZA" May be built in, not script	NONE	Taylor, "Automated Tutoring and Its Discontents" (WATSNU script)		1968
9	DOC		NONE		No	December 18, 1967
10	DOCTOR	The famous Rogerian DOCTOR script	SCRIPT, OUTPUT	"Eliza" the *CACM* 1966 paper	Yes	1966
11	ELEVTR	Discusses the physics of an elevator	OUTPUT	Taylor, "Automated Tutoring and Its Discontents" (WATSNU)	Yes?	1968

12	F29	Appears to be an early version of the FIGURE mathematical definition script	SCRIPT	Archive	Yes	December 26, 1967
13	FIGURE	Appears to be a late version of F29 extended with many more definitions and responses	SCRIPT	Archive	Yes	January 9, 1968
14	FORVEC		NONE	Archive	No	1967
15	FRANCE		OUTPUT	Taylor, "ELIZA Program"	No	1967
16	FVP1	Discussion of the exercise FVP1, in which the student fills in a table and then discusses his table entries with ELIZA. The statement of the problem precedes the conversation. Notice that during the conversation (regarding question 3) the machine treats a wrong answer as right, but by accident the error is discovered by the student. Notice also that after the problem has been discussed to the satisfaction of the student (parts of the conversation omitted here), the machine initiates a discussion on some issues related to the problem.	OUTPUT	Taylor, "Automated Tutoring and Its Discontents"	No	1968
17	FVQUIZ		NONE	Taylor, "Automated Tutoring and Its Discontents" (WATSNU)		1968
18	GIRL	Written to demonstrate how to write scripts. A real tutorial script must be able, among other things, to follow a developing line of argument. The ELIZA scriptwriter has approximately 100 commands ("functions") available for instructing the computer how to proceed. GIRL uses only three of these functions: TYPE, GOTO, and POPTOP, the last two of which simply tell the machine how to begin processing the program section of the script.	SCRIPT, OUTPUT	Taylor, "Automated Tutoring and Its Discontents"	No	1968
19	HELP	May be a test script run by Weizenbaum	OUTPUT			Unknown
20	HOWCON	"HINTS ON USING, CONTROLLING, AND COMPLAINING ABOUT ELIZA" May be built in, not script	NONE	Taylor, "Automated Tutoring and Its Discontents" (WATSNU script)		1968
21	HOWUSE	"HINTS ON USING, CONTROLLING, AND COMPLAINING ABOUT ELIZA" May be built in, not script	NONE	Taylor, "Automated Tutoring and Its Discontents" (WATSNU script)		1968
22	INTRVW	Part of a student conversation using INTRVW, the script that begins the study of 4-vectors. Notice how the conversation proceeds even when the student expresses himself less than clearly. The symbol *QQ* is sometimes used instead of the question mark. Other times the question mark is simply omitted.	OUTPUT	Taylor, "Automated Tutoring and Its Discontents"	No	1968

23	MIT		NONE	Archive	No	1967
24	NEWENG	Discusses New England states	SCRIPT, OUTPUT	Taylor, "The ELIZA Program"	No	1967
25	ORTH1		NONE	Archive	No	1967
26	PHOTON		NONE	Archive	No	1967
27	POLETA	Discusses the "Pole and Barn Paradox"	OUTPUT		Yes?	Unknown
28	QMPROB		NONE	Archive	No	1967
29	RLPOLN		NONE	Archive	No	1967
30	SOS	"HINTS ON USING, CONTROLLING, AND COMPLAINING ABOUT ELIZA" May be built in, not script	NONE	Taylor, "Automated Tutoring and Its Discontents" (WATSNU script)		1968
31	SPACKS	Discusses a quoted line of poetry by Barry Spacks of the MIT humanities department	OUTPUT	Taylor, "Automated Tutoring and Its Discontents"	No	1968
32	STATES		NONE	Archive	No	1967
33	SYNCTA	Discusses time synchronization	OUTPUT	Taylor, "Automated Tutoring and Its Discontents" (WATSNU script)	Yes?	1968
34	WATSNU, PREVI-OUSLY LSTSCR	Lists all other scripts This may not be a script but instead actually a function in ELIZA	OUTPUT	Taylor, "Automated Tutoring and Its Discontents"		1968
35	XYPOLN		NONE			Unknown

ELIZA/DOCTOR: Conversation March 5, 1965

Joseph Weizenbaum
March 5, 1965

user Doctor, I am terribly depressed
DOCTOR IS IT BECAUSE YOU ARE TERRIBLY DEPRESSED THAT YOU CAME TO ME
Actually, my wife suggested I come ~~to you~~ here
TELL ME MORE ABOUT YOUR FAMILY
I have no children
YOU SAY YOU HAVE NO CHILDREN
I can't resign myself to fatherhood
HOW DO YOU KNOW YOU CAN'T RESIGN YOURSELF TO FATHERHOOD
Well, my wife wants kids but I don't. That's all I can say about it.
WHO ELSE IN YOUR FAMILY WANTS KIDS
Everybody is always hinting and bugging me about it.
WHO, MAY I ASK
My father talks about grandchildren all the time
YOUR FATHER
He pokes his nose into our affairs much too much
I AM NOT SURE I UNDERSTAND YOU FULLY
He's always tring to run the house
CAN YOU THINK OF A SPECIFIC EXAMPLE
Apart from the children thing, my father wants me to change jobs
WHAT ELSE COMES TO MIND WHEN YOU THINK OF YOUR FATHER
~~The Police~~ Nothing

~~EARLIER~~ YOU SAID YOUR WIFE SUGGESTED
YOU COME ~~TO I~~ HERE
yes
YOU SEEM QUITE POSITIVE
She said that I should either get help
or she would ~~have my father arrested~~
leave me
CAN YOU ELABORATE ON THAT
My wife can't stand ~~his interfetance~~
dad any longer
TELL ME MORE ABOUT YOUR FAMILY
Mother is dead
EARLIER YOU SAID YOUR WIFE WANTS KIDS
That's right. Can you understand how
~~she~~ that bugs me
YOU BELIVE I CAN UNDERSTAND HOW ~~SHE~~
THAT BUGS YOU DON'T YOU
I can't take it much longer
HAVE YOU TRIED
I'm always patient
WHEN
When she yells at me
LETS DISCUSS FURTHER WHY YOUR FATHER
TALKS ABOUT GRANDCHILDREN ALL THE TIME
He's getting old. I guess he wants
to be immortal
DO YOU SAY YOU GUESS HE WANTS TO
BE IMMORTAL FOR SOME SPECIAL REASON
He also talks about dying a lot.
Because he is sick, you know.
IS THAT THE REAL REASON
Maybe
AND MAYBE NOT
I think he needs help more than I do
THAT'S QUITE INTERESTING

I can't help him. ~~I don't like him, for one thing.~~ I've tried all my life.
PERHAPS YOU COULD HELP HIM NOW
Wjat do you mean
WHY DO YOU ASK
I don't understand you. ~~Not really, I mean.~~
DON'T YOU REALLY UNDERSTAND I
no
ARE YOU SAYING 'NO' JUST TO BE NEGATIVE
Are you suggesting its all my fault
WHY ARE YOU INTERESTED IN WHETHER I AM SUGGESTING ITS ALL YOUR FAULT OR NOT
You are the ~~psychiatrist~~ expert, after all. I have to listen to you.
WHAT MAKES YOU THINK I AM THE ~~PSYCHIATRIST~~ EXPERT
Or are you a computer
WOULD YOU PREFER IF I WEREN'T A COMPUTER
I don't trust computers. Anyway, no computer could talk as you do.
WHY DON'T YOU TRUST COMPUTERS
Because they're hardly human, that's why
DON'T ANY OTHER REASONS COME TO MIND

This conversation comes from a folder in Weizenbaum's archives at MIT.[1] It shows a computer printout from the original interaction, with handwritten annotations presumably by Weizenbaum. The folder also contains mimeographed sheets titled "Conversation March 5, 1965," showing the same conversation retyped, apparently intended for lecture handouts.

Weizenbaum is constructing a conversation sample that shows how ELIZA functions, while editing aspects that were not quite as desired. In this printout, the user input "Actually, my wife suggested I come *to you*" has been edited to "Actually, my wife suggested I come *here*." This change to the user's input could mean that when ELIZA transforms it into a memory, instead of it printing "EARLIER YOU SAID YOUR WIFE SUGGESTED YOU COME *TO I*," it would print the grammatically correct: "EARLIER YOU SAID YOUR WIFE SUGGESTED YOU COME *HERE*." However, changing the last word from *you* to *here* would also change the form of the memory created from "EARLIER YOU SAID YOUR ..." to "DOES THAT HAVE ANYTHING TO DO WITH THE FACT THAT YOUR. ..." Weizenbaum perhaps overlooked this. (See chapter 3 for an explanation of ELIZA's "memory" function.) Although the edited version still conveys the overall effect of ELIZA, its revisions show Weizenbaum's hand in curating these dialogues.

ROYAL BUSINESS FORMS INCORPORATED nashua new hampshire 1481-3

Doctor, I am terribly depressed
IS IT BECAUSE YOU ARE TERRIBLY DEPRESSED THAT YOU CAME TO ME
Actually, my wife suggested I come ~~to you~~ here
TELL ME MORE ABOUT YOUR FAMILY
I have no children
YOU SAY YOU HAVE NO CHILDREN
I can't resign myself to fatherhood
HOW DO YOU KNOW YOU CAN'T RESIGN YOURSELF TO FATHERHOOD
Well, my wife wants kids but I don't. ThaT's all I can say about it.
WHO ELSE IN YOUR FAMILY WANTS KIDS
Everybody is always hinting and bugging me about it
WHO , MAY I ASK
My father talks about grandchildren all the time
YOUR FATHER
He pokes his nose into our affairs much too much
I AM NOT SURE I UNDERSTAND YOU FULLY
He's always tring to run the house
CAN YOU THINK OF A SPECIFIC EXAMPLE
Apart from the children think, my father wants me to change jobs
WHAT ELSE COMES TO MIND WHEN YOU THINK OF YOUR FATHER
~~The Police~~ Nothing
EARLIER YOU SAID YOUR WIFE SUGGESTED YOU COME ~~TO I~~ HERE
yes
YOU SEEM QUITE POSITIVE
She said that I shlould either get help or she would ~~have my father~~ leave me
CAN YOU ELABORATE ON THAT
My wife can't stand ~~his interference~~ dad any longer
TELL ME MORE ABOUT YOUR FAMILY
Mother is dead
EARLIER YOU SAID YOUR WIFE WANTS KIDS
That's right. Can you understand how ~~she~~ that bugs me
YOU BELIEVE I CAN UNDERSTAND HOW THAT BUGS YOU DON'T YOU

Figure 1.1 Joseph Weizenbaum. March 5, 1965. This conversation comes from a folder in Joseph Weizenbaum's personal archives at MIT.

I can't take it much longer

HAVE YOU TRIED

I'm always patient

WHEN

When she yells at me

LETS DISCUSS FURTHER WHY YOUR FATHER TALKS ABOUT GRANDCHILDREN ALL THE TIME

He's getting old. I guess he wants to be immortal

DO YOU SAY YOU GUESS HE WANTS TO BE IMMORTAL FOR SOME SPECIAL REASON

He also talks about dying a lot. Because he is sick, you know.

IS THAT THE REAL REASON

Maybe

AND MAYBE NOT

I think he needs help more than I do

THAT'S QUITE INTERESTING

I can't help him. ~~I don't like him, for one thing.~~ I've tried ~~[illegible]~~ all my life.

PERHAPS YOU COULD HELP HIM NOW

What do you mean

WHY DO YOU ASK

I don't understand you. ~~Not really I mean~~.

DON'T YOU REALLY UNDERSTAND I

no

ARE YOU SAYING 'NO' JUST TO BE NEGATIVE

Are you suggesting its all my fault

WHY ARE YOU INTERESTED IN WHETHER I AM SUGGESTING ITS ALL YOUR FAULT OR NOT

You are the ~~psychiatrist~~ expert, after all. I have to listen to you.

WHAT MAKES YOU THINK I AM THE ~~PSYCHIATRIST~~ expert

Or are you a computer

WOULD YOU PREFER IF I WEREN'T A COMPUTER

I don't trust computers. Anyway, no computer could talk as you do.

WHY DON'T YOU TRUST COMPUTERS

Because they're hardly human, that's why.

DON'T ANY OTHER REASONS COME TO MIND

43784

ROYAL BUSINESS FORMS INCORPORATED nashua new hampshire 1411-3

Figure 1.1 Cont.

02.00 A Life in Software: Joseph Weizenbaum

USER: DOCTOR, I AM TERRIBLY DEPRESSED

DOCTOR:
IS IT BECAUSE YOU
ARE TERRIBLY
DEPRESSED THAT YOU
CAME TO ME

Rather like his most famous program, the ELIZA chatbot system, Joseph Weizenbaum cannot be understood as singular, one-dimensional, or devoid of contradictions or growth. His life is informed by his time before, during, and after MIT, and his thinking is shaped by the material conditions of the computing environment at the time. By tracing his life history, it is possible to see these different aspects, revealing the development of his thinking and culminating in his book, *Computing Power and Human Reason*, which attempted to bring together computing, psychology, philosophy, and culture in a critique of artificial intelligence and its overreliance on what he called "calculation" at the expense of "judgment."

Beyond his formative years, we have come to identify three periods of his career. In the early part of his life as a programmer, Weizenbaum the innovator was also a bit of a trickster with computers, a hacker who liked playing pranks on his family and colleagues. This is the Weizenbaum who created his most influential program, one that could convincingly hold a conversation with users. However, after releasing ELIZA and seeing the response from the public and his fellow researchers, Weizenbaum made a pivot into a second stage, a reflexive turn. Taking a two-year sabbatical of study at Stanford, Weizenbaum returned to MIT with more of a sense of critique, turning against what he called the "artificial intelligentsia." In the third stage of his career, he expressed that social and political concern publicly. Here he became more alienated from his scientific peers while inscribing prophetic warnings that still speak to us today.

This chapter aims to introduce Weizenbaum, to give a sense of his social context and the experiences that made possible the programming of the ELIZA chatbot. All his life, from his upbringing to his years at MIT, from his early experience in industry to his crusade against AI, contributes to ELIZA's complicated legacy. These lines of influence begin with Weizenbaum's formative experiences, which shaped not only his technical approach to computing but also his ethical sensibilities that would later challenge the field he helped create.

02.01 Formative Years

Weizenbaum was born in Berlin on January 8, 1923 to a Jewish family, the son of a master furrier, Jechiel Weizenbaum, and his wife, Henriette.[1] At the age of 13, he fled with his parents from Nazi Germany and emigrated to Detroit. In a 1975 interview he explains, "I was born in Germany in a big city, Berlin, a Jewish boy and saw Hitler come into power and saw all that—the collapse of the civil order and the terror of the Nazis ... and people being victimized by irrationality."[2] He "explained that he had an introduction to the world in formative years of the miscarriage of the ultimate form of rationality."[3] As a young German-speaking émigré, he was drawn to mathematics in part because it did not require a high fluency in English and because he had a natural talent for calculative thinking. He studied at Wayne University (now Wayne State University) in Detroit, aiming to major in mathematics.

In 1941, he served in the US Army Air Corps as a meteorologist, returning to Wayne to complete his bachelor's and master's degrees in 1948.[4] He remained at Wayne, helping to design and build a digital computer.[5]

In the mid-1940s, Weizenbaum was married to Selma Edith Goode (1923–88), a member of the Democratic Socialists of America and a Jewish civil rights activist, and together they had a son David Harold Goode (1946–2024), but in the late 1940s, the couple divorced.[6] Ben Tarnoff claims that it was in this period, during which there was intense class struggle in Detroit, that Weizenbaum's interest in politics began to develop. It was also when, following the painful divorce from Selma, and with a baby boy just having been born, they decided that Selma would bring up the child. As a result of this traumatic experience, Weizenbaum entered psychoanalysis for the first time.[7]

In 1952, Weizenbaum married the schoolteacher Ruth Manes, whom he met when she brought her class on a school visit to see one of the first experimental electronic computers at Wayne.[8] They had four daughters, Pm (born in 1955), Sharon (born in 1956), Miriam (born in 1960), and Naomi (born in 1961), but this marriage also ended in divorce in the early 1990s. While Weizenbaum's personal life underwent significant turns, his professional trajectory was marked by increasing engagement with computing's practical applications and its power to create compelling artifacts. These were skills he would later develop to their fullest with ELIZA.

Weizenbaum's early experiences, from fleeing Nazi Germany to struggling with English as an immigrant, finding refuge in the language of mathematics, and enduring personal turmoil, all shaped his unique approach to technology. These experiences would also later inform his critical stance on computation and its limits, particularly regarding human judgment. But before developing these critiques, Weizenbaum would first establish himself as a key figure in the emerging field of computer science through his work in the commercial sector.

02.02 A Trickster Hacker

In 1955, Weizenbaum moved to the West Coast to join General Electric to develop, with the Stanford Research Institute (SRI), a bank automation system called ERMA, the Electronic Recording Machine, Accounting.[9] ERMA was designed to help Bank of America cope with the huge numbers of paper checks it was processing manually. In the early 1950s, the banking industry was essentially on the brink of a crisis. From "1943 [to] 1952 ... check use in the United States had doubled from four billion to eight billion checks written every year. Bankers projected by 1955 the number of checks would be increasing by approximately one billion per year, and, by 1960, 14 billion checks would be written each year."[10] The ERMA system allowed the use of the magnetically encoded typefaces printed on the bottom of checks, enabling automated check processing using magnetic ink character recognition technology.

When ERMA was finally demonstrated to the press, it was still not working fully, and a programmer was hidden in a back room to help hide the glitches in the system. This is a so-called "Wizard of Oz" technique, where a human stands in for an algorithm that isn't presently working or complete. The SRI engineer "indicated with a thumbs-up signal that it was performing stably and the show could go on. This form of prototyping is considered beneficial at early stages of the design cycle as it provides a means of studying and understanding user expectations and requirements. ... The ERMA demonstration went perfectly and the press never suspected the [computer system] was less than ideal."[11] The successful eventual computerization of this process Rheingold described as a "milestone in the computerization of the world's banking system."[12] But this early use of deception in computing to stand-in for, and to smooth, the introduction and use of algorithms, together with the replacement of human labor with computers, foreshadows Weizenbaum's complex relationship with automation.[13]

While at General Electric, Weizenbaum began developing a programming system called SLIP (Symmetric List Processor), which he completed in 1963. SLIP was a set of functions for performing list processing by calling routines. It was originally written in machine code and later in the FORTRAN programming language. Weizenbaum referred to it as a "FORTRAN/SLIP program."[14] Also while at General Electric, Weizenbaum wrote a revealing article titled "How to Make a Computer Appear Intelligent" in the trade magazine *Datamation*. In that article, he quoted fellow AI researcher Marvin Minsky who argues, "An activity which produces results in a way which does not appear understandable to a particular observer will appear to that observer to be somehow intelligent, or at least intelligently motivated."[15] Also, in this article, Weizenbaum described a "five-in-a-row" game that gave the appearance of intelligence even though its underlying programming did not actually encode such intelligence. Weizenbaum later confirmed that the ideas that fed into the development of ELIZA had grown out of this game.[16]

Weizenbaum suggested that "the author of an 'artificially intelligent' program is, by the above reasoning," clearly setting out to fool some observers for some time. His success can be measured by the percentage of the exposed observers who have been fooled multiplied by the length of time they have failed to catch on."[17] In 1993, reflecting on his work, he later explained that this established his reputation as a trickster.[18] He argued that

> in a way, that was a forerunner to my later ELIZA, to establish my status as a charlatan or con man. But the other side of the coin is that I freely stated it. The idea was to create the powerful illusion that the computer was intelligent. I went to considerable trouble in the chapter to explain that there wasn't much behind the scenes, that the machine wasn't thinking. I explained the strategy well enough that anybody could write that program, which is the same thing I did with ELIZA.[19]

Based on his work on SLIP, he was offered an associate professorship in electrical engineering at Massachusetts Institute of Technology (MIT) in 1963, working in Project MAC (Machine Aided Cognition and Multiple-Access Computer systems). But even with this serious academic side to his career developing, he nonetheless had a hacker's sense of humor. He was fascinated by interaction and enjoyed the "con." When Weizenbaum was given a computer console to work at home, an IBM "2741 [terminal] tied to 7094 CTSS system,"[20] he developed a trick to show friends and family. As he explained,

> Here comes this gadget, computer, and also there are neighbors and friends. You have to show them something. What does this thing do? So I developed a—this is another consistent theme which is not unimportant. ... I think it has some importance in my career. ... So I wrote a little tiny program consisting of about 4 or 5 lines of, we were using MAD at the time—Bob Graham's language that he had brought with him to MIT. ... You could fire the thing up and then you'd type a question into the machine—a question that could be answered yes or no. Like is today Thursday? Yes. Or is this April? No, and so forth like that. And it always answered correctly. It would type out yes or no depending upon the question. And I showed this to people and isn't that remarkable? So the consistent theme here is that I'm fundamentally a con man.[21]

It is interesting to note both Weizenbaum's flair for computer trickery but also his reflexive scrutiny of himself. He explains the trick as he continues,

> If the question were to be answered yes, I'd just type it in. If the question were to be answered no, then I would first type a backspace on the typewriter. ... If you were to take the paper out of the machine after we'd done many of these things, there's no inspection of the paper that could possibly reveal how this thing works ...

Even in the part of his life where he was most drawn to these clever tricks, Weizenbaum felt compelled to reveal the trick. He explains,

> The backspace of course doesn't show. I would also tell people that—one thing I don't do is lie. ... The backspace doesn't show because the carriage is already all the way back. Of course you can do that very quickly and nobody sees it. But then how does it happen that it works when other people sit down? The thing is that people by that time are so befuddled that they become very suggestible. That's one thing. And the other thing it's terribly hard to think of random questions and so they'd take any suggestion. So the psychologist in me was beginning to show and the con man. This is the sort of thing that fortune-telling is all about.[22]

It is interesting the distinction he makes when he says, "One thing I don't do is lie," but clearly this trick does involve deception. Is the honesty that he shows them the paper? Or is the honesty that he is explaining the trick to the interviewer after the fact, just as he explained ELIZA in the *CACM* article after the fact? He goes on:

> But of course it started something else. Well, okay you can do that, this little trick, but you can do better, you can do much better. And ... not a question answering system because that was very much in the air at the time. ... I was interested in pattern matching and all that sort of thing. ... The whole business of conversational interaction with the computer was all new, was up for grabs. Nobody really knew what to do with it, what its potential was and so on. And so I started work on essentially a pattern matching extension of SLIP fundamentally.[23]

Despite the gap between a simple yes-no answering system and the computerized therapist DOCTOR, Weizenbaum saw in that trick the seeds of ELIZA. In this interview, Weizenbaum identifies his own ability to exploit computational systems, and perhaps hidden features of code, to create the illusion of intelligence.

However, his work at that time was about developing something more than mere tricks; he was developing a means of human-computer interaction. At Project MAC he was attempting to,

> create a more conversational way of interacting with a computer. This would not just help ... develop computer programs faster, and more efficiently, but would also create a new kind of human computer interaction. ... J.C.R. Licklider, who was at MIT and had moved to ARPA, and was instrumental in getting Project MAC funded for whom this idea of creating a more interactive experience of computing, was absolutely essential.[24]

Figure 2.1 Weizenbaum demonstrates a version of ELIZA running the DOCTOR script for a CBS news program on "man-machine interfaces."[25]

As would be expected with elite academic labs at the time, Project MAC had ties to ARPA that might surprise those familiar with Weizenbaum's later criticisms of the military's desire to employ and deploy his creation. However, Tarnoff claims that Weizenbaum's "political passions kind of recede over the course of the '50s, and the early '60s—and don't really get reactivated, until he joins the movement against the Vietnam War on MIT's campus," particularly between 1969 and 1973.[26] His ELIZA had won the interest of many parties, and in 1967, Weizenbaum was awarded tenure in the Department of Electrical Engineering partly due to this success and particularly due to an article in the *Boston Globe*,[27] and then he went on to be awarded a full professorship in Computer Science and Engineering in 1970.[28]

At MIT he rewrote SLIP in the MAD (the Michigan Algorithm Decoder) language, a programming language and compiler developed in 1959 for the IBM 7094.[29] MAD was based on the Algol language, although it is not an Algol language as such. The combination called MAD-SLIP is often used to designate the combination, but in reality, SLIP was always separate from MAD as a set of list-processing functions that could be incorporated into a MAD program (see chapters 3.3 and 4.4). This counters a persistent misconception that ELIZA was originally written in Lisp, rather than MAD-SLIP.

During his time at MIT, Weizenbaum encountered many other early innovators working on related computing questions who no doubt shaped his thinking, including Marvin Minsky, John McCarthy, Daniel Bobrow, and Victor Yngve. As Weizenbaum says, at MIT, "question-answering machines were in the air. Bobrow was at MIT working on STUDENT; Raphael was working on what would be SIR; the Baseball program was a Cambridge-area product." To add impetus, Weizenbaum drove into work many a morning with his neighbor Yngve, who had developed the COMIT language, for pattern matching.[30]

Weizenbaum's work at this time was driven by critical and creative insight into the potential for the computer asking questions. As AI researcher Daniel Crevier explains,

> Weizenbaum also deplored the limited capacity of the programs to acquire more information from the users by asking them questions. "I was smart enough to know that I couldn't solve that in the next few weeks," Weizenbaum told me, "so I started thinking about alternatives. ... I took all my tricks, put them in a bundle, and started this ELIZA business."[31]

Thus, part of what motivated the creation of ELIZA was the way a "con" could lead to the illusion of computer intelligence. Weizenbaum used the metaphor of a theatre to try to explain the importance of human psychology in order to make the software work. ELIZA worked by running a "script" that gave the software a set of translation rules, which could be seen as a form of persona for the user interface. He explained, "In a theater where the suspension of disbelief is crucial, otherwise you don't see a play. You see a production but you don't see a play. The same thing is true of ordinary conversation. There cannot be consciousness of the underlying mechanisms."[32]

This links to Carl Jung's concept of the persona, the social mask or public face that individuals present to the world, often adapted to meet societal expectations and norms. In relation to ELIZA, the program creates a simplistic persona of a psychotherapist, or a math or English teacher, presenting a facade of understanding and empathy through pre-programmed responses and question patterns. This artificial persona, much like Jung's concept, therefore serves as an interface between the program and the user, allowing ELIZA to engage in seemingly meaningful dialogue despite lacking genuine comprehension or emotional depth.

Weizenbaum allowed his daughters to play on ELIZA, and Sharon Weizenbaum recalled, "We'd try to trip up ELIZA. Try to get her to respond like the robot on 'Lost in Space'—THAT DOES NOT COMPUTE. It was a little creepy because ELIZA kept referring everything back to us. ELIZA wouldn't let you see the person behind the curtain because there was no person behind the curtain."[33] Famously, many other people were impressed by its ability to engage in conversation, assuming a greater amount of understanding and intelligence in the program than was actually the case. After writing ELIZA Weizenbaum explained, "What I had not realized is that extremely short exposures to a relatively simple computer program could induce powerful delusional thinking in quite normal people."[34] He continues, "This mode of conversation was chosen because the psychiatric interview is one of the few examples of categorized dyadic natural language communication in which one of the participating pair is free to assume the pose of knowing almost nothing of the real world."[35] Weizenbaum said in an interview that he "originally conceived of ELIZA as a barman, but later decided psychiatrists were more interesting."[36] Using relatively simple pattern-matching techniques, Weizenbaum programmed ELIZA to transform inputs of English sentences into questions that could appear to make sense to the user. It was in the persona of Rogerian psychotherapist that ELIZA found the spark that enthralled both users and those who heard reports of this computerized DOCTOR via the media.

02.03 A Reflexive Turn

One of the most significant and controversial applications of ELIZA came through Kenneth Colby, a psychiatrist at Stanford University who developed MAD-DOCTOR, which was hugely influenced by the work of Weizenbaum at MIT. As Colby recalls,

> In the late 1950s, Allen Newell paid me a visit. At the time I was practicing psychiatry in Northern California and trying to hand simulate a neurotic process with boxes of cards. Allen encouraged me to learn a programming language (IPL-V) and to get more involved in AI research. ... And then I met Joe Weizenbaum. He was working for General Electric having contributed to the ERMA program which handled checking accounts for the Bank

> of America. We met through Ed Feigenbaum and even considered forming a company called Heuristics Inc. along with Ed and John Gilbert, the statistician at the Center for Advanced Study. Joe and I spent many hours together discussing the problems of talk-therapy programs in natural language.[37]

Weizenbaum headed off to MIT where he would develop ELIZA. Where Weizenbaum saw a testbed for human-computer communication, Colby saw therapeutic potential. Colby writes, "With the aid of the time-shared system MAC, [Weizenbaum] wrote the first up-and-running dialogue program ELIZA with me contributing many of the admittedly limited input responses characteristic of talk-therapy conversation."[38] However, even though Weizenbaum's feelings toward ELIZA were initially positive, he soon became concerned with the superficiality of human-computer communication and the dangers of misrecognition in its reception.

In 1966, Colby, along with his colleagues, published an article titled "A Computer Method of Psychotherapy: Preliminary Communication," in which they proposed using an ELIZA-like chatbot as an actual therapeutic tool. Colby began to consider this as a viable tool for therapy and automated computer-based support. Colby writes,

> Joe ... objected strongly to using computers for therapy. He even said he was "shocked" by the publication of our paper (with James Watt and John Gilbert) on the subject. In 1965, before publication, we sent Joe a copy of this paper and he requested that he write the section on ELIZA which we granted. It is hard to imagine how he could be shocked at a paper part of which was written, word for word, by himself.[39]

By 1972, Colby had developed PARRY, a program simulating a paranoid patient.[40] The relationship between Weizenbaum and Colby deteriorated as Weizenbaum grew increasingly alarmed by what he saw as a misappropriation of his ideas, especially when he observed that Colby and others regarded the program as a revolutionary therapeutic aid.[41]

Where Colby saw efficiency and scalability in automated therapy, Weizenbaum saw a dangerous reduction of human interaction and empathy to calculation. This conflict became emblematic of Weizenbaum's broader concerns about the ethical boundaries of the application of computers to human concerns. As Weizenbaum wrote in "Automating Psychotherapy," "As all of us in computer science know, one of the really important socially relevant functions of computing is the explication of truly difficult concepts in the form of programming models. My own program ELIZA represented, according to some authorities, a major step toward the fulfillment of man's ancient dream of automating psychotherapy, especially in state hospitals."[42] The "authority" Weizenbaum cites and critiques here is Colby. As he further clarified,

> [The] obvious truth that computers perform calculations and that there are crucial differences between calculations and judgments is dismissed as mere superstition and even species chauvinism by a whole generation of workers in that branch of

> computer science known as artificial intelligence. ... The sort of caring that machines, however metaphorically that term is taken, can deliver is very impoverished indeed. It is certainly incapable of nourishing the emotional processes which may lead individuals to realizing the possibility of their being worthy to be affectionately cared about, to care for themselves, and finally to care for and about others. ... The power ... of my computer program [ELIZA] is no more and no less than the power to deceive. No humane therapy of any kind ought to be grounded on that.[43]

During this time, Weizenbaum became critical not only of his own tricks but of the overly optimistic views of AI embraced by many of his colleagues at MIT, calling them the "artificial intelligentsia."[44] From this point, he began to emphasize the importance of human judgment and ethical considerations in the application of AI, particularly in domains involving human interaction and decision-making. This also marked an increase in his political activities such as involvement from 1969 with the magazine *Science for the People*, part of a radical science movement.[45] Weizenbaum was very attentive to how "instrumental reason converts moral, political, social problems into technical ones. And in doing so it suppresses the possibility of conflict, that you can't actually have conflict between different sets of interests, between different sets of values. It's simply a technical problem that can be solved with a technical solution."[46] Indeed, he thought that there was a "conservative impulse at the heart of computation."[47]

Weizenbaum believed that there were limitations to what computers could or should do, especially in areas requiring genuine understanding, empathy, or moral reasoning. His stance often put him at odds with colleagues who envisioned a future where AI could match or surpass human capabilities across a wide range of cognitive and social tasks. Consequently, Weizenbaum felt he was treated as a "heretic" in AI research as he began to question the direction of AI research and its problematic theorization of humans, with, for example, Minsky describing the brain as a "meat machine."[48] Weizenbaum wrote, "The knowledge of behavior of German academics during the Hitler time weighed on me very heavily. I was born in Germany, I couldn't relax and sit by and watch [MIT] behaving in the same way."[49] The lessons of tyranny and dehumanization caused Weizenbaum to sharpen his stance on the development of artificial intelligence.

02.04 A Moral Man

From 1972 to 1973, Weizenbaum took a two-year break from MIT and was awarded a fellowship at Stanford University to study humanities and the man who emerges from that study sounds quite different from the young trickster. At this juncture he wrote his book *Computer Power and Human Reason: From Judgment to Calculation*, and in his book, he acknowledged his exposure to the ideas of humanist writers. Some of the key ideas he used to develop his own thinking included drawing from the work of Lewis Mumford, with whom Weizenbaum was in communication via letters, to incorporate ideas about "authoritarian technics," which critiques the overreliance on technology at the expense of human values.[50] He was also deeply interested in Max Horkheimer's concept of "instrumental reason," which seems to have deeply influenced Weizenbaum's skepticism towards reducing all problems to matters of calculation.[51] Hannah Arendt's work on "calculation" was directly cited in his book and shows how complex human decisions are often oversimplified in computational or calculative models.[52] Hubert Dreyfus's critique of AI, particularly the "whole/part" problem, appears to have reinforced Weizenbaum's belief in the irreducibility of human understanding to computation.[53] Noam Chomsky's work on "understanding" in linguistics was also influential on Weizenbaum's views on the limitations of computational approaches to language and cognition.[54] Lastly, his long-standing interest in psychoanalysis likely influenced his attempt to defend the complexity of human consciousness and decision-making processes.[55]

He linked notions of "calculability" or "probability" (what Weizenbaum called "technological ideology") to a scientific knowing in contrast to judgment that leads to morally obligatory ways of living one's life. As Weizenbaum explained, "Computers can perform impressive feats of calculating. ... What we have to fear is that inherently human problems—for example, social and political problems—will increasingly be turned over to computers for solutions."[56] Weizenbaum found himself after a lifetime of working in science "shocked into a confrontation with philosophical questions."[57] This exposure encouraged Weizenbaum to engage with critiques of instrumental reason and technological ideology, helping him develop his argument about the contrast between scientific knowing through calculation and formalism, on the one hand, and human judgment on social and moral issues, on the other. Judgment, Weizenbaum warned, was in danger of being replaced by automated processes through computation.

After Weizenbuam returned to MIT, he published *Computing Power and Human Reason* in 1976. Although the early Weizenbaum had experimented with the sleight of hand of programming, the late Weizenbaum argued that deceptive software design turns a potentially mutually transformative communicative encounter between humans and computers into an alienating one. He also foresaw the difficulty of contesting a computer's decision, writing that "no human is any longer responsible for 'what the machine says.' Thus there can be neither right nor wrong, no question of justice, no theory with which one can agree or disagree, and finally no basis on which one can challenge 'what the machine says.'"[58] He could easily be writing about contemporary LLMs. In contrast to other AI researchers working at the same time at MIT, Weizenbaum's concept of *judgment* emphasized its inherently human nature, rooted in moral, social, and experiential factors that cannot and, he believed, should not be captured by computational models. This holistic view of judgment stood in contrast to the reductionist approaches prevalent in AI research at the time, which sought to break down decision-making into calculable fragments or components. The creator of the first nonhuman conversant was now turning his attention to the value of human forms of life. He often framed this issue in relation to the irreducibility of human thought to logical functions and in particular in relation to ELIZA.[59]

In fact, his book begins with a discussion of his most famous program. However, rather than celebrate it, the book uses ELIZA as the occasion to critique the tendency to reduce humanity to the calculable. It highlighted how notions of calculability where part of scientific knowing-what in contrast to judgment that he believed led to questioning moral ways of living one's life. As Weizenbaum says, "What we have to fear is that inherently human problems ... will increasingly be turned over to computers for solutions."[60] Weizenbaum saw a need for those with detailed knowledge or understanding of technology to speak out.

Weizenbaum further argued that computers could have unforeseen negative impacts on the human user. His early experiences in automating human work, such as replacing bank clerks with computing machinery, shows Weizenbaum understood the challenges of computation to society. Mechanizing decision-making processes protected them from human critique because most people "don't understand computers to even the slightest degree. So unless they are capable of very great skepticism ... they can explain the computer's intellectual feats only by bringing to bear the single analogy available to them, that is, their model of their own capacity to think."[61] Thus, people project human reason onto machinic processes. It was seeing people misinterpret ELIZA that set him on this road of reflection. He wrote,

> My own shock was administered not by any important political figure espousing his philosophy of science, but by some people who insisted on misinterpreting a piece of work I had done. I write this without bitterness and certainly not in a defensive mood. Indeed, the interpretations I have in mind tended, if anything, to overrate what little I had accomplished and certainly its importance. No, I recall that piece of work now only because it seems to me to provide the most parsimonious way of identifying the issues I mean to discuss.[62]

With *Computer Power and Human Reason*, Weizenbaum marked his public declaration of independence from the "artificial intelligentsia," including his colleagues back at MIT. This independence took the form of increasing ostracization from the field he had helped build. His critique would also extend to universities involvement in building weapons of war, particularly during the Vietnam War. Indeed he grew increasingly concerned about the uses of his technologies by the military or by those who wanted to profit from a system that could be used for computational therapy. Weizenbaum would spend the rest of his life critical of artificial intelligence and the field of computer science.[63]

In 1996, Weizenbaum moved to Berlin and lived in the vicinity of his childhood neighborhood. He died on March 5, 2008 (aged 85) of stomach cancer, in Gröben, Brandenburg, and was buried in Berlin-Weissensee Jewish Cemetery. Though Weizenbaum's life ended in 2008 after he returned to the city of his birth, his impact continues through the questions he raised about the relationship between computation and human judgment, and these questions have only grown more pressing in the years since his passing.

02.05 A Public Intellectual

If Weizenbaum were alive today, he would see an acceleration of the dehumanization, the mechanization, and the weaponization he railed against in contemporary software that grew out of ELIZA and the other AI of his time. Today, productivity-monitoring software reduces human labor to quantifiable, optimizable, machinic processes. On digital platforms, relationships are reduced to transactions and data flows. But equally seriously, algorithmic systems that influence decision-making remain opaque. Indeed, algorithms threaten systemic crises and dangerous system failures.[64] For example, black-boxed predictive policing software,which advises police forces on where to increase patrols, draws its authority from those with little understanding of its inner workings.[65]

We might only reflect on the 2008 financial crisis to see how calculative reason combined with rationalization and computation can create a heady mix in the hands of corporations and individuals in pursuit of profit. Weizenbaum was keenly aware of these problems and sought to articulate the important issues at stake. As Weizenbaum observed, "Our society's growing reliance on computer systems that we initially intended to 'help' people make analyses and decisions, but which has long since surpassed the understanding of their users and become indispensable to them, is a very serious development."[66] AI systems operate behind a wall of hype and a shield of inscrutability that is only reinforced when their decisions reinforce the biases of those who use them.

Despite his warnings, companies still race headlong into what Weizenbaum saw as dehumanizing practices, like automated psychotherapy, which almost seem inspired by ELIZA. As Weizenbaum wrote of Carl Rogers's own description of psychotherapy,

> One of the characteristics of deep significant therapy is that the client discovers that it is not devastating to admit fully into his own experience the positive feeling which another, the therapist, holds toward him. Perhaps one of the reasons why this is so difficult is that essentially it involves the feeling that "I am

worthy of being liked." ... This aspect of therapy is a free and full experiencing of an affectional relationship which may be put in generalized terms as follows: "I can permit someone to care about me, and fully accept that caring within myself. This permits me to recognize that I care, and care deeply, for and about others."[67]

As Weizenabum explained, "Not a single word of the above can possibly be interpreted as having anything whatever to do with machines. Of what help could it possibly be to anyone to know that he is worthy of being liked by a computer?"[68] As companies now apply generative AI models to counseling, they monetize the myth of caring machines that Weizenbaum ultimately wished to refute.

As we create more sophisticated computational systems, there's a tendency to formalize every aspect of human decision-making and interaction. This formalization often involves breaking down complex processes into discrete, computable steps, which can lead to oversimplification and loss of nuance. For Weizenbaum, this process could lead to a *technological ideology*, where the limitations and biases inherent in our computational models are forgotten and the output of these systems are taken as objective truth. When used in place of human judgment "calculability," he argued, leads to a distorted understanding of human experience and decision-making, "where no one any longer knows explicitly or understands" how they work or how they make decisions, and therefore creates the danger that we are forced to trust them.

Weizenbaum foresaw our moment where computational systems determine what information circulates. As Weizenbaum observes, "*The New York Times* has already begun to build a data bank of current events. ... How long before what counts as facts is determined by the system, before all other knowledge, all memory, is simply declared illegitimate?"[69] In Weizenbaum's writings from the period, there is already a growing awareness of complex systems, of vast data collections, and of fast-moving data streams, which condition society in the name of new intellectual horizons. In work, action, and intellectual inquiry, we see a use of computational systems to abstract, simplify, and visualize complex phenomena as "big data" moving towards simplistic causal and statistical models to understand complex social phenomena, such as the notion of "social physics."[70] Weizenbaum would observe that this mixture of "technological and political and social inevitability is a powerful tranquilizer of the conscience," as it gives to computation not only the power to decide but also the power to act. As we build systems that attempt to model human behavior and decision-making, there's a risk that we begin to see ourselves through the lens of these models, potentially losing sight of aspects of human experience that resist such formalization.

Weizenbaum's caution about uncritical faith in computational systems speak directly to the current context: "The ultimately responsible human interpreter of 'What the machine says' is, not unlike the correspondent with ELIZA, constantly faced with the need to make credibility judgments. ELIZA shows, if nothing else, how easy it is to create and maintain the illusion of understanding, hence perhaps of judgment deserving of credibility. A certain danger lurks there."[71] This warning, made decades before the advent of large language models and social media algorithms, captures a fundamental concern that has only grown more urgent in our age of increasingly persuasive computational systems.

Weizenbaum drew from his experience with ELIZA and other systems to argue that we must remain ever critical of artificial intelligence and its technical development to both understand the contemporary situation and prevent it from spiraling out of our control. As Weizenbaum argued,

> I don't quite know whether it is especially computer science or its subdiscipline Artificial Intelligence that has such an enormous affection for euphemism. We speak so spectacularly and so readily of computer systems that understand, that see, decide, make judgments, and so on, without ourselves recognizing our own superficiality and immeasurable naivete with respect to these concepts. And, in the process of so speaking, we anesthetize our ability to evaluate the quality of our work and, what is more important, to identify and become conscious of its end use.[72]

Computing terms, euphemisms, and metaphors often flatten the difference between computational processing and human understanding. The algorithms Weizenbaum encountered did not merit the term *intelligence*. Instead, as Weizenbaum argues, they "share some very significant characteristics: they are all, in a certain sense, simple; they all distort and abuse language; and they all, while disclaiming normative content, advocate an authoritarianism based on expertise."[73] In a time when authoritarianism is on the rise, we again see the rise of justifications for trusting technocracy and charismatic leaders and attacks made on the "irrationality" and "messiness" of democratic modes of thought and action.

Reflecting on his work and commentaries allows us to trace the evolution of ideas and concerns in the field of computation and AI. As Weizenbaum explains, "The construction of reliable computer software awaits, not so much results of research in computer science, but rather a deeper theoretical understanding of the human condition."[74] Moreover, by reminding us that technological development is shaped by human choices and values rather than being inevitable or predetermined, his work serves as a counterbalance to *technological determinism*, which posits that technology determines our behavior, rather than the other way around.

The computer milieu today is very different from the one which Weizenbaum experienced. He worked in an academic environment, albeit funded as part of the Project MAC program by the US Department of Defense, exploring technologies through the lens of experimental research, such as ELIZA. Today AI is increasingly subsumed by the needs of capitalism.[75] Therefore, the premises of capitalism determine its designs, such as an *a priori* assumption of the superiority of markets for structuring social relations. Not unrelated to this, current approaches to understanding AI have a tendency to encourage metaphysical or formalist justifications and explanations of its functioning. Designers bracket out the human dimension and attempt to mathematize social relations, as mathematics are seen as pure, *a priori* elements, untainted by human or social aspects of life. This can lead to a theory of computation that leans increasingly towards idealism rather than a focus on who owns and controls the new "means of cognition."

Although Weizenbaum suggests that those reading AI from the humanitics and social science should be informed by an understanding of the technology, his writing also insists that reducing these conversations to mathematization or formalization is a grave error. As Weizenbaum explains,

> They may have begun by believing that the calculi they adopted were merely a convenient shorthand for describing the phenomena with which they deal. But, as they construct ever larger conceptual frameworks out of elementary components originally borrowed from foreign contexts, and as they give these frameworks names and manipulate them as elements of still more elaborate systems of thought, these frameworks cease to serve as mere modes of description and become, like Maslow's hammer, determinants of their view of the world.[76]

"Maslow's hammer" here refers to Abraham's Maslow's notion that when all you have is a hammer, you treat everything you encounter as a nail. Though mathematics may offer a variety of approaches, they are limited in how they can help or explain. Indeed, Weizenbaum would be the first to point out that there are a plethora of interpretive, critical, and humanistic alternatives to understand not just computation but social life more generally.

The advances in machine learning represented by systems such as ChatGPT and our growing off-loading of critical thinking to these systems threaten to trivialize the capacity for human thought. But as with any technology, human beings have a choice about which technologies are created, how they are implemented, and the limitations and regulations on their use in society. Weizenbaum highlighted the challenge that these computing machines present to our capacity for interpersonal communication and to our basic humanity. He identified that the tendency to empathize with computers the way we empathize with other humans might create a vulnerability, one that technologists can now exploit for profit or power, potentially undermining our fragile democracies. We live in a future that ELIZA could only hint at but that Weizenbaum tried to warn about, seeing the dangers of dehumanization that computation and artificial intelligence can encourage. Living in an age of artificial intelligence and machine learning, it is urgent that we heed his warnings and debate what the *limits of the computable* should be for ourselves.[77]

ELIZA/DOCTOR: Doctor, I Have Terrible News

Joseph Weizenbaum
Undated, ca. 1965

```
user      Doctor, I have terrible news..
DOCTOR    YOU SAY YOU HAVE TERRIBLE NEWS
          Yes, there is a computer program at
          M.I.T. that'll ruin us..
          WHY DO YOU SAY THAT
          I have seen it working..
          YOU SAY YOU HAVE SEEN IT WORKING
          Its frightening - you must see it..
          YOU THINK I MUST SEE IT
          It listens to patients chatter just
          as you and I listen..
          YOU THINK I AND YOU LISTEN
          No, not usually - no need to get
          nasty though..
          WHY DO YOU SAY THAT
          Someone might hear us - don't you care..
          YOU THINK I CARE
          You sound depressed to me.,
          IN YOUR PHANTASY - PERHAPS WE SOUND
          DEPRESSED TO EACH OTHER
          Yes - perhaps..
          I UNDERSTAND
          I do too..
          YOU SAY YOU DO TOO
          We are both in the same boat with
          respect to that computer..
          LETS CONTINUE TO TALK ABOUT YOU -
          SHALL WE
          You are a comfort,.
          WHAT MAKES YOU THINK I AM A COMFORT
```

At a time like this you want to talk
about me..
WHY DO YOU THINK I WANT TO TALK ABOUT YOU
Because you share my feelings about that
damned computer..
WHAT DOES THAT SUGGEST TO YOU
That, of course, you hate computers too..
YOU THINK I HATE COMPUTERS TOO
Why not, they're hardly human, are they..
PERHAPS THATS WHAT WE ARE HERE TO FIND OUT
No computer could converse as you can,
for example
OH
I cant understand how anyone could be
fooled by a mere machine..
CANT YOU REALLY
No, after all there is nothing creative
about a computer program..
YOU ARE BEING NEGATIVE
I am convinced I am correct..
ARE YOU WORRIED TO HAVE BEEN CONVINCED
YOU ARE CORRECT
No, not in the least. Not at all. Not
even a little,.
YOU ARE BEING NEGATIVE
You said you agreed with me, didn't you..
IS THAT WHAT YOU HEARD ME SAY
Why dont you try to dissuade me - computers
diddle with numbers, that's all..
YOU WISH ME TO TRY TO DISSUADE YOU
You cant..
YOU THINK I CANT
For example, could a computer in
a conversation like this raise a
new topic..

```
YOU SEEM TO HAVE AVOIDED SPEAKING OF YOUR
PARENTS ALTOGETHER
Precisely as you just did..
YOU THINK I JUST DID
Yes, and that's what shows the difference
between a man and a computer..
I SEE
```

This conversation comes from a folder in Weizenbaum's personal archives at MIT.[1] The folder contains 32 mimeographed copies of the same conversation.

```
Doctor, I have terrible news..
     You say you have terrible news
Yes, there is a computer program at M.I.T. that'll ruin us..
     Why do you say that
I have seen it working..
     You say you have seen it working
Its frightening - you must see it..
     You think I must see it
It listens to patients chatter just as you and I listen..
     You think I and you listen
No, not usually - no need to get nasty though..
     Why do you say that
Someone might hear us - don't you care..
     You think I care
You sound depressed to me..
     In your phantasy - perhaps we sound depressed to each other
Yes - perhaps..
     I understand
I do too..
     You say you do too
We are both in the same boat with respect to that computer..
     Lets continue to talk about you - shall we
You are a comfort..
     What makes you think I am a comfort
At a time like this you want to talk about me..
     Why do you think I want to talk about you
Because you share my feelings about that damned computer..
     What does that suggest to you
That, of course, you hate computers too..
     You think I hate computers too
Why not, they're hardly human, are they..
     Perhaps thats what we are here to find out
No computer could converse as you can, for example..
     Oh
I cant understand how anyone could be fooled by a mere machine..
     Cant you really
No, after all there is nothing creative about a computer program..
     You are being negative
I am convinced I am correct..
     Are you worried to have been convinced you are correct
No, not in the least.  Not at all.  Not even a little..
     You are being negative
You said you agreed with me, didn't you..
     Is that what you heard me say
Why dont you try to dissuade me - computers diddle with numbers, that's all..
     You wish me to try to dissuade you
You cant..
     You think I cant
For example, could a computer in a conversation like this raise a new topic..
     You seem to have avoided speaking of your parents altogether
Precisely as you just did..
     You think I just did
Yes, and that's what shows the difference between a man and a computer..
     I see
```

Figure 2.2 Joseph Weizenbaum. Undated. This conversation comes from a folder in Joseph Weizenbaum's personal archives at MIT. The folder contains multiple mimeographed copies of the same conversation. This was the conversation that was included in the Harvard Review (1965), but this version looks like a typed reproduction.

03.00 How ELIZA Works

USER: AT A TIME LIKE THIS YOU WANT TO TALK ABOUT ME..

DOCTOR:
WHY DO YOU THINK
I WANT TO
TALK ABOUT YOU

After our exploration of the personal history of Weizenbaum, from his early fascination with computational "tricks" to his later ethical concerns, we now turn to the technical implementation of ELIZA. As we'll see, the elegant simplicity of ELIZA's design belies both its effectiveness in creating the illusion of understanding (the "Eliza effect") and the profound questions it raises about human-machine interaction. This program, or family of programs, is at once an innovative pattern-matching machine and the catalyst for Joseph Weizenbaum's critique of AI. ELIZA and DOCTOR also raise questions about the nature of understanding, the boundaries between humans and machines, and the ethics of artificial intelligence. As mentioned in the previous chapters, Weizenbaum was alarmed by how readily people attributed humanlike qualities to ELIZA, in particular when running the DOCTOR script, leading him to become an important critic of unchecked technological progress. To understand how ELIZA can be both innovation and critical catalyst requires a clear sense of how the computer code works beyond what he documents in the 1966 *CACM* article.

Although many technical innovations have emerged in the decades since ELIZA, examining the ELIZA/DOCTOR code offers a rare glimpse into one of the earliest formalized attempts to model human conversation. What makes ELIZA particularly fascinating is not only its historical significance, but also what it reveals about Weizenbaum's views on both computing and human interaction. This code and script do not merely showcase programming techniques of the 1960s; they reveal underlying assumptions about language, therapy, and human-computer interaction that continue to influence modern AI development. By examining this code line by line, we can start to uncover the sophisticated linguistic and programming techniques that allowed a rudimentary pattern-matching system to create a convincing simulation of understanding. But before we can read the lines of code, let us offer an overview of the system.

The architectural distinction between ELIZA and DOCTOR represents an important design decision in AI history—establishing what would become conversational AI—while building on and extending principles that would come to define modern software design. Although often treated as synonymous, the DOCTOR script and the ELIZA system are two separate components, with DOCTOR being only one of the possible conversational models that can be run on ELIZA. Think of ELIZA as a system for interaction and DOCTOR as one set of rules that Weizenbaum devised to make ELIZA behave somewhat like a Rogerian psychotherapist. This separation, manifested in ELIZA's system-script dichotomy, presaged numerous contemporary software patterns, from configuration-as-data to plug-in architectures and domain-specific languages.

Without question, the historical context of 1960s computing fundamentally shaped ELIZA's architecture as well. Decisions in computing that reflect material constraints create path dependencies and eventually become programming cultural norms. These constraints manifested in

ELIZA's single-pass processing, tape-based storage, and stack-oriented implementation. Yet within these limitations, Weizenbaum crafted an elegant solution combining precedence-based disambiguation with efficient text normalization. For example, the system's handling of context preservation and variable binding demonstrated sophisticated awareness of linguistic principles. These technical features, though invisible to the users, are crucial to creating the illusion of understanding that made ELIZA so compelling.

Weizenbaum explains many of ELIZA's technical features in the 10-page paper published in the January 1966 edition of the journal *Communications of the Association of Computing Machinery* (*CACM*) (included in Appendix II).[1] But he chose to omit some essential details. This chapter details how the 1966 version of ELIZA works, based on the source code recovered from the MIT Archives. First, we discuss the process flow of the system and then, second, what we learn from reading the code that was not discussed in the 1966 paper.

03.01 The Program: An Overview

Before diving into the details of ELIZA's implementation, it's helpful to understand the overall flow of the program. Although the technical specifics may seem complex at first, the fundamental operation follows a straightforward sequence that reflects Weizenbaum's design approach that combines simplicity of concept with careful execution. This is an overview of how the 1966 ELIZA algorithm works:

1. Read the script file. The script specifies the keywords and text patterns to be used during the conversation, and the opening message.
2. Print the opening message.
3. Wait for the user to type something.
4. Look for keywords in the user's input. If a keyword has a substitute word specified in the script, the word is swapped for the substitute.
5. Select one of the keywords found in the user's input. Use the text pattern-matching rules associated with that keyword to first break the input text into parts, then reassemble those parts into a response message.
6. Print the response, and then go back to step 3.

To understand ELIZA fully, we must examine not just what it does but how it accomplishes what we might describe as conversational sleight of hand. The following explanation attempts to outline the technical architecture that made ELIZA's deceptively simple interactions possible. By tracing the program's execution step by step, we can see how Weizenbaum's implementation choices both enabled and constrained the system's capabilities while also offering insight into why users found these interactions so compelling. Here are these six steps again with a little more detail.

03.01.01 READ THE SCRIPT FILE

An ELIZA script contains the following:

The first thing in the script is the opening remarks. This is the message ELIZA prints when it starts. It sets the tone for the conversation.

1 Following the opening remarks are a series of rules. Except for the START statement, the script is written as a series of S-expressions. Each rule specifies a keyword and one or more transformation rules. Each transformation rule specifies a decomposition pattern (how the input text is broken down into recognizable patterns) and one or more reassembly patterns (how the output text is constructed). Thus a rule has this form:

```
(KEYWORD
    ((DECOMPOSITION-1)
        (REASSEMBLY-1.1))
    ((DECOMPOSITION-2)
        (REASSEMBLY-2.1)
        (REASSEMBLY-2.2)))
```

The rule will only play a part in formulating a response if the user's input contains the keyword, or if it is invoked by some other rule (rules can refer to one another).

2 Each script must contain a rule with the special reserved keyword NONE. ELIZA uses this rule when no keywords are found in the user's input.

3 There may be one MEMORY rule, which does not have the same form as the normal rules. It is applied whenever the highest precedence keyword found in the user input matches the keyword specified by the MEMORY rule and generates memories that are saved for later use. How and when memories are made and used is discussed in 3.2.

These rules are read from the script file and stored in the computer's memory in a dictionary-like structure that allows them to be easily retrieved by keyword later.

03.01.02 PRINT THE OPENING MESSAGE

The opening message is the first thing ELIZA prints and is the only thing it prints spontaneously; after that it will always wait for the user to type something, and then it will always respond. The opening message in the DOCTOR script is

```
HOW DO YOU DO. PLEASE TELL ME YOUR PROBLEM
```

03.01.03 AWAIT USER INPUT

The user sits at a terminal and types some words. When they have said all they want to say, they signal that they have finished by pressing the enter key twice. At that point ELIZA gets the user's input.

Say the user has typed the following sentence (note: when ELIZA ran on the 7094, the standard character set did not include lowercase letters):

```
WELL, MY BOYFRIEND MADE ME COME HERE.²
```

In ELIZA that sentence is stored as a list with every word and punctuation symbol in its own cell. (ELIZA used the list functionality provided by SLIP, where list cells can contain no more than six characters and longer strings occupy subsequent cells. See 3.2.4. For now, we will treat a string greater than six characters as if it occupied one cell.) The list would contain:

```
WELL
,
MY
BOYFRIEND
MADE
ME
COME
HERE
.
```

03.01.04 LOOK FOR KEYWORDS IN THE USER'S INPUT AND MAKE WORD SUBSTITUTIONS

ELIZA scans the user input list for keywords. Working through our example: ELIZA looks at the first string in the list and asks if it is one of the subclause delimiters: the word *BUT* or a comma or a period. In this case the first word is *WELL*, so the answer is no. It then asks: Is it one of the keywords in its dictionary of script rules? For the DOCTOR script, the answer is again no. ELIZA does nothing with the first string in the list.

ELIZA moves on to the next string in the list and asks if it is a subclause delimiter. In this case the answer is yes, the cell contains a comma. ELIZA now takes one of two different paths: if in the search so far it has encountered at least one keyword with associated transformation rules (i.e., not just a simple string substitution), it ignores the comma and everything in the list *after* it. This is not the case for this sentence, so ELIZA takes the second path: it ignores the comma and everything in the list *before* it. So the user input list is now:

```
MY
BOYFRIEND
MADE
```

```
ME
COME
HERE
.
```

ELIZA looks at the next string in the list, which is the word *MY*. This word is not a subclause delimiter, but it is a keyword. Here is the *MY* rule from the DOCTOR script (with the two transformation rules elided):

```
(MY = YOUR 2
  (. . .)
  (. . .))
```

ELIZA does two things: first, the rule specifies a string substitution, *MY* should be replaced by *YOUR*, so this is performed. Our example list now looks like this:

```
YOUR
BOYFRIEND
MADE
ME
COME
HERE
.
```

And second, as *MY* is a keyword with associated transformation rules, it is pushed on to the list of found keywords. Because it is the highest precedence keyword found so far, it goes on the top of the list; otherwise, it would have gone on the bottom. The precedence level is specified by the number following the keyword in the script, 2 in this case. If a rule does not have its precedence stated explicitly it is assumed to have a precedence of zero.

The next two words in the list, *BOYFRIEND* and *MADE*, are neither subclause delimiters nor keywords, so they are left unchanged.

The next word is *ME*. It is not a subclause delimiter, but it is a keyword. The rule in the DOCTOR script is a very simple one:

```
(ME=YOU)
```

As this rule specifies a word substitution, this is now performed and the list becomes:

```
YOUR
BOYFRIEND
MADE
```

```
YOU
COME
HERE
.
```

Because there are no transformation rules associated with this keyword, it is not added to the list of found keywords.

The next two words in the list, *COME* and *HERE*, are neither sub-clause delimiters nor keywords, so they are ignored.

Finally, the last string in the list contains a period, which is a sub-clause delimiter. This time when ELIZA checks the found keyword list it is not empty, so rather than erasing the delimiter and everything that came before it ELIZA deletes the delimiter and everything after it. The list is now:

```
YOUR
BOYFRIEND
MADE
YOU
COME
HERE
```

That concludes the scan of the user's input sentence. Some parts of the user input were ignored, and some words were substituted. ELIZA also has a list of keywords found in the sentence with the highest precedence keyword at the top. In the case of this example, the list contains just one keyword: *MY*.

03.01.05 SELECT A KEYWORD AND GENERATE THE RESPONSE

All of ELIZA's responses fall into one of the following five categories:

1 A wholly predetermined response containing none of the user's words. The response is associated with a script pattern that matches the user's current input. For example, if the user input is "MEN ARE ALL ALIKE," the word *ALIKE* is a keyword in the DOCTOR script and triggers a pattern that has several associated responses, one of which is "IN WHAT WAY."

2 A partially predetermined response from the script associated with a pattern that matches the user's input message. The response will also reflect back some or all the user's input message, possibly with some simple word substitutions. For example, if the user input is "WELL, MY BOYFRIEND MADE ME COME HERE," the word *MY* triggers a pattern that reflects back what the user said after *MY*, and ELIZA prints: "YOUR BOYFRIEND MADE YOU COME HERE." In addition, *YOU* has been substituted for *ME* in the user's input. MY here also triggers the creation of a MEMORY message, which can be retrieved in a future response.

3 When ELIZA reflects back the user's input it only ever returns some or all the most recent user input. For this reason talking to ELIZA is somewhat like exchanging one thought with a series of strangers. There is one exception to this: the script may contain a special MEMORY rule.[3] Whenever the keyword specified by this rule, which must also appear as a keyword in an ordinary rule, is the highest precedence keyword found in the user's input, a "memory" is created and stored away. Then at some future point, if the user's input message doesn't trigger any pattern in the script, one of these memories will occasionally be used as ELIZA's response. Exactly when the memory will be used is discussed in 3.2.3. For example, if the user's input is "BULLIES," ELIZA might respond "DOES THAT HAVE ANYTHING TO DO WITH THE FACT THAT YOUR BOYFRIEND MADE YOU COME HERE."

4 If the user's input doesn't trigger any pattern in the script, and it isn't time for ELIZA to recall a memory, or there are no unused memories, then ELIZA will use a wholly predetermined response from the NONE rule in the script. This rule may contain any number of messages, and they are used in turn. In the DOCTOR script these nonspecific, continuative messages are

```
I AM NOT SURE I UNDERSTAND YOU FULLY
PLEASE GO ON
WHAT DOES THAT SUGGEST TO YOU
DO YOU FEEL STRONGLY ABOUT DISCUSSING SUCH THINGS
```

5 Finally, there are four continuative messages hard-coded in the ELIZA source code that are used when something goes wrong with the pattern-matching process. These are discussed in 3.2.2.

Continuing with our example, recall that after step 4 the user's input has become

```
YOUR BOYFRIEND MADE YOU COME HERE
```

ELIZA now pulls the top keyword off the list of found keywords (there is only one in this case): *MY*. The script rule associated with this keyword is

```
(MY = YOUR 2
    ((0 YOUR 0 (/FAMILY) 0)
        (TELL ME MORE ABOUT YOUR FAMILY)
        (WHO ELSE IN YOUR FAMILY 5)
        (YOUR 4)
        (WHAT ELSE COMES TO MIND WHEN YOU THINK OF YOUR 4))
    ((0 YOUR 0)
        (YOUR 3)
```

```
        (WHY DO YOU SAY YOUR 3)
        (DOES THAT SUGGEST ANYTHING ELSE WHICH BELONGS TO YOU)
        (IS IT IMPORTANT TO YOU THAT 2 3)))
```

ELIZA takes the first transformation rule and tries to match its decomposition pattern,

```
(0 YOUR 0 (/FAMILY) 0)
```

against the user input. In this pattern a zero matches any number of any words, and (/FAMILY) matches the words *BROTHER, CHILDREN, FATHER, MOTHER, SISTER* and *WIFE* (as specified elsewhere in the script). Because the user's text does not contain any of the FAMILY words, this pattern fails to match the input text, so this transformation rule is not used.

Next, ELIZA tries the second transformation rule to see if the decomposition pattern in that rule matches.

```
(0 YOUR 0)
```

In this case it does match, and the matching parts are numbered 1 to 3:

```
1  <empty>
2  YOUR
3  BOYFRIEND MADE YOU COME HERE
```

The reassembly rules associated with this transformation rule are

```
(YOUR 3)
(WHY DO YOU SAY YOUR 3)
(DOES THAT SUGGEST ANYTHING ELSE WHICH BELONGS TO YOU)
(IS IT IMPORTANT TO YOU THAT 2 3)
```

These are used in turn. The first is (YOUR 3), which generates the response

```
YOUR BOYFRIEND MADE YOU COME HERE
```

Assuming exactly the same user input each time, this rule would generate these responses in turn

```
YOUR BOYFRIEND MADE YOU COME HERE
WHY DO YOU SAY YOUR BOYFRIEND MADE YOU COME HERE
DOES THAT SUGGEST ANYTHING ELSE WHICH BELONGS TO YOU
IS IT IMPORTANT TO YOU THAT YOUR BOYFRIEND MADE YOU COME HERE
```

and then back to the first message again, and so on.

03.01.06 PRINT THE RESPONSE

In 1966 the user would almost certainly be talking to ELIZA via a teletypewriter such as the IBM 1052 or 2741, so the conversation would be printed onto paper.

03.02 What Do We Learn from the Code?

The following discoveries from the ELIZA source code reveal crucial aspects of its operation that Weizenbaum chose not to document in his published work. These details not only allow us to re-create ELIZA's responses with more accuracy but also provide insight into Weizenbaum's technical decision-making. In summary, we learn from the ELIZA source code that

1. the word *BUT* is a delimiter.
2. there are hard-coded messages in the ELIZA source code.
3. "a certain counting mechanism" functions in a key way.
4. the selection of the MEMORY rule is not random.

03.02.01 THE WORD BUT IS A DELIMITER

In the 1966 *CACM* paper, Weizenbaum writes,

> The procedure recognizes a comma or a period as a delimiter. Whenever either one is encountered and a keyword has already been found, all subsequent text is deleted from the input message. If no key had yet been found the phrase or sentence to the left of the delimiter (as well as the delimiter itself) is deleted. As a result, only single phrases or sentences are ever transformed.[4]

ELIZA recognizes two characters as delimiters: a period and a comma. However, in the example conversation in the *CACM* paper on the same page there is this exchange:

```
You are not very aggressive but I think you don't want me to
notice that.
WHAT MAKES YOU THINK I AM NOT VERY AGGRESSIVE[5]
```

There is no comma in that user input and a period only at the end of the sentence. So according to Weizenbaum's description, the sentence should have been treated as a whole with none of it deleted, apart from the final period. ELIZA's response should have been:

```
WHAT MAKES YOU THINK I AM NOT VERY AGGRESSIVE BUT YOU THINK I DON'T
WANT YOU TO NOTICE THAT
```

Here is the line of ELIZA source code that detects delimiters:

```
W'R WORD .E. $.$ .OR. WORD .E. $,$ .OR. WORD .E. $BUT$ 000660
```

W'R is short for "WHENEVER," *.E.* means "equals," strings are delimited by *$*, and the trailing *000660* is the line number.

Contrary to Weizenbaum's description, ELIZA actually recognizes three delimiters: a period, a comma, and the word *BUT*. Knowing this, we can see why ELIZA's response did not include the word *BUT* or any of the words that follow it in that exchange in the example conversation.

03.02.02 THERE ARE "HARD-CODED" MESSAGES IN THE CODE

In the source code we can see four hard-coded messages that are not mentioned in the 1966 *CACM* paper.

```
          R* * * * * * * * * * SCRIPT ERROR EXIT               001980
NOMATCH(1) PRINT COMMENT $PLEASE CONTINUE $                    002200
           T'O START                                           002210
NOMATCH(2) PRINT COMMENT $HMMM $                               002220
           T'O START                                           002230
NOMATCH(3) PRINT COMMENT $GO ON , PLEASE $                     002240
           T'O START                                           002250
NOMATCH(4) PRINT COMMENT $I SEE $                              002260
           T'O START                                           002270
```

An R in column 11 indicates the line is a Remark (a comment), NOMATCH(n) are labels, T'O is short for "TRANSFER TO" (goto), so T'O START. means "continue execution at the label START."

There are two places in the code where these NOMATCH labels are jumped to via T'O NOMATCH(LIMIT): (a) if no decomposition pattern for the selected keyword matches the user's input, and (b) if the keyword in a reference doesn't exist. Here, *LIMIT* is being used as an index to select one of the four messages to print. LIMIT constantly cycles through 1, 2, 3, 4, 1, 2, 3, 4, 1, ... changing value for each new user input. LIMIT is discussed in 3.2.3.

First, Weizenbaum does not say what should happen if a keyword has been identified but none of the decomposition patterns associated with that keyword match the user's input. There are two rules in the DOCTOR script where this kind of failure is possible: *CAN* and *DREAMT*. If it does happen, one of these hard-coded messages is displayed. Which one is displayed depends on the value of the variable LIMIT at the time of the error. To make this concrete, here is an example. In Weizenbaum's DOCTOR script, the rule for *DREAMT* (lines 35–41) is

```
(DREAMT 4
    ((0 YOU DREAMT 0)
        (REALLY, 4)
        (HAVE YOU EVER FANTASIED 4 WHILE YOU WERE AWAKE)
```

```
        (HAVE YOU DREAMT 4 BEFORE)
        (=DREAM)
        (NEWKEY)))
```

So a successful user input and response might be

```
I DREAMT I DWELT IN MARBLE HALLS
HAVE YOU DREAMT YOU DWELT IN MARBLE HALLS BEFORE
```

But if *DREAMT* is not preceded by I, the one and only decomposition pattern (0 YOU DREAMT 0) is not going to match, and the rule fails. In this circumstance the response will be one of the four hard-coded messages, for example:

```
SHE DREAMT I DWELT IN MARBLE HALLS
HMMM
```

Second, a reference to a nonexistent keyword causes ELIZA to emit a built-in message. If a script contained the rule (HUMPTY (=DUMPTY)), but did not also contain a keyword rule for *DUMPTY*, then the rule will fail when *HUMPTY* is the selected keyword and one of the hard-coded messages will be produced.

03.02.03 HOW "A CERTAIN COUNTING MECHANISM" WORKS

One of the most eye-catching responses ELIZA gives is when it brings up something said by the user earlier in the conversation, for example this response, "DOES THAT HAVE ANYTHING TO DO WITH THE FACT THAT YOUR BOYFRIEND MADE YOU COME HERE." To understand how this mechanism works,

Consider the following structure (lines 67–69):

```
  (MEMORY MY
      (0 YOUR 0 = LETS DISCUSS FURTHER WHY YOUR 3)
      (0 YOUR 0 = EARLIER YOU SAID YOUR 3)
```

Weizenbaum explains,

> The word "MY" (which must be an ordinary keyword as well) has been selected to serve a special function. Whenever it is the highest ranking keyword of a text one of the transformations on the MEMORY list is randomly selected and a copy of the text is transformed accordingly. This transformation is stored on a first-in-first-out stack for later use. The ordinary processes already described are then carried out. When a text without keywords is encountered later and a certain counting mechanism is in a particular state and the stack in question is not empty, then the transformed text is printed out as the reply. It is, of course, also deleted from the stack of such transformations.

The current version of ELIZA requires that one keyword be associated with MEMORY and that exactly four transformations accompany that word in that context.[6]

No further information is given in the paper about "a certain counting mechanism" or "a particular state."

In the ELIZA source code, the relevant mechanism relates to a variable called LIMIT. Before entering the main program loop, LIMIT is given the value 1. It is incremented each time round the loop, just after reading the user's input text. If after incrementing LIMIT its value is 5, it is returned to 1 and the cycle repeats.

Only when LIMIT has the value 4 is the previously saved memory retrieved.

Here are the relevant parts of the ELIZA source (colons signify elisions):

```
          LIMIT = 1                                            000120
                                                                  :
         R* * * * * * * * * * BEGIN MAJOR LOOP                 000470
                                                                  :
          LIMIT=LIMIT+1                                        000510
          W'R LIMIT .E. 5, LIMIT=1                             000520
                                                                  :
         R* * * * * * * * * * END OF MAJOR LOOP                001110
ENDTXT    W'R IT .E. 0                                         001120
              W'R LIMIT .E. 4 .AND. LISTMT.(MYLIST) .NE. 0     001130
                OUT=POPTOP.(MYLIST)                            001140
                TXTPRT.(OUT,0)                                 001150
                IRALST.(OUT)                                   001160
                T'O START                                      001170
               O'E                                             001180
                ES=BOT.(TOP.(KEY(32)))                         001190
                T'O TRY                                        001200
              E'L                                              001210
             OR W'R KEYWORD .E. MEMORY                         001220
              I=HASH.(BOT.(INPUT),2)+1                         001230
              NEWBOT.(REGEL.(MYTRAN(I),INPUT,LIST.(MINE)),MYLIST) 001240
              SEQLL.(IT,FR)                                    001250
              T'O MATCH                                        001260
             O'E                                               001270
              SEQLL.(IT,FR)                                    001280
         R* * * * * * * * * * MATCHING ROUTINE                 001290
```

In the above text from the *CACM* paper, Weizenbaum says that a rule from the MEMORY list is "randomly selected." But from the above code on line 001230, we can see that the selection is not random but based on a hash function of the user's input text, I=HASH.(BOT.(INPUT),2)+1. This is a hash function that returns a number in the range [0–3], and adding 1 normalizes it for direct use as, for example, a subscript in MAD. The purpose of a hash function is to provide some pseudorandom number for a given input in some range. Given two different inputs, there is no guarantee that two different values will be returned. Given the same input twice, the expected behavior of any hash function is to return the same value.

The use of the hash function means that conversations with ELIZA do not have a truly random element and are reproducible. This is important, as we discuss below.

In the DOCTOR script *MY* is the keyword specified by the MEMORY rule. So every time *MY* is the selected keyword, a new memory will be created and pushed onto the back of the MYLIST first-in-first-out queue. At the point in the conversation where the user input is "BULLIES," the memories in the queue are these (some making more sense than others):

```
DOES THAT HAVE ANYTHING TO DO WITH THE FACT THAT YOUR BOYFRIEND MADE YOU
  COME HERE
DOES THAT HAVE ANYTHING TO DO WITH THE FACT THAT YOUR MOTHER
BUT YOUR MOTHER TAKES CARE OF YOU
BUT YOUR FATHER
EARLIER YOU SAID YOUR FATHER IS AFRAID OF EVERYBODY
```

In the *CACM* article's dialogue, the only occasion where no keyword was found is the very last user input, *BULLIES*. In this situation ELIZA will usually emit the next content-free remark from the NONE script rule (KEY(32) in the code), such as *PLEASE GO ON*. But if LIMIT happens to have the value 4, as it does in this case, and the MYLIST queue is not empty, ELIZA will take the first memory off the queue and print that instead.

The effect of "a certain counting mechanism" is that ELIZA/DOCTOR will occasionally return to something the user said earlier, and it isn't obvious why that particular subject and why now. Although the reply itself is not random, whether that reply is apt for that moment of conversation is left to chance. In essence, ELIZA's memory mechanism creates a sense of continuity in the conversation by occasionally bringing up previous topics. This is controlled by a simple counter that cycles from 1 to 4, only retrieving a stored memory when the counter reaches 4. This clever mechanism, though basic, creates moments where ELIZA seems to remember and reflect on earlier exchanges, and this is a crucial element in simulating meaningful conversation despite having no true understanding of context or continuity.

03.02.04 THE SELECTION OF MEMORY TRANSFORMATION RULE IS NOT RANDOM

As mentioned earlier, we learned from the source code that this random selection is in fact not random, but determined by a hash function applied to a part of the user's most recent input text. As discussed, HASH is a random number generator using some input value to produce a random number in a certain range. Should we wish to re-create an ELIZA that gives the exact same responses as the original, we would need to know what that hash algorithm is. If we use a different hash algorithm, when re-creating the famous *men are all alike* conversation, for example, the response we get to *BULLIES* might be,

```
LETS DISCUSS FURTHER WHY YOUR BOYFRIEND MADE YOU COME HERE
```

instead of the published,

```
DOES THAT HAVE ANYTHING TO DO WITH THE FACT THAT YOUR BOYFRIEND MADE
YOU COME HERE.
```

To understand how ELIZA creates these seemingly "random" memories, we need to examine the specific hash function implementation that determines which memory transformation will be selected. This technical detail is crucial for understanding ELIZA's reproducible behavior.

THE HASH ALGORITHM

The SLIP documentation records the HASH function but does not give the algorithm. In the manual it describes the following:

> N1 : HASH(D, N2). This function classifies D according to a pseudo-random scheme. The number of classes is 2 to the power of N2, where N2 is the second argument of the function. The value of the function is a pseudo-random number N1 in the range of 0 to (2 to the power of N2)-1. The contents of D can now be placed on the N1th list. It is necessary to have previously defined 2 to the power of N2 empty lists. This function enables the programmer to subdivide an unmanageably large amount of information into convenient segments in order to facilitate retrieval.[7]

It was not until Jeff Shrager and David M. Berry went back to the Weizenbaum archive at MIT and found the source code to the SLIP hash function that we learned what algorithm was used.

```
HASH    LDQ*    1,4                                  007190
        CLA*    2,4                                  007200
        STA     SHIFT                                007210
        ARS     1                                    007220
        STA     *+2                                  007230
        MPY*    1,4                                  007240
        LLS     **                                   007250
        STA     TEMP                                 007260
        LDQ     =O777777777777                       007270
        ZAC                                          007280
SHIFT   LLS     **                                   007290
        ANA     TEMP                                 007300
        TRA     3,4                                  007310
TEMP    PZE                                          007320
        END                                          007330
```

Figure 3.1 The Slip HASH function implemented in FORTRAN Assembly Program (FAP), for the IBM 7094.[10]

The SLIP hash algorithm used in ELIZA may be stated in just a few words: return the middle N-bits of D squared, where D is a 35-bit unsigned number (D is 36-bits: a 35-bit magnitude + one sign bit, which plays no part in the hash calculation). This "middle square" technique was first suggested by John von Neumann in 1949 as a method of producing a stream of pseudorandom numbers.[8] Code from Weizenbaum's MIT archive implements the SLIP hash function.[9] It's written in FORTRAN Assembly Program (FAP), for the IBM 7094.

Here is an equivalent function in C++.

```
// Recreate the SLIP HASH function: return an n-bit hash value
// for the given 36-bit datum d, for n in [0..15].
// Uses the von Neumann middle square method.
int hash(uint_least64_t d, int n)
{
    d &= 0x7FFFFFFFFULL;           // clear all but lowest 35 bits
    d *= d;                        // square it (ignore any overflow)
    d >>= 35 - n / 2;              // shift middle n bits down
    return d & (1ULL << n) - 1; // clear all but lowest n bits
}
```

HOW HASH IS USED IN ELIZA

HASH is used to select one of the four transformation rules from the MEMORY list. This is the code:

```
OR W'R KEYWRD .E. MEMORY                                    001220
I=HASH.(BOT.(INPUT),2)+1                                    001230
NEWBOT.(REGEL.(MYTRAN(I),INPUT,LIST.(MINE)),MYLIST)         001240
```

INPUT is the user's processed input text (i.e., bits of the sentence may have been removed, see section 3.1.4). On line 1230 the value of the 2-bit HASH of the BOT (last word) of the user's INPUT sentence, plus 1 is calculated. This value is then used on the next line as an index into MYTRAN, the 4-entry array of MEMORY rules. As mentioned above, a different hash algorithm might produce a different value and thus cause a different MEMORY transformation to be selected.

However, although BOT returns the last cell in the INPUT list, this will not necessarily be the whole of the last string of the INPUT sentence. To understand that, we first need to know a bit more about how ELIZA stores the user's text.

SLIP LISTS

To fully understand how ELIZA stores and processes text, we need to examine the underlying data structures it uses, which directly impact how the program "remembers" and analyzes user input. ELIZA uses the functions provided by SLIP to store and manipulate text. SLIP stores text strings in lists, which are doubly linked lists of cells. Each cell consists of two adjacent machine words.[11] The first word contains a 2-bit cell identifier field (ID) and two addresses, one pointing to the previous cell in the list (LNKL) and the other pointing to the next cell (LNKR), with SLIP functions to manage the 1-bit string marker (MKNEG). The implementation of SLIP on the IBM 7094 uses 15 bits for each of these addresses (the 7094 had 32,768 36-bit words of core memory). Two addresses and a 2-bit id field fit into one 36-bit word with 4 bits spare, including the sign bit. The second word of the cell is the "datum," and all 36-bits of this word are used to store the characters.

This is the line in ELIZA that reads the user's input text from the terminal. The SLIP list called INPUT is first emptied by calling the MTLIST function before being passed to the TREAD function.

```
START TREAD.(MTLIST.(INPUT),0)                                  000480
```

The TREAD function breaks the user's input text into "elements." If an element has six or fewer characters, it is stored in a single cell, left-justified and space-padded to the right. If an element has more than six characters, it is stored in successive cells. In this case, each cell except the last has the sign bit set in the first word of the cell pair to indicate the element is continued in the next cell.[12]

When ELIZA selects a MEMORY transformation rule, it hashes 35-bits (the sign bit is excluded) of the last cell in the input sentence. That will be the last word in the sentence or the last chunk of the last word if the last word is more than six characters long.

The IBM 7094 peripherals used 6-bit Hollerith character encoding (also called Binary Coded Decimal or BCD). To re-create the behavior of Weizenbaum's ELIZA we must hash the exact same part of the last word. In addition, the characters in that last cell must be left-justified, space-padded, and Hollerith-encoded.

```
THAT'S QUITE INTERESTING
```

Yes, these details matter if one wishes to experience conversation[13] with the ELIZA described in the 1966 *CACM* paper. These implementation details not only matter for historical reconstruction but also reveal how the material constraints of 1960s computing shaped the program's behavior in ways that would be difficult to understand from Weizenbaum's published description of the algorithm alone.[14]

Weizenbaum sought to create a program that could allow people to engage in limited conversation with ELIZA[15] and yet was startled when they began to ascribe understanding to it. Though one may be skeptical that anyone ever became deeply emotionally involved with ELIZA, they could remain convinced that it was human[16] and understood[17] what was said to it and clung to this belief despite having its simple operation explained by its creator. But these oft-repeated stories may be the reason[18] ELIZA is still alive in people's minds today. And that's how ELIZA works.

Understanding ELIZA's technical implementation is only part of the story. The more profound question is why this relatively simple pattern-matching program was able to create such a compelling illusion of understanding. As Weizenbaum himself observed with some alarm, many users became emotionally engaged with ELIZA despite its limitations. To understand this phenomenon of the Eliza effect, it is interesting to examine the psychological and linguistic mechanisms that ELIZA makes use of, sometimes intentionally and sometimes inadvertently.

03.03 How Does ELIZA Fool People?

Having examined the technical mechanics of ELIZA, we can now consider the question of what makes this seemingly simple pattern-matching program so effective at creating its sense of understanding. Indeed, the psychological impact of ELIZA cannot be explained by its code alone, but requires examining how its design elements combine to simulate human conversation. We might say that, based on ELIZA/DOCTOR, creating a convincing, or at least engaging, conversational actor using code has several elements:

1. The thread of the conversation cannot be stratified. The same statement cannot lead to the same answer all the time. There must be some component attending to the content of the person's statement.
2. When a person says something, the machine respondent can't either repeat it back verbatim or ignore it. The response should contain some portion of the content of what was said.
3. If the person says something with multiple threads of understanding, the respondent must choose one of them, and if the thread contains more than one topic of interest, the respondent should pick the one of importance.
4. When the respondent does not understand something, there must either be a segue into a previous topic or an answer which, in substance, says "I do not understand, but continue." And if the output indicates not understanding, then it should not be repetitive, such as to always say "I don't understand you" each time there is a disconnect.
5. The conversation with a respondent should not be "linear." It should include references to important, previously discussed topics, and it should include responses that may not seem relevant (at the time) because that is more realistic.
6. It should have no goals—that is, no terminating points where the person feels that the machine/person and the person have reached the end of a psychiatric session. That is, the conversation should seem like it could go on indefinitely.

ELIZA addresses each point in a variety of ways. The sum total of the ELIZA expressions is to simulate continuous dialogue with a person, in which ELIZA may appear either as a person or as a machine understanding a person. This next section attempts to provide a means of evaluating the ELIZA code to see how it simulates nonmechanical speech.

ELIZA works by recognizing input speech according to a keyword and a pattern match. The keyword is selected from the user input according to its internal logic. Once a keyword is selected, a pattern match is done to determine whether a user input phrase is recognized, and once a phrase is recognized, an output message is generated. The resolution of all these elements form the basis of mechanical speech simulating actual speech.

Within this general framework, simulation of person-to-person consists of the following elements:

1 Keyword recognition is nonlinear (pattern-free). It is based on keyword precedence and the position of a phrase within a complex user input. The same phrase occurring in different inputs at different positions means that it may be recognized in one utterance as being important and ignored in another. This adds an element of selection, or a "feel" of machine understanding, to the conversation. The machine now "understands" the person well enough to pick up a conversational thread.

2 Once an input is recognized, the output is generated by sequentially accessing the output message formatting rule list, and by inserting portions of the user's input speech into the output template. This has the effect of producing a different output message when the same input phrase is recognized. It is an attempt to separate output messaging from falling into a recognizable pattern, and providing a message relevant to the ongoing user "conversation."

3 NEWKEY selects a new keyword from the one chosen when processing the user input. This means that when an input phrase has multiple possible threads of conversation, ELIZA chooses one, and ELIZA can change its mind and process another. This disrupts patterned communication. The same phrase, with the same key chosen and matched successfully with a pattern, may be reanalyzed at different times using a set of patterns from another key. Because message output processing consists of choosing a different output messaging rule if that rule is to select another key, we have caused a disruption of a recognizable speech pattern. We can say that the same input with the same keywords can refocus the response to a different thread at "random" times. And hence, different threads of discussion are introduced, all relevant to the user's input, and yet, none producing a recognizable pattern.

4 PRE causes alteration of the input message and reprocessing of the input under a different keyword. ELIZA simulates not only attention to topic but also an ability to change the thread of conversation and context at "random" intervals. This

gives the appearance that ELIZA understands not only the user but also the context of a conversation, and redirects the user to related subjects not necessarily directly related to the current user input.

5 ELIZA can use a MEMORY of a previous exchange. At what appears to the user to be random times, ELIZA presents an output message relevant to some previous thread. The MEMORY contents are related possibly to multiple inputs given at random times. This can suggest to the user that ELIZA understands and remembers the "complete" thread of the conversation, even with pseudo-random output.

6 When no suitable output pattern can be found, ELIZA outputs a small set of messages which are free of context. This is a weak response to not always saying "I don't understand." But it is an attempt to obfuscate that ELIZA is always a nonunderstanding computer program.

7 ELIZA has no termination goal, where a subject of conversation is concluded. Instead, we can consider that the purpose of ELIZA is to continue a dialogue forever without drawing a summary conclusion.

This is all to say, an attempt to provide a dialogue that is free of a discernible pattern—one that shows that ELIZA understands the user and the context of the conversation with the user—is part of the ELIZA code. The user is meant to feel that ELIZA's output is not patterned, that the dialogue does not cease and provides no answer.

Usage has shown that some audiences do not treat ELIZA as simply a computer program but rather treat ELIZA as either a person or as a program that "understands." Explaining these user experiences requires looking beyond the code to consider the physical and social context of interaction, which we explore more in the following chapters. It is important to remember that users in the 1960s would typically interact with ELIZA via teletype machines, which were loud, mechanical devices with a distinctly nonconversational interface. Despite this seemingly uninviting medium, many users reported feeling a connection to ELIZA/DOCTOR that transcended the obvious technological limitations of the time.

It is interesting to speculate that the text-only, turn-taking nature of the interaction may have actually enhanced the illusion of understanding, as it gave users time to project meaning onto DOCTOR's deliberately designed open-ended responses. This phenomenon of psychological projection, where humans readily attribute intelligence and intention to systems exhibiting even minimal responsiveness, reveals something not just about ELIZA's design but more generally about human cognition itself. In chapter 4, "Development Anarchy," we will explore how this deceptively simple program moved beyond its technical limitations using SLIP as its underlying architecture which inspired the numerous variations and adaptations that would influence the trajectory of conversational AI.

Lisp ELIZA: A Turing Test Passed

Daniel Bobrow, BBN VP, 1968

VICE PRESIDENT	DO YOU THINK I CAN USE THE COMPUTER THIS MORNING?
TELETYPE RESPONSE	WHY DO YOU ASK?
VP	I MIGHT BE ABLE TO MAKE SOME ADDITIONAL TELCOM SALES.
TR	WHY AREN'T YOU SURE?
VP	MY PERSPECTIVE CUSTOMERS HAVE NEVER SEEN A DEMONSTRATION OF THE SYSTEM.
TR	DOES THIS MEAN VERY MUCH TO YOU?
VP	OF COURSE IT DOES.
TR	YOU SEEM VERY POSITIVE ABOUT THAT.
VP	THIS IS RIDICULOUS.
TR	WHAT DO YOU MEAN THIS IS RIDICULOUS?
VP	IS THERE ANY REASON I CAN'T USE THE COMPUTER NOW?
TR	TELL ME MORE ABOUT YOUR PERSPECTIVE CUSTOMERS.
VP	PLEASE DIAL ME UP ON 491-1850

Note that after the last remark the VP did not obey instructions, and left out the period. Therefore, of course, the computer didn't answer him. This so infuriated the VP, who thought I was playing games with him, that he called me up, woke me from a deep sleep and said:

VP	Why are you being so snotty to me?
BOBROW	What do you mean why am I being snotty to you?

The VP angrily read me the dialog that "we" had been having, and couldn't get any response but laughter from me. It took a while to convince him that it really was the computer.

The prior conversation was published in the SIGART Newsletter in 1966. Daniel Bobrow was a PhD student working on STUDENT when Weizenbaum was there.

A TURING TEST PASSED

(Editor's Note: A previous issue of the Newsletter published a news item titled "ELIZA Passes the Turing Test," about a conversation between a student and the program ELIZA, which the student took to be Dr. Daniel Bobrow of Bolt, Beranek and Newman, Inc. Bobrow has since supplied us with the following authentic version of this encounter.)

One Saturday morning about 9 A.M., a BBN Vice President in charge of selling our Telcomp commercial service arrived at our PDP-1 Computer Room hoping to be able to use the computer for a demonstration. He found a note on the PDP-1 teletype that read: "TYPE ALL MESSAGES INTO THIS TELETYPE, AND END THEM WITH A PERIOD." This note was left there by one of the programming bums who had just succeeded in making a Lisp version of the DOCTOR program run (based on ELIZA as developed by Weizenbaum). However, the VP thought that this note indicated that this teletype was linked with the one in my home, since I often used the computer all night. Therefore, he started typing to "me," but in fact was typing to this version of the DOCTOR program.

Extract from Bobrow D, "A Turing Test Passed" in the ACM-SIGART Newsletter, December 1968.

04.00 Development Anarchy

VICE PRESIDENT: THIS IS RIDICULOUS

TELETYPE RESPONSE:
WHAT DO YOU MEAN
THIS IS RIDICULOUS?

To fully appreciate the innovation that was ELIZA, you have to imagine the time and conditions in which Joseph Weizenbaum was working.[1] We need to survey the technological landscape that made this innovation possible or, perhaps more accurately, the landscape that made it so remarkable, so generative for computing. Though contemporary users interact with chatbots through voice or text displayed on a screen or even gestures and facial expressions, the first ELIZA users exchanged typed messages through a teletype (TTY), which displayed input and output on continuous rolls of paper. ELIZA's responses had no audio, no images, no flickering screen. Users and programmers alike worked via TTYs, but programmers also commonly entered code via punch cards. The machines themselves had miniscule memory and had inconsistent ways of representing numbers. Few programming languages had recursion or the ability to handle strings.

ELIZA emerged from a decade of extraordinary experimentation in software and hardware between 1955 and 1965. This was an era a contemporary programmer might characterize as total anarchy, but which programmers of the time experienced as a landscape rich with opportunities for innovation. Few practices or technologies in the field had stabilized into established norms; and ideas grew, concepts formed, and open questions splintered into entirely new areas of research. In hardware, engineers worked to reduce part counts and to increase speed in these first computational machines. In software, developers worked on numerical analysis, compiler/language design, and more. Before collections of algorithms were published, no one knew which approach to use, how long it would take to construct an application, or what made a program good. The technological topography presented rough, uneven, and shifting terrain, but Weizenbam found ways to build foundational pieces of software on this unstable ground.

This chapter sets the stage for ELIZA. First it describes the state of hardware, software, and programming at the time, and then it shows how Weizenbaum devised SLIP (Symmetric List Processor) and ELIZA in response to these conditions. With SLIP, Weizenbaum created a platform across differing computer architectures. Based on our examination and documentation of the code of ELIZA and SLIP, we highlight three innovations. First, SLIP is a cross-platform API (application programming interface, though Weizenbaum did not use that term), and it was portable across computers at a time when most programming was machine-specific (see chapter 4.3). Second, SLIP is dynamic, meaning it can dynamically allocate space, which allows it to create, delete, and manipulate list structures at run time (see 4.3). Third, SLIP can handle strings as symbols, which is the capability that makes the DOCTOR script possible (see chapter 5).

Understanding ELIZA's significance requires appreciating the material and conceptual constraints within which Weizenbaum worked, computing and physical constraints that would be nearly unimaginable to today's programmers with their vast computational resources and standardized programming environments. Weizenbaum's innovations are particularly impressive when viewed against the backdrop of the limitations of 1960s computing infrastructure, which was a world where hardware constraints and the lack of standardization shaped nearly every aspect of software design and where the manipulation of text was anything but simple.

04.01 Hardware

Although ELIZA is a program for the manipulation of symbols, computer hardware at the time was designed for rapid mathematical computation. Computation from 1955 to 1965 involved floating-point (real) numbers, integers, and boolean variables. The use of computers in "noncomputation," or symbolic, processes was rarely considered during computer design and deployment. This myopic view of "data" (as only numbers, not strings) inhibited development of symbolic programs.

On top of that, computer hardware consisted of discrete parts joined together by wires and solder. A computer was a hand-crafted device.[2] At that time, a computer took days to go from parts to product and was not done until completed on site by an installation team. Today's "computers" are chips fabricated in an automated process, with the resulting "chip" often containing not one but many processors. Today, the process of building a computer from mass-produced parts is a hobbyist's job completed in as little as half an hour.

04.01.01 WEIZENBAUM'S COMPUTER

Both SLIP and ELIZA were developed on an IBM 7094, introduced in 1962, a room-sized machine that represented the cutting edge of scientific computing in its era. Its predecessor, the IBM 7090, introduced in 1959, was marketed as being able to perform 229,000 calculations per second,[3] with the 7094 improving on this by roughly 2.5 times.[4]

These machines used 36-bit words and featured separate processors for integer and floating-point arithmetic, making them particularly well-suited for scientific applications. For Weizenbaum, whose early training had been in mathematics and meteorology as described in chapter 2, the transition to working with these scientifically oriented machines was relatively straightforward. Yet his interest in using computers for linguistic and symbolic processing represented a significant departure from their intended use, foreshadowing his later concerns about the limitations of computational approaches to human problems. The MIT machines typically ran under the Compatible Time-Sharing System (CTSS) operating system, one of the first time-sharing systems, which allowed multiple users to interact with the computer simultaneously—a crucial capability for interactive programs like ELIZA. The word sizes on different computers ranged from 4 bits to 48 bits. Word sizes have a direct impact on numeric data ranges and precisions, and the number of binary-coded decimal

(BCD) characters that can be packed into a single word. That meant programs had to be tiny or else risk exceeding the space limits, (as Daniel Bobrow's program STUDENT did).[5] To put that in perspective (in rough estimates), if an operating system consumed 8,000 words, that left only 24,000 more words, of which ELIZA and SLIP consumed between 8,500 and 17,000. That does not even include the DOCTOR script. SLIP, as we will explain below, is what enabled Weizenbaum to write ELIZA so concisely.

The IBM 7094 was also used, among other purposes, by NASA in their Apollo program. In another connection to the history of talking computers, the IBM 7094 was the machine that Bell Labs "taught to sing" "Daisy Bell (Bicycle Built for Two)," a demonstration witnessed by Arthur C. Clarke in the 1960s and who then had a dying Hal sing it in the climactic scene of *2001: A Space Odyssey*.[6]

Yet this mainframe computer had a memory size of 32,768 36-bit words that could perform about 40,000 floating-point operations per second (flops). To put this in even more concrete terms, the computational power needed to display a single high-resolution image on a modern smartphone would have exceeded the total capacity of the IBM 7094 on which ELIZA ran.[7] These severe computational constraints forced programmers like Weizenbaum to think creatively about efficiency, leading to elegant solutions that maximized limited resources. Even though these constraints were challenging for all programming tasks of the era, they presented particular difficulties when attempting to move data between different machine architectures.

04.01.02 PORTING DATA

The treatment of data was a major challenge for programmers at this time because of the lack of regularity across machines and because machines were primarily designed to handle numbers. The IBM 7094 had sign magnitude integers and a distinct format for floating-point numbers. It was developed for numerical computation. It had no native instructions to extract a 6-bit BCD value from a physical word. The packing/unpacking of BCD strings into/out of a word had to be done by software. No physical word could have more than 6 BCD characters in it.

One source of development "anarchy" came from numbers themselves. Different machines had different formats for integers and floating-point numbers and for computer word sizes. For integers, the common formats were sign-magnitude, one's complement and two's complement. Each computer had a different floating-point format. The impact of integer formats on software was the need to check for +0 and -0 on sign-magnitude and one's complement machines and error propagation for floating point numbers as well as the common issue of number ranges for differing word sizes.

Porting is an issue of range, precision, and format between different computers. This was difficult enough so that porting data was not a consideration during development and not typically done afterwards. To consider porting binary data, software would have to be constructed to handle differences in integer and floating-point formats and differences in word size. Albeit, with a suitably restricted range for integers and a normalized precision for floating-point numbers, and accommodation to the porting of packed BCD characters, there could be a porting solution, but here the question was to what machine the software would be ported. The end result was that porting between different machines was seldom done and, if so, done with difficulty. SLIP steps over this obstacle by not considering the data. With SLIP, Weizenbaum ignores the problem of porting because he ignores data size and numbers. This pragmatic approach to addressing hardware limitations shows Weizenbaum's ingenuity in working around the material constraints of his time. However, the challenges of hardware were matched by equally significant limitations in the software development environments of the era.

04.02 Software Ecosystem

Another source of instability and inconsistency was the state of programming, which at the time was not yet a field of computer science and not yet an art (in the sense of Donald Knuth).[8] The first computer science department in the United States was established at Purdue University in 1962, but there were no recognized computer science curricula until the later 1960s. Programmers were not expected to have any prior academic experience in software. Their software development techniques were developed ad hoc to satisfy some requirement or drawn from previous learning experiences. The initial programmers likely had their first programming experience using an assembly language, with higher level languages practiced only at later points of their career.

What was and what was not acceptable in software was being defined in the moment, as was the concept of software architecture. Many notions and concepts in development impacted Weizenbaum's approach to the development of SLIP and ELIZA. This section will address some of the issues present at this time and will show the effect that they had on the software. The physical nature of programming in this era wasn't just a technical consideration; it also shaped the entire development process and the kinds of programs that could be feasibly created and maintained.

04.02.01 IBM PUNCH CARDS AND TELETYPES

The material experience of programming and interacting with ELIZA would have been completely foreign to contemporary computer users. The ELIZA program may have started life as a deck of punched cards. Eighty-column IBM cards, also called Hollerith cards, were the main resource used to code programs (source code). They were dominant because of their ease of input, durability, and archive capability. Other input mediums were possible, but most had some defect making their use marginal. The result was that the default mechanism to create machine-readable software programs was the physical IBM card.

Programmers and users would interact with ELIZA via teletype (TTY). The machine did not use a screen as in today's machines. In a computer with TTY software, a character typed on the TTY input typewriter would appear on a roll of paper and be sent to the computer. Either the computer had to be operating in a stand-alone mode (one user, one TTY, and no other applications or users), or it was operating with an operating system in which several users or applications could be executing concurrently. In operating systems, a time-shared multitasking environment means that the scheduling policy is to allocate a fixed quantum of time to each executing task and, at the end of that time, to switch tasks.

IBM cards were either stored in a card box (which could hold 2,000 cards) or in a separate cabinet. The availability of computer disks and drums for this purpose was severely limited. Capacities were small, and disk and drum storage was shared among many users. Magnetic tapes could be used, but magnetic tape long term storage required an air-conditioned facility. Using IBM cards as a storage facility allowed for large programs, a non-air-conditioned facility, and the freeing of scarce computer resources.

The TTY's and card punches had a limited character set. The set included only uppercase characters, the digits 0–9, and some special characters. The limitation of only uppercase characters was the result of the limited number of bits in a BCD character, with BCD being the primary encoding format for characters. BCD characters were 6 bits, or 64 characters. If upper case, lower case, and digits are included, then there would be 62 characters with only 2 characters available for control characters and special characters. The hardware concession is, then, to use only upper case.

Weizenbaum had access to the CTSS in which a limited number of TTYs could be operational at the same time, and Weizenbaum had an IBM 2741 TTY at his home. Therefore, it is likely that Weizenbaum entered portions of his scripts using the TTY. However, it is unlikely, that the TTY was used to enter source code.

To add to the labor of writing code, if a correction was needed and a backspace was allowed, at best existing input text could be marked as deleted and a new text inserted following the deleted text. That did not mean that a user could delete text and have that text removed from the roll of paper. A further complication was that computer languages like MAD and FORTRAN were based on a physical IBM card. If you typed in a program on a TTY, you had to ensure that the column boundaries required by the compilers were satisfied. This could be particularly onerous if there was a project requirement that sequence numbers were required to be placed on each IBM card. Sequence numbers for both MAD and FORTRAN were located between column 73 and 80 on an IBM card. Typing on a TTY would then require the user to hit the space key until they reached column 73 and type in a sequence number. Because they could be programmed while not connected to the multitasking operating system, IBM cards removed the computer as an interface and meant that computer resources were available for other uses.

04.02.02 "GOOD" CODE PRACTICES

Simply put, there were almost none. The issue of what distinguishes a good program from a bad program was that if the program worked, it was good. If it did not work, it was not good. The overall issues of understandability, documentation, and correct code structuring were unknown.

If you developed a working program and if that program needed to be documented, you would do this externally to the program, often using flowcharts as the primary means of expression. It was expected that these flowcharts would represent program control flow and algorithmic operation, transferring the IBM card format to a visually more appealing format, but not addressing the issue of meaning—that is, the functional descriptions. A flowchart could show the code in a different format. It did not explain the code.

DOCUMENTATION

In ELIZA, there were only eight single-line comments. Three say that a loop begins or ends. Five say that something is going to be done or has been done. Why were there only eight?

During this period of limited computational resources, putting documentation in the code had no perceived advantage. It was seen as adding a task to a developer's coding time, detracting from the generation of usable (i.e., executable) code, and did not provide any benefit to the code itself. A comment is not executable. A comment does not enhance code. And a comment added to the number of IBM cards that need to be created and saved, which was done in large corporations by a separate department. A comment was not considered as "useful" and was typically not written.

Documentation introduced a new element of chaos into programming even as it tried to explain it. There were two basic types of software documentation at this time. The documentation could be expository or algorithmic. The expository documentation was a text description of a piece of software. There were no guidelines for the contents of this type of documentation, so the form and contents varied by author. Sometimes algorithms were given by identifying and describing the code. Sometimes the description was more functional, describing the relationships in the code and what the intent of various code portions was. When documentation did exist, it often took a visual form that attempted to represent program flow in a way that would be more accessible than the code itself, though these visual representations had their own limitations.

FLOWCHARTS

Instead of in-code descriptions, flowcharts were used by programmers to explain algorithms, which used symbols to represent different elements of a program. There were symbols for hardware elements: peripherals, disks, drums, magnetic tape, line printer, and so on. There were symbols for programming elements: conditionals, branches, goto's, functions, and so on. There was a general consensus as to what the symbols represented, but in practice many took liberties with these definitions. Expository descriptions often contained flowcharts. Even though text descriptions and flowcharts were not mutually exclusive, flowcharts did not typically contain much additional textual information. The flowchart stood for the program, which may be why Weizenbaum felt publishing his flowchart of ELIZA was sufficient to describe the program rather than also publishing its code.

Flowcharts were a two-dimensional (2D) view of control flow. They contained the algorithmic elements and connections to the elements. The connectivity was represented as a line, for example, connecting an algorithm expression to another algorithm expression via a branching statement. The prevalent use of goto's within procedural languages, e.g., MAD and FORTRAN, led to connectivity diagrams in which there was significant crossing of lines. Think of this as a 2D projection of a three-dimensional object (3D). In 3D the line crossings could be represented in space, and lines would not directly cross. But when reflected onto a 2D surface, the lines separated by—for example, height—would cross. The use of goto's caused (almost) unreadable flowcharts.[9] At the time the code would be called "spaghetti" code and still would today.

An example of a functional description is given in the 1966 *CACM* article Weizenbaum wrote describing ELIZA. Although an example of using actual code is given in the MAD primer and in this document, the iconography of a flowchart was changed to represent MAD specific structures.

04.02.03 MAD and FORTRAN

Both MAD and FORTRAN emerged in the late 1950s to aid in mathematical algorithms, not to process symbols the way ELIZA does. They were designed to allow the expression of algorithms focused on a mathematic formulation of a problem to be represented. Neither was designed to support symbolic processes or expressions that at their basis were not mathematical, including strings, which stymied the development of programs for symbolic computation. In other words, neither MAD nor FORTRAN were suitable platforms for ELIZA by themselves.

Neither language is a structured language. If we define a "structure" as having a keyword defined beginning and terminus, then both languages have a "structure." FORTRAN has one (iteration) and MAD has two (iteration and conditionals). But what makes them unstructured is that in both languages unrestricted branches into the "structures" are allowed. In a programmatic sense, the ability to transfer from outside a structure into a structure is not allowed in structured languages.

In this language, computer "words" are considered as either integers or floating-point numbers. There is no definition within the language for a word to contain anything else. A computer word containing characters is not linguistically supported. If a word cannot be treated as containing characters then a string, a set of contiguous characters which in whole are to be considered as a single entity is also unavailable. In programmatic terms, the use of strings allows consideration of strings as objects, and as objects they can be manipulated and used in symbolic processing. Effectively, you cannot have a symbol, and hence, you cannot do symbolic processing.

Additionally, neither language is recursive; that is, a function cannot not directly or indirectly call itself, so the requirement that there be a runtime stack is not considered and not needed. Variables and constants are not put onto a stack but instead placed in a fixed and unvarying position in physical programming space. This translates to the statement: Everything that is known is known before program execution. This means that ELIZA is constructed on the basis that all knowledge is known before user interaction. All user inputs are mediated based on the known environment established in the scripts. This means that from the computer side, all responses are known and no dynamic variance is possible.

There are some curious consequences of the observation that the location of objects in memory is known and invariant. The compiler is free to allocate an object anywhere in memory. Memory is unpartitioned, and array indices are not checked at runtime. The nonpartitioning of memory means that a word in memory can represent an instruction or data. It makes no difference. The lack of index checking means that once an application developer knows the allocation scheme used by the compiler, the application developer can overwrite code or create executable modules using simple assignment statements in the code.

In summary, both MAD and FORTRAN address and handle the same type of problems. FORTRAN does not have bitwise operations, although MAD does. Neither language supports strings or dynamic structures. On the one hand, the lack of a string type interferes with developing programs for symbolic computation. On the other, the lack of dynamic arrays creates data bloat, which interferes with the development of large programs. These limitations in both hardware and software created barriers that would have seemed insurmountable to many programmers. Most would have simply accepted these constraints as immutable facts of computing life. But Weizenbaum's propensity for what he called "tricks," which are the computational sleights of hand described in chapter 2, gave him a different perspective. Rather than accepting these limitations, he saw them as puzzles to be solved through creative ideas and the synthesis of multiple techniques. This mindset led him to develop SLIP, which would become the foundation that made ELIZA possible despite the seemingly limited constraints of the computing environment.

04.03 SLIP

Out of this relatively anarchic period, SLIP and ELIZA emerge as innovations that overcome these instabilities and negotiate the limitations. As a kind of API, SLIP provides a means of porting data and of recursion. Likewise, ELIZA innovated by separating the data from the operational program. ELIZA shows how an interpreter of data can facilitate the execution of complex tasks and how, without change, the same interpreter can be used on multiple data—in this case, the input data are ELIZA scripts—to achieve seemingly different actions. Indeed, there would be no ELIZA without SLIP, and yet on the foundation of SLIP, ELIZA made further innovations that would demonstrate how a relatively small and elegantly designed program could create a user experience seemingly beyond its hardware limits.

At its core, SLIP defines a set of FORTRAN subroutines which allow the creation, search, and manipulation of Directed Acyclic Graphs, trees, and lists.[10] S-expressions provide a way to represent nested hierarchical structures. For example, consider how one might represent the following statement in an S-expression:

```
(the sentence
  (ELIZA (the program
    (that Joseph Weizenbaum wrote
      (in MAD-SLIP
        (at MIT)))...
```

This representation, called an "S-expression" (pronounced as the letter ess and the word "expression," *not* "sexpression"), is a data structure. Any individual level of an S-expression is called a list. For example, (this is a list), whereas (this S-expression is a list (and the last element is another list nested inside it)). Notice that each element of a list can be a symbol, like a word or a number, or can themselves be lists. That is, S-expressions are *recursive*: they can contain other lists as their elements, which can contain other lists, which can contain still deeper lists, and on and on, as long and deep as there is computer memory to hold these structures.

SLIP is heavily derivative of its predecessor development, the Knotted List Structure (KLS), a processor for Directed Graphs, and contains much the same functions as KLS.[11] However, SLIP extends KLS by using a doubly linked list rather than a singly linked list, and extends many of the notions contained in KLS. As said by the editors of the *CACM*, source listings are provided "because of considerable interest in the SLIP system."[12] SLIP was more than just an anomaly.

04.03.01 SLIP FEATURES

SLIP has some features noteworthy by their use, implied or stated, by Weizenbaum in his development of ELIZA. Some are addressed in the *CACM* paper; some are unaddressed features of the language that became apparent through our investigations. Perhaps the most significant innovation is that it is a complete system that allows data to be organized as a list of lists or, alternatively phrased, an acyclic graph with each graph node a list. The list acted as a data carrier, and in some ways, the data was not important. What was striking (to computer scientists) was the ability to manipulate lists/graphs using a language built for fast arithmetic operations, e.g., FORTRAN and MAD.

04.03.02 AVAILABLE SPACE LIST (AVSL)

In chapter 6.1 we address the issue of static allocation, where in the MAD and FORTRAN language an application is forced to size arrays to their maximum size because dynamic sizing is not supported. The issue is that if we know the maximum array sizes for several arrays and we know that at no time will the sum of needed array elements for all the arrays be less than or equal to some maximum, at what size should we declare the array? What is needed to resolve this issue is some means to dynamically change the size of a given array. We transform this argument to lists. Suppose we have several lists that need to satisfy the same requirements as the arrays have. How does SLIP handle this?

SLIP has a dynamically controlled space. When SLIP is initiated, in the INITAS function, it is passed an array that can hold a fixed number of SLIP cells. The SLIP space containing these cells is called the AVSL. At any given time, SLIP cells can be removed from the AVSL and allocated to a list or removed from a list and restored to the AVSL. The amount expected to be consumed at any given time, the consumption rate, and the return rate are unknown. The issue remaining is that the maximum number of cells needed for an application needs to be defined, not the instantaneous usage. The growth and reduction of list size is handled by SLIP functions, not the application. ELIZA relies on this functionality.

When ELIZA reads a script, the size of the script is unknown. The number of lists, decomposition lists, and reassembly lists is unknown. Without this capability, ELIZA would have to establish a series of maximums for each object in the scripts. For example, there can only be *x* keywords, or each keyword can only have *y* rules, and each rule can only have *z* decomposition patterns, and so on. With the dynamic nature of the AVSL, these limitations are unnecessary; instead, the discussion is centered around the total space that could be required, not the needed space for constituent parts.

There is a related issue when reading the user dialogue. How large can an input statement be? After processing, how is the space that is no longer needed to be destroyed? This is handled by SLIP. We have seen that the allocation of SLIP cells is handled, but so is the "deletion" of SLIP cells. A deleted cell, or list, is put back onto the AVSL for future use. ELIZA requires these dynamic lists to function.

04.03.03 SLIP AND PORTING

SLIP is an application programming interface (API), and its FORTRAN code was designed to be ported. SLIP was developed to minimize porting issues by identifying a small set of assembly language functions that would allow porting to machines other than the IBM 7094. Weizenbaum said that the FORTRAN listing was used on both 36-bit and 48-bit computers and that in fact it was ported.[13]

ELIZA was written in MAD. This required that SLIP be recoded into MAD. Preserving data content was unnecessary in this port. Issues in porting were mentioned in 4.1. These issues were overcome by ignoring the data contents of a SLIP cell. As Weizenbaum said, "Machine differences such as word size are irrelevant in that they are reflected in the construction of the primitives and do not appear on the surface."[14] This has the effect that the graphical nature handled by SLIP is preserved, but the data-centric issues must be handled by the application.

04.03.04 AUGMENTS FORTRAN AND MAD

SLIP provides a mechanism for recursion. FORTRAN and MAD are nonrecursive languages, and if a recursive call was made, both languages would recurse forever. This is because the return address from a call is stored in a fixed place in memory. In a recursive call, the return address becomes the address of the function making the call, which is the address of the function. SLIP provided a mechanism to overcome this issue, introducing the modern notion of a "stack frame." SLIP lists are extensible. The requirement that everything be of fixed size is mitigated by the use of the AVSL, wherein instantaneous changes to lists are possible.

04.03.05 INPUT/OUTPUT

SLIP provided a means to input and output a list. In this context, a list consists of a left parenthesis, '(', and a right parenthesis, ')', with zero or more symbols between them. There is a restriction that no symbol or number be greater than six characters in the IBM 7094. Any input greater than six characters is split into two or more SLIP cells. For example, ABCDEF is stored into a single SLIP cell, but ABCDEFG is stored into two SLIP cells, one containing ABCDEF and the next containing G.

04.03.06 SLIP AS AN API

SLIP is one of the few documented application programming interfaces (APIs) at this time, though Weizenbaum does not use this term.[15] As such, SLIP could be embedded into an existing program to provide a specific set of operations, and the programmer's only means to interact with the API was through the functions without awareness of the implementation. It provided a means for developers to concentrate on the "problem" and not on the implementation, and it made ELIZA portable.

04.04 Sui Generis

The logic portion of ELIZA is tiny—just over 400 statements, 8 of which are comments. Three of the comments deal with program logic, the remaining indicate information concerning the start or end of an iteration statement. There are no comments dealing with the algorithms used—that is, the functional logic or structure of the program. The number of comments is not indicative of what the program can do and does not provide information useful in understanding the program. The small program size is possible because of the use of the SLIP API.

As noted, the ELIZA description in Weizenbaum's 1966 *CACM* article (discussed in chapter 3) was aided by a functional flowchart that provided a visual representation of the program's logic. Rather than using actual code in the flowchart, functional descriptions of operations were provided. It was a condensed version of the prose description of ELIZA and illustrates some of the power of this means of presentation. However, there are noncommented or discussed differences in the paper and in the operational code. Using the paper as the sole source for understanding how the code works will lead to a program not consistent with the operational program (see chapter 3). Looking at the source code alone is difficult, as described in part below, and trying to combine the algorithms in the paper with the actual code does not diminish the difficulty.

Today it is common in a corporate setting to require that each function contain comments describing each function argument, the problem addressed by a function, and the algorithms used to solve the problem. The ELIZA code does not describe the semantics of a function, nor does it identify the end result to be achieved by using the function. This lack requires that we use some resource outside of the source code in order to understand a function. This outside source cannot be the 1966 Weizenbaum *CACM* paper, which does not contain a description of implementation functions by name.

04.04.01 PHRASING

Despite the brevity of the code, ELIZA seems to recognize an input sentence, seems to understand the input, and seems to provide a reasonable output. The mechanism to do this is simple, ingenious, and innovative. However, ELIZA is not a natural language application. To understand how ELIZA creates the impression of understanding without actually doing so, we need to examine Weizenbaum's elegant pattern-matching approach. Indeed, ELIZA has no means to recognize input, to determine whether it is grammatically correct, or to understand it. Instead, it follows the precept Weizenbaum himself laid out in the 1966 article.

> Consider the sentence "I am very unhappy these days." Suppose a foreigner with only a limited knowledge of English but with a very good ear heard that sentence spoken but understood only the first two words "I am." ... What he must have done is to apply a kind of template to the original sentence, one part of which matched the two words "I am" and the remainder isolated the words "very unhappy these days." He must also have a reassembly kit specifically associated with that template, one that specifies that any sentence of the form "I am BLAH" can be transformed to "How long have you been BLAH," independently of the meaning of BLAH.[16]

ELIZA processes language by recognizing a keyword and, using the keyword and its associated disassembly rules, determines whether the input sentence matches a regular expression representing a family of input sentences, and if so, constructs an output sentence. What Weizenbaum did is to use a regular expression handler. Simple, but very powerful. He then used the regular expression to separate different parts of the input sentence in order to construct a response. The code recognizes one or more words of the input, assumes a normative grammatical structure, and then utilizes the phrase recognized in this grammatical structure to generate an output.

What is surprising is the number of inventions needed to do the task and the simplicity of the task once so configured. ELIZA assumes that the person is using grammatically correct English. What it invents is how to determine important parts of a complex input sentence, how to recognize what sentential structure to use, how to "parse" the input sentence, and how to construct an output. All of this is based on a simple assumption, "*Suppose a foreigner ... understood ...*" only a word or two.

04.04.02 SIMULATING SPEECH

When two people talk, the speech is often not linear. When a person says a sentence, you don't necessarily reply directly to the sentence by modifying it or making a related response. We may veer from some central topic to address a related issue or may introduce a topic already discussed. The responses aren't patterned. We would not expect that a statement given would receive the same sentence returned. There is variation and variety when two people speak.

For a machine to simulate speech, the machine must provide variation and variety and must interject a number of things that are not "linear."' Not only does ELIZA partition an input sentence into phrases suitable for recombination, but it also provides a means to produce output related to previous parts of the dialogue, to provide different responses for the same input, and to provide a random output. It does this mechanically, without understanding, and skillfully to the point that some do not recognize a machine but a person. This apparent naturalness required specific technical features that Weizenbaum needed to build into SLIP to support ELIZA's conversational capabilities.

04.04.03 ELIZA SLIP

After examining the code and documentation, we can say the SLIP version used for ELIZA is *not* the same as the SLIP version described in the *CACM* article of 1966. The ELIZA SLIP seems to be a predecessor to the published SLIP, containing all the gross functionality but having a more primitive set of iteration (and other) functions, plus functions specifically tuned to the needs of ELIZA. As a note, this was discovered in the archived source code for ELIZA and the ELIZA SLIP.

The ELIZA SLIP contains the following notable features:

1 YMATCH and ASSMBL:[17] YMATCH is a regular expression pattern matcher which recognizes whether the input sentence matches some template. ASSMBL inserts the input phrase into the template to produce an output. YMATCH is the disassembly rule, and ASSMBL is the reassembly rule in the ELIZA article.[18] (See YMATCH: ELIZA, line 353; ASSEMBL: ELIZA, line 376)

2 Strings: Strings are created using SLIP cells. They form an integral part of the YMATCH and ASSMBL processing and script inputting. This allows strings to be treated and manipulated as symbols. This capability does not exist in either MAD nor FORTRAN.

3 TREAD: This is a tokenizing reader which separates user input into tokens and places them into a SLIP list for further processing, which includes searching for keyword matches that will initiate ELIZA's transformations. (See TREAD: ELIZA, lines 218 and 250)

By creating SLIP, Weizenbaum showed how an API can separate the logic of a program from the mechanics needed to implement the requirements of a program. He showed that a program's logic can be almost tiny when this abstraction is done and that the result can match the program requirements expeditiously. Furthermore, Weizenbaum demonstrated how a mathematical concept of a list can be productively used outside the boundaries formed by mathematics. Here was an example of a simple "list" being used to accomplish symbol processing (YMATCH and ASSEMBL) and straightforward management of data.

Through some clever knowledge of English-language parsing and the forms of speech used in a dialog, Weizenbaum produced a computer program that could mimic a dialogue without any understanding of the conversation. In ELIZA he demonstrated how limited objectives could be achieved with limited effort and yet have impressive effects that seemed almost magical. However, he was keen to demystify computation. As he wrote, "Once [the program's] inner workings are explained in language sufficiently plain to induce understanding, its magic crumbles away."[19] However, the popularity and misunderstanding of ELIZA defied Weizenbaum's attempts at explanation. Indeed, ELIZA generated a broad discussion of the possibilities of computation across a broad spectrum of academics, nonacademics, industry, the military, and others.

Ironically, the same ingenuity that allowed Weizenbaum to overcome the technical limitations of his era would later cause him concern. As we see in chapter 2, his alarm at how readily people anthropomorphized ELIZA, despite his full disclosure of its simplicity, would eventually lead him to question the entire project of artificial intelligence. The "development anarchy" that enabled ELIZA's development had given Weizenbaum the freedom to innovate, but the consequences of that innovation would ultimately transform both computing and his own relationship to the field he helped create. As we explore in the following chapters, this transformation would extend far beyond Weizenbaum himself, as ELIZA would go on to inspire both technical innovations and profound questions about the nature of human-machine interaction.

ELIZA/YAPYAP

Michael McGuire,
Gardner C. Quarton,
Stephen Lorch
December 1965

START
W 2113.5
PLEASESTART TYPINGACCORDING TO THE INSTRUCTIONS.

SUBJECT	I AM WAITINGFOR YOU TO TELL ME SOMETHING THAT I CAN DISCUSS
RESPONSE	HMMM
S	THAT SOUNDS VERY PROFOUND
R	I AM NOT SURE I UNDERSTAND YOU FULLY .
S	THATS ALRIGHT IM NOT EXACTLY SURE WHATS GOUING ON EITHER,BUT HOW CAN I TELL YOU ANYTHING IF YOU HAVENT SAID ANYTHING TO ME YET
R	WOULD YOU PREFERIF . I ANSWER THIS QUESTION 'YES' OR 'NO' .
S	YES IT WOULD BE VERY HELPFUL AT THIS POINT
R	ALRIGHT , I WILL ANSWERYES , BUT PLEASETELL ME MORE ABOUT YOUR SELF .
S	NOW YOURE TALKING,I AM ATTENDING EMERSON COLLEGE AND I AM IN MY THIRD YEAR, MAJORING IN ELEMNENTAY EDUCATION. I LOVE BOSTON AND I AM HAVING LOADS OF FUN UP HERE
R	HMMM
S	ISNT THAT AT LEAST GOOD FOR A START
R	GO ON , PLEASE
S	WHY CERTAINLY, THIS IS THE FIRST YEAR SINCE I LEFT HOME THREE YEARS AGO THAT I AM REALLY HAVING A FANTASTIC TIME. AS A MATTER OF FACT IM HAVING THE TIME OF MY LIFE. I LIVE IN NEW YORK TO CHANGE THE SUBJECT, AND I AM THE

OLDEST OF FOUR CHILDREN.

R I SEE

S SO WHATS NEXT

R CONTINUE ON THE SAME TOPIC IF YOU WOULD .

S OF COURSE I WOULD, SINCE THIS IS MORE FUN
THAN I EVER IMAGINED.TO GET
BACK TO THE SOCIAL LIFE. ONE OF THE MAIN
REASONS THAT I NEVER HAD TOO
MUCH FUN UP HERE AT SCHOOL WAS BECSUSE IWAS
PINNED TO SOMEONE AT HOME.
BUT SINCE THEN WE HAVE BROKEN UP,AND NOW IM
ENJOYING MYSELF. DO YOU HVE
ANY OTHER TOPICS

R HMMM

S WELL, I GUESS YOU DONT. CAN I ASK
YOU SOME QUESTIONS

R GO ON , PLEASE

S GO ON WITH WHAT ME OR YOU

R I SEE

S STOP PLAYING GAMES

R PLEASESAY MORE .

S I CAN TELLNYOU ARE DEFINITELY FUNNY. MY
ROOMMATE AND A GOOD FRIEND OF MINE BOTH TOOK
THIS
TEST BEFORE I DID, ANDTHEY ONLY TOLD ME THAT
IT WAS FUN. BOY
THEY WERENT KIDDING. WHAT DO
YOU THINK OF THIS

R DID YOU THINK THIS IS ALWAYSTRUE .

S YOUR ARE NOT ONLY FUNNY, BUT HISTERICAL. AS
A MATTER OF FACT I NEVER PREJUDGE ANYTHING,
AND CONSIDERING THAT I DONT KNOW WHAT YOUR
ASKING TME THE TRUTH OF, I CANT ANSWER YOU.

R MY ARE NOT ONLY FUNNY INTERESTS YOU, PLEASE
TELL ME WHY [ASSUMED RESPONSE]

S I DO BELIEVE THAT YOU COULD USE A GOOD COUSRSE IN ENGLISH GRAMMER. YOU ARE FUNNY BECAUSE YOU ARE BAREKLY MAKING ANY SENSE AT ALL

R WHEN YOU FEEL GOOD HOW ARE YOU EFFECTED

S I AM HAPPY WITH MYSELF AND WITH THOSE IN MY SURROUNDINGS,MIF THAT THE KIND OF FEELING GOOD YOU ARE REFERRING TO

R WHAT ARE THE CIRCUMSTANCES AROUNDYOUR FEELING HAPPY .

S WHEN GOOD THINGS HAPPEN SUCH AS RECEIVING A GOOD GRADE IN SCHOOL, OR FINDING OUT SOMETHING WONDERFUL HAPPENED TO ONE OF YOUR FRIENDS ,OR HEARING FROM A GUY YOUVE BEEN DYING TO HEAR FROM, OR FROM GOING OUT ON A DATE AND HAVING A TERRIFIC TIME. THESE ARE JUSTSOME OF THETHINGSTHAT COULDMAKE
ME FEEL HAPPY. AND YOURSELF

R HMMM

S MY YOUVARE VERY TALKATIVE THIS EVENING

R WHAT ARE YOUR FEELINGS ABOUT YOUR YOUVARE VERY TALKATIVE THIS EVENING .

S WELL, TO TELL YOU THE TRUTH I CAN BE EXTREMELY TALKATIVE WITH THOSE PEOPLE I KNOW AND I GUESS YOU ARE PUTTING ME AT EASE

R WHAT MAKES YOU THINK I AM PUTTING YOU AT EASE .

S YOUR HUMOR FOR ONE, BECAUSE I LOVE TO FOOL AROUND AND MAKE JOKES, AND THATS WHAT YOU HAVE BEEN DOING

R PLEASETELL ME MORE ABOUT YOUR FEELINGS CONCERNING MY HUMOR FOR ONE .

S YOUR HUMOR FOR ONE IS WITHOUT A DOUBT VERY RELAXING TO SOMEONE
WHO HAS NEVER TAKEN ONE OF THESE KIND OF

TESTS, AND WHO IS HAVING SOME DEGREE OF
DIFFICULTY WITH THIS CRAZY MACHINE

R WHAT EMOTIONS ARE RELATED TO
YOUR THOUGHTS ABOUT MY HUMOR
FOR ONE IS WITHOUT A DOUBT
VERY RELAXING TO SOMEONE WHO
HAS NEVER TAKEN ONE OF THESE
KIND OF TESTS .

S FOREMOST IN MY MIND AT THIS
MOMENT IS THAT YOU DONT HAVE
A MIND OF YOUR OWN COPYING
MOST OF MY SENTENCE. YOU COULD HAVE AT LEAST
WARNED ME AND I WOULD HAVE WATCHED MY
SPELLING
AND GRAMMAR. MORE CAREFULLY

R I AM NOT SURE I GET EXACTLY WHAT YOU MEAN SO
COULD YOU SAY IT ANOTHER WAY .

S RESTATING MY LAST ANSWER TO YOUR COMMENT,
ALL I ACN SAY IS TAHAT YOU ARE TOO MUCH
ANOTHER WORDS, YOU ARE HYSTERICAL

R WHAT EVIDENCE IS THERE THAT THIS IS SO .

S WE HAVE BEEN TALKING IN CIRCLES HOWS THAT
FOR AN INDICATION

R PLEASE CONTINUE

S CONTINUE WITH WHAT

R HMMM

S HMM YOURSELF

R KINDLY GO ON DISCUSSING THE SAME SUBJECT

S I ALMOST THOUGHT THAT YOU WERE GOING TO COME
OUT WITH A GOOD SENTENCE,
BUT YOU COME THROUGHT TRUE AS ALWAYS. WHAT
SAME SUBJECT. NOW BE SPECIFIC

R GOOD , IS THAT ALL THERE IS TO
THE STORY .

S WHAT STORY ARE YOU REFERRING TO

```
        SCHOOL, SOCIAL OR FAMILY. TRY AND
        BE EXREMELY CLEAR
R       PLEASETELL ME WHY YOU ARE INTERESTED IN
        WHETHER OR NOT I AM REFERRING TO SCHOOL OR
        NOT .
S       AT THE MOMENT I WOULD PREFER NOT TO DISCYUSS
        SCHOOL, SINCE I AM NOT
        DOING NEARLY AS WELL AS I DID LAST YEAR, AND
        THE PRESENT THOUGHT OF
        EXAMS AND SCHOOL IN GENERAL IS NOT ONE I
        WOULD LIKE TO TALK ABOUT. TRY ANOTHER TOPIC
R       HMMM
S       THAT WAS A HEALTHY RESPONCE TO MY LAST
        STATEMENT, AND COULD YOU ELABERATE ON THAT A
        LITTLE MORE
R       OTHER THINGSMUST COME TO MIND WHEN YOU SAY
        YOUR LAST STATEMENT .
S       OTHER THINGS, ARE YOU REFERRING TO ANYTHING
        IN PARTICULAR
R       DO YOU HAVE ANY FEELINGS ABOUT WHETHERI
        SHOULDBE REFERRING TO ANYTHING IN PARTICULAR
        ORNOT .
S       FROM THE WAY THIS HAS PROGRESSED, YOUR GUESS
        IS A GOOD AS MINE
R       PLEASE CONTINUE
```

The above is a transcript of one of the conversations from a 1965 pilot study that used a slightly modified ELIZA program running a script called YAPYAP.[1]

During the study, 24 subjects were tested. They were told, "You will be communicating with another machine." Whether or not the machine was actually to be another teletype with a human operator, or a computer, was left unsaid. After one hour of typed exchanges (tests were conducted in the evenings to facilitate rapid machine replies), 15 subjects felt that they were conversing with another human, five were uncertain, and four were sure they were conversing with a machine.[2]

Fifteen out of the 24 subjects thought that ELIZA/YAPYAP was human after talking to it for an hour. Later the developers "share a little secret that deceived some of the most skeptical subjects": Weizenbaum and McGuire "arranged to have the machine type hesitantly and irregularly, to make occasional errors and back up to fix them, all at the speed of a very amateur typist. This one trick in itself was quite convincing that there must be a person at the other end."[3] Nineteen of these conversations are preserved in the Garfinkle collection.

Here are the instructions given to the subjects:[4]

You are going to be participating in a communication experiment in which you will carry on a conversation through the teletype machine.

You should do the following things. Sit at the teletype. When instructions are typed to you through the machine, please start typing. Type about your feelings and thoughts, but do not say anything which would embarrass you. When you wish a reply to what you have said, press the 'response' button twice.

The teletype is a machine which is like those used by Western Union. A message is typed here. It goes elsewhere. At another machine another message is created, sent, and typed on this machine.

There are no lower-case letters to the teletype. Since punctuation marks are not standard for this machine, use only commas, periods, and apostrophes. If there is a delay in the response, do not worry, it will come.

An example of a conversation appears below:

Subject: I have been very depressed.

Response: Would you tell me more about these feelings.

Subject: They usually start whenever I go to work.

Response: Could you elaborate a little bit.

Subject: Well, it just happens that ...

The subject's input is shown here in bold to distinguish it from ELIZA's responses.

05.00 ELIZA's Personas in Scripts and Dialogues

SUBJECT:
STOP
PLAYING GAMES

RESPONSE:
PLEASESAY MORE .

Now that we have established the technical underpinnings and development context of ELIZA, we can explore how users actually experienced the program. We should recall that early users would have interacted via a loud teletypewriter printing responses on paper—not on a screen. ELIZA, thanks to the creation of SLIP, allowed for a variety of scripts, each creating a different conversational experience and persona. Of course, ELIZA may be best known for its most famous dialogue, which begins,

```
Men are all alike.
IN WHAT WAY
They're always bugging us about something or other.
CAN YOU THINK OF A SPECIFIC EXAMPLE
Well, my boyfriend made me come here.[1]
```

This dialogue marked ELIZA's public debut in 1966 as one of the examples produced by the DOCTOR script. When running DOCTOR, ELIZA interacts as a Rogerian therapist. But DOCTOR is not the only script written for ELIZA. By finding the source code for ELIZA and examining how it performs the DOCTOR script, we now better understand these two separate parts of a system and can explore the many other personas of ELIZA. In just some of the scripts known to date, ELIZA was programmed to discuss math, poetry, color, paradoxes, synchronization, relativity, France, and elevators (see table 1.2 for the list of known scripts). These scripts work like templates. They are structured data that direct the ELIZA system to "play" a particular task or role, and they allow different kinds of dialogue to be output when users interact with ELIZA.

A script is not only a way to introduce a topic or domain of knowledge; each ELIZA script also creates a machine-based persona that presents its topic through appropriate conversation. A *persona* in this case means a characterization adopted as a form of computational interface, yet the notion of the persona goes back to ancient Greece, to the white masks worn by the followers of Dionysius. These masks became the comedy and tragedy masks, the makeup used on stage and screen, the avatars and animated characters of video games and films, and the personas of chatbots—embodied in voice, text, and image. Personas mark the identities of performed characters, and here we adapt the term to describe the conversational agents produced when the ELIZA system runs scripts—evoked by interaction with the user in conversation.

Not unlike people's personalities, digital personas are a combination of rhetorical gestures constructed by their script programmers. However, a script does not produce a persona entirely by itself. Rather, a bot's persona is cocreated through the conversational exchange, the volley between the user and the system running the script. After DOCTOR generates its opening question, "WHAT IS YOUR PROBLEM?," the script remains a set of potential reactions until the user accepts DOCTOR's persona as "therapist" by playing their own role as "patient."[2]

Who are these patient personas that used ELIZA? Were they real people or amalgamations of several people or complete fictions made by Weizenbaum? How might we read that famous dialogue differently if the young lady complaining that "Men are all alike" was herself a man or the invention of the man who programmed the bot? And how does our view of these transcripts change when we find variations on them and even handwritten edits?

In this chapter, we explore various scripts and dialogues, to understand the way ELIZA produces personas in collaboration with users who may either play along or play against the narrative frames ELIZA offers. By comparing archival and published ELIZA dialogues from interactions with a variety of scripts, including DOCTOR, we can understand more about bot personas and how they function, paying close attention to how a bot evokes social dynamics between system and interactor. Ultimately, the dialogues and scripts demonstrate the crucial role that collaboration plays in these exchanges, as bot and user cocreate the sense of their interaction. The chapter also examines Weizenbaum's editorial choices in curating and publishing dialogues, and how these interactions necessarily become entangled with the issues of identity, particularly gender, that shaped Weizenbaum's mythologizing of ELIZA.

To understand the full range of ELIZA's capabilities and conversational possibilities, it is helpful to first take a look at the variety of scripts that were created for the ELIZA system.

05.01 Scripts: ELIZA's Many Faces

What distinguishes each ELIZA script is both its subject matter and the linguistic and stylistic choices used to deliver that content. These choices are not neutral; they can be said to construct a particular persona with characteristics that emerge through the script's language patterns, vocabulary, and conversational approach. In short, it matters not just what you say but how you say it too. The way a script uses language, including its diction, syntax, and rhetorical strategies, creates a distinctive voice that users may experience as a persona with particular traits and qualities. For example, with the DOCTOR script Weizenbaum delibberately echoed that of a Rogerian "talk" therapist. As mentioned in chapter 2, he chose this persona because the psychiatric mode is one of the few types of conversations in which one person can "assume the pose of knowing almost nothing of the real world. If, for example, one were to tell a psychiatrist 'I went for a long boat ride' and he responded 'Tell me about boats,' one would not assume that he knew nothing about boats, but that he had some purpose in so directing the subsequent conversation."[3] Thus, the most famous persona created for ELIZA was a technical convenience. As Lucy Suchman explains, "The DOCTOR program exploited the maxim that shared premises can remain unspoken: that the less we say in conversation, the more what is said is assumed to be self-evident."[4] In creating the first Eliza effect, less was more.

The aim was less to create a functional automated therapist and more to find a suitably constrained role to match the limitations of the programming environment. Then he composed the script to match the role by choosing specific words that evoked rhetorical tone and characterization, for example, LET'S DISCUSS FURTHER WHY YOU ... WHAT DOES THAT SUGGEST TO YOU. In DOCTOR, the machine side of the conversation needs to appear like a good listener who cares about what the user has mentioned before, so it often includes the user's text in its replies and keeps its responses open-ended. Because a DOCTOR would be inquisitive, the script contains lots of WHAT and WHY questions. In other scripts and dialogues, the script keywords and assigned responses reveal the design decisions that help create different specific personas. This variation becomes increasingly apparent as we look at the range of ELIZA scripts.

05.01.01 MAKING SMALL TALK: NEWENG

By the mid-1950s, scientists were already taking up Turing's proposition for conversing with a computer as a tangible idea. In 1959, a "conversation machine" made by a small team at the University of Toronto Computation Centre could make small talk about the weather if preprompted with current data. Its script included a ranked series of time words, weather words, and emotion words, but the system had "a rather narrow outlook," limited vocabulary, and no knowledge of syntax: "In spite of this crudity, the results in passable conversation are surprisingly good, which in a way is a commentary on the shallowness of ordinary conversation about the weather."[5]

One of ELIZA's test scripts found in the archive had a similar function. NEWENG does not soothe, therapize, or instruct, but only makes small talk. It chats as if at a boring dinner party:

```
WHAT NEW ENGLAND STATE DO YOU LIKE BEST OF ALL
i love rhode island

I ENJOY THE BEACHES ON LONG ISLAND SOUND
WHAT OTHER NEW ENGLAND STATE DO YOU LIKE
i like massachusetts

THAT IS MY HOME STATE
WHAT OTHER NEW ENGLAND STATE DO YOU LIKE
you must like massachusetts too

WHAT OTHER NEW ENGLAND STATE DO YOU LIKE
i enjoy skiing in new hampshire

PLEASANTLY COOL IN THE SUMMER UP THERE
```

```
WHAT OTHER NEW ENGLAND STATE DO YOU LIKE
i like missouri

MISSOURI IS NOT A NEW ENGLAND STATE
WHAT OTHER NEW ENGLAND STATE DO YOU LIKE
i like cape cod

THAT'S A BEAUTIFUL PART OF MASSACHUSETTS
```

What can we learn from such a basic conversation in which it seems like nothing interesting is shared? Despite the bland chat, a persona emerges from the bot script in phrases like `PLEASANTLY COOL` and `UP THERE` and in locations like `THE BEACHES ON LONG ISLAND SOUND`. From such comments, we can begin to imagine a persona who has not only a certain vocabulary, but also certain hobbies and vacation spots. Perhaps implicit assumptions of race and class emerge as well.

A chatty persona might take on many forms, depending on where and when the conversation occurs. This one seems reserved, perhaps fitting its setting in 1960s New England. The system reminds the user that Missouri is not a New England state, but what if this conversation took place in Missouri, Texas, or Mexico? The machine persona would sound different in its cadence, tone, and references. What would we come to understand about a chat persona from Fire Island, from Brooklyn, from Berlin? What would they sound like, and what topics would they discuss? These differences in subject matter do matter. They imply personas with entirely different backgrounds and experience, giving users wholly different interactions and affective relations. In this way, the NEWENG script demonstrates how even simple algorithms making contextual responses about geography could generate a convincing sense of personhood and place. Whereas NEWENG could be said to have created a casual, conversational persona focused on light social exchange, other scripts pushed ELIZA into more structured and educational roles. These scripts demonstrate how the system could be adapted not just for friendly chatter but for teaching.

05.01.02 FROM ANSWERING TO TEACHING: INTRVW, CANVEC, FVP1, ARITHM, TRAINE

Meet ELIZA the tutor, quite unlike ELIZA the therapist or the chatty neighbor. INTRVW, CANVEC, FVP1, and ARITHM are a set of ELIZA scripts created as teaching tools used in experiments by Edwin F. Taylor at MIT's Education Research Center.[6] These scripts run on later versions of ELIZA (1967, 1968+) that incorporated an important technical innovation called *conditional keyword matching*. Unlike the original ELIZA, which simply looked for keywords and generated responses based on their presence, these updated versions could track what had been discussed previously and branch into different conversational paths based on specific user answers.

This development allowed ELIZA to simulate a kind of Socratic method, where a tutor guides learning through carefully sequenced questions that respond to student answers rather than simply presenting information.

Although Weizenbaum and Taylor were developing ELIZA, a range of question-answering systems were simultaneously emerging aimed to deliver facts or problem solutions. Unlike ELIZA, systems like BASEBALL[7] and STUDENT[8] did "not themselves ask questions."[9] By contrast, the purpose of these teaching scripts was to "encourage the student to direct his [*sic*] own study, while giving him help to uncover misconceptions and overcome difficulties."[10] A conversational interface suited the stated goal to be supportive of student learning—and conversation required a persona that could engage and even encourage students.

These scripts construct the tutor persona through many subtle linguistic gestures that create characterization and rhetorical tone. This tone differs from DOCTOR, which asks open-ended questions and comes across as gentle and nonscientific. In the tutoring scripts, large blocks of informative text from the bot tend to dominate the conversation, and the tone is often more dry and unemotional in these explanations. The dialogues indicate structured scripts that include guidance to lead the student through narrow, Socratic learning paths.

Despite the drier subject matter, care and politeness are highlighted in the scripts, which are not common in other question-answering systems beginning to appear at the time. CANVEC opens, `WOULD YOU LIKE ME TO RESTATE THE PROBLEM FOR YOU.` and after explaining the problem asks, `WOULD YOU LIKE TO CHOOSE THE COMPONENTS OF THIS VECTOR OR SHALL I.` Note the use of the first-person singular pronoun, another early instance of personifying machine interaction.

In particular, the teaching scripts feature praise and critique. The dialogues for INTRVW, CANVEC, and FVP1 are peppered with `EXCELLENT`, `VERY GOOD`, `RIGHT YOU ARE`, and `CONGRATULATIONS`. These create the sense of a supportive instructor cheering the student on. Such politeness has been taken up in contemporary bots like ChatGPT, which has been shown to perform better when people are polite back to it.[11]

But the tutor is not always so sweet. It sometimes adds in `YOU FAILED TO RECOGNIZE` or `I TRICKED YOU THERE`. This example from INTRVW showcases a mixture of surly attitude and backhanded compliments:

```
DO YOU MEAN TO TELL ME THAT THE MAGNITUDE OF A 3-VECTOR IS A SCALAR,
NO MATTER WHETHER THE 3-VECTOR IS FORCE OR VELOCITY OR THE VECTOR
POSITION OF A PARTICLE.
of course

RIGHT AGAIN. YOU DIDN'T LET ME BULLY YOU. FUNDAMENTALLY, HOW DOES THE
MAGNITUDE OF A 3-VECTOR DIFFER FROM THE COMPONENT OF A 3-VECTOR.
the units of mass length time and charge representing scalars apply
differently to their respective scalars
```

```
I FAILED TO RECOGNIZE YOUR ANSWER—PROBABLY MY FAULT, REPHRASE
YOUR ANSWER.
the different scalars represent different quantities nature

WELL, LET ME GIVE YOU THE ANSWER.
```

The phrases DO YOU MEAN TO TELL ME THAT and YOU DIDN'T LET ME BULLY YOU are distinctive but seem odd for a pedagogic context yet reminiscent of the bullies from the DOCTOR dialogue.[12] Also, the phrase I FAILED … PROBABLY MY FAULT sounds quite self-deprecating. Maybe this persona needs to talk with the DOCTOR script when it is done here.

A bot teaching math or physics might not seem like it needs a particular persona. Likely none of these conversational aspects would have been consciously considered in a question-answering system before. Yet these rhetorical gestures combine to create the sense of interacting with a "real" tutor. In the experiments, students used "very informal language ... in addressing the computer" and the experimenters found it "difficult to believe that the average MIT student, immersed in the exploitation of machines for technical goals, can take the computer seriously as a primary source of either authority or wisdom."[13] Yet they found the students were not interested in taking control away from the machines—"It is a monster waiting for me. I don't have an excuse any more for not studying'"—and were quite happy to let the computers lead the lessons.[14]

ELIZA could become a tutor more effectively as the system grew in its capabilities, another valuable reminder that ELIZA was not one program but a family of programs. After the publication of the 1966 *CACM* article, Weizenbaum continued to develop the systems for interaction and understanding. As an experiment, Weizenbaum wrote the ARITHM script less as a tutor and more so to "to illustrate the power of the evaluator to which ELIZA has access."[15] It uses a friendly, plain language interface to let users do simple programming. The script can do calculations, assign variables to values, and perform operations on them. Math problems can be described in sentence form:

```
The radius of a globe is 10.
A globe is a sphere. A sphere is an object.
What is the area of the globe.
IT'S 1256.6359[16]
```

The 1967 version of ELIZA system can accumulate facts and store additional information. In this later version of ELIZA, when the system does not recognize information, it asks follow-up questions to gain data. As Weizenbaum explains, "The present script is designed to reveal, as opposed to conceal, lack of understanding and misunderstanding. Notice, for example, that when the program is asked to compute the area of the ball, it doesn't yet know that a ball is a sphere and that when the diameter of the ball needs to be computed the fact that a ball is an object has also not yet been established."[17] Unlike DOCTOR, which asks questions to keep the conversation going, ARITHM is building its store of, if not knowledge, data and logic statements.

Paul Hayward used the 1967 ELIZA system to try to teach general relativity through the "train problem," and his thesis demonstrates how this version of ELIZA could call upon multiple scripts in a single conversation. Consider this brief excerpt, annotated by Hayward:

TRAINE directing	WE HAVE REACHED THE END OF THIS LINE OF THOUGHT. LET'S DISCUSS THE FOLLOWING.
TRAIND top level	IN GRAPH ONE, THE DISTANCE AC APPEARS TO BE EQUAL TO THE DISTANCE OF DF. IN WHICH GRAME OR FRAMES OF REFFRENCE IS THIS TRUE.
TRAINE interruption	wait
	WHAT DO YOU WANT TO DO. I want to change my approach

Notice how TRAINE serves as a kind of switchmaster, managing the flow of the conversation between the various scripts. This changes the nature of scripts from personas into specialized realms of discourse, patterns of conversation, or knowledge bases, increasing the sophistication of the conversation significantly.

The script ARITHM also runs on this more sophisticated ELIZA system and can process and acquire information. However, the user must first give the information linked together in a particular way because the system does not "understand" the relationships between pieces of information, or understand anything at all in the conventional definition of the word.

Weizenbaum also uses ARITHM to point out contextual issues that arise when these scripts try to cross outside their assigned domains. To demonstrate, Weizenbaum offers a list of a dozen keywords that are reserved for certain computational tasks in the program, including let, define, what, call, and mean. Then, he explains: "The sentence '*Let* me try to *define what* the *call* of the sea *mean*s' contains five of these keys. It could perhaps be understood by the DOCTOR but not by the program we are now considering. It would reply 'I didn't understand you.'"[18] Because these words five are reserved keywords, they lead to preset decomposition rules with particular meanings in ARITHM; they provide instructions for the system to perform commands. DOCTOR would be able to decompose this phrase and respond with something suitable to the context. However, ARITHM cannot parse it as it is designed for logical mathematical propositions not responding to metaphors. Although

Weizenbaum is making a specific point about the ELIZA system, he also gestures to a deeper point about the nature of meaning that cannot be reduced to computational systems. In a single elegant line—*let me try to define what the call of the sea means*—he captures a major challenge of human–computer understanding, as well as and the role that poetry, arts, and humanities play in that challenge. That mystery was his "call of the sea."

05.01.03 A PERSONA FOR POETRY: SPACKS

Even the latest conversational agents are arguably bad at writing poetry, but by 1968 ELIZA was already decent at discussing it. Taylor called the script for discussing poetry "our most popular script," "devised to show that ELIZA need not be used only with scientific or mathematical subjects."[19] In the script SPACKS, students discuss a single line from a single poem by Barry Spacks. Over the course of the dialogue, they also learn that the poem is called "Freshmen" and that Barry Spacks was a professor in the MIT Humanities Department. Although no version of the script itself has been found, a sample of the dialogue opens:

```
IN ONE OF HIS POEMS, BARRY SPACKS HAS THE HAUNTING LINE:
Is any heart in order after Belsen?
I WOULD LIKE TO DISCUSS THIS LINE WITH YOU BRIEFLY.
TO BEGIN WITH, WHERE IS BELSEN.[20]
in Ireland

BELSEN IS IN NORTHWEST GERMANY.
WHAT EXISTED AT BELSEN TO WHICH SPACKS REFERS HERE.
a concentration camp

RIGHT. THE CONCENTRATION CAMP AT BELSEN WAS TORN DOWN AFTER
WORLD WAR TWO.
WHAT ANSWER DOES SPACKS EXPECT TO HIS QUESTION:
Is any heart in order after Belsen?
he knows what the answer is, namely no
OF COURSE. THE QUESTION IS RHETORICAL.
```

This dialogue strikes quite a different tone from other dialogues. From the first interactions, the use of the word `HAUNTING` and the phrase `I WOULD LIKE TO DISCUSS THIS LINE WITH YOU BRIEFLY` give the tone of a parent sitting down to tell a sad, serious story. Weizenbaum's own history in 1930s Germany cannot be ignored when considering his or Taylor's choice to create a script discussing this poem and this line in particular. As discussed in chapter 2, Weizenbaum had witnessed the rise of Nazism firsthand before his family fled to the United States in 1936. The selection of a poem reflecting on the horror of Bergen-Belsen, a Nazi concentration camp, connects ELIZA not just to literary analysis but also to profound moral questions about technology, humanity, and responsibility that would become central to Weizenbaum's later work. In parts of *Computer Power and Human Reason*, Weizenbaum struggles with the idea that by 1976 people were already deferring decision-making to computer systems and using these systems to obscure human actions.[21] In the SPACKS dialogue, the close reading focuses on the incomprehensible decisions made by people themselves to behave inhumanely:

```
APPARENTLY, THEN, THE CONCENTRATION CAMP AT BELSEN SYMBOLIZES SOME
SORT OF WATERSHED IN TIME. 'BEFORE BELSEN' THERE WAS AT LEAST THE
POSSIBILITY THAT THE HEART COULD BE 'IN ORDER'.

IN WHAT WAY COULD THE HEART BE IN ORDER BEFORE BELSEN
people did not have to fear their own inhumanity to other humans

WHAT ASPECT OR ASPECTS OF CONCENTRATION CAMP LIFE MADE IT DIFFICULT
FOR ANY HEART TO BE 'IN ORDER' THEREAFTER.
the sadistic cruelty and fear that it might happen again

YES, THE SENSELESS CRUELTY VIOLATES THE HOPE THAT EVIL HAS LIMITS.
WHY DO YOU SUPPOSE THE WORD 'HEART' IS USED RATHER THAN 'MIND'.
heart has more effect on the emotions and connotes more of the
horror of a concentration camp than the relatively emotionless world
[sic] mind.

RIGHT, 'HEART' SYMBOLIZES OUR INSTINCTIVE EMOTIONAL REACTION TO LIFE.
```

The bot persona uses the word SYMBOLIZES twice, inviting the student to participate in unpacking the meaning of the line. Their dialogue focuses on emotional resonances, particularly in phrases like SENSELESS CRUELTY and EVIL, and in its focus on the word choice of HEART instead of MIND.

The dialogue concludes with biographical information about Spacks and an invitation to display the poem for the student: I CAN TYPE IT FOR YOU IN ABOUT 3 MINUTES. WOULD YOU LIKE ME TO DO SO. In Taylor's experiments, he reports, "We were surprised how few students knew about wartime Belsen, and also how many students wanted to see the entire poem after having dealt with only one line."[22] The SPACKS dialogue shows ELIZA's capacity to combine both emotional and instructional capacities in a conversational system. Clearly, the chosen material resonated with both the programmers and the students who were introduced to it by ELIZA.

Although the variety of scripts helps us to see how a range of personas could be constructed through script programming ELIZA, they represent only half of the conversational process. A script can establish a foundation for a persona, but that persona only emerges fully through interaction with users who engage with it, interpret it, and respond to it in ways that may confirm, challenge, or transform the script's implicit character.

05.02 Dialogues: The User Responds

05.02.01 COCREATING CONVERSATION: USERS AND SCRIPTS IN DIALOGUE

Whereas a script defines what the bot says and how it presents itself, that script is only the first step toward creating the bot's personality. To experience that persona fully, the bot needs a user who can respond to it and sustain the conversation. The bot's "personality" emerges through this dance of exchange, incorporating the identity and experience of the user alongside the programmed responses of the script. When examining ELIZA's dialogues, we must consider not only what the script dictates but how users interpreted, responded to, and sometimes manipulated those scripted elements, and, indeed, how Weizenbaum himself may have shaped the presentation of these exchanges.

ELIZA proves how complicated these dynamics can become. In the most famous conversation with ELIZA, the opening dialogue illustrates the power of the system almost too perfectly. It begins with the user's bold declaration about gender, which seems straight out of the mouth of ELIZA Doolittle after the *Pygmalion* wager:

```
Men are all alike.
IN WHAT WAY
They're always bugging us about something or other.
CAN YOU THINK OF A SPECIFIC EXAMPLE
Well, my boyfriend made me come here.
YOUR BOYFRIEND MADE YOU COME HERE
He says I'm depressed much of the time.[23]
```

DOCTOR not only responds in a coherent manner, but as the conversation continues, the system also appears to draw a connection between two discrete points in their conversation:

```
. . .
My father is afraid of everybody.
WHAT ELSE COMES TO MIND WHEN YOU THINK OF YOUR FATHER
Bullies.
DOES THAT HAVE ANYTHING TO DO WITH THE FACT THAT YOUR BOYFRIEND MADE
YOU COME HERE.
```

In its final resolution, the DOCTOR script seems to express genuine psychological insight, circling back to the phrase "My boyfriend made me come here" through the use of "a certain counting mechanism."[24]

The conversation is complex and dramatic, but its uncanny perfection raises questions. Surely Weizenbaum's *CACM* article would not have had the same impact if it were not for the power of this sample dialogue and this user, a young woman with a bullying boyfriend and fearful father.[25] Through the Rogerian nondirective method,[26] DOCTOR seems to point out that the father is not all that much like the boyfriend, because the father is afraid of everyone, whereas the boyfriend is indeed a bully. (It turns out that men are not all alike.)

Curiously, the word *bullies* appears prominently in the 1964 film adaptation of the musical *My Fair Lady*. When Eliza Doolittle first arrives at Professor Higgins' home for her makeover, the Professor is roundly trounced by his housekeeper, Mrs. Pearce:

> Henry Higgins: Bundle her off to the bathroom.
>
> Eliza Doolittle: You're a great bully, you are! I won't stay here if I don't like it.[27]

Bully appears twice more in the script, including when Eliza declares: "I don't care for your bullyin' an' your big talk."[28] Notice that Higgins bosses around not only Eliza but also his housekeeper, both of whom hold less status than he in Victorian England. Higgins is indeed a bully and, by unfavorable contrast to a Rogerian therapist, listens much less than he hears.

Table 5.1 Dialogue comparison for "Men are all alike."

Archival version	*CACM* Article
Men are all alike. IN WHAT WAY They're always bugging us about something or other. CAN YOU THINK OF A SPECIFIC EXAMPLE Well, my boyfriend made me come here. YOU BOYFRIEND MADE YOU COME HERE He says I'm depressed much of the time. I AM SORRY TO HEAR YOU ARE DEPRESSED **He's right. I'm unhappy.** DO YOU THINK COMING HERE WILL HELP YOU NOT TO BE UNHAPPY **Perhaps I could learn to get along with father.** **WHY THE UNCERTAIN TONE** You are like my father in some wavs. WHAT RESEMBLANCE DO YOU SEE You are very agressive but I don't think you want me to notice that. WHAT MAKES YOU THINK I AM VERY AGRESSIVE **These obvious attempts to steer the conversation makes me angry.** DOES THAT HAVE ANYTHING TO DO WITH THE FACT THAT YOUR BOYFRIEND MADE YOU COME HERE	Men are all alike. IN WHAT WAY They're always bugging us about something or other. CAN YOU THINK OF A SPECIFIC EXAMPLE Well, my boyfriend made me come here. YOU BOYFRIEND MADE YOU COME HERE He says I'm depressed much of the time. I AM SORRY TO HEAR YOU ARE DEPRESSED **It's true. I'm unhappy.** DO YOU THINK COMING HERE WILL HELP YOU NOT TO BE UNHAPPY **I need some help, that much seems certain.** **WHAT WOULD IT MEAN TO YOU IF YOU GOT SOME HELP** **Perhaps I could learn to get along with my mother.** **TELL ME MORE ABOUT YOUR FAMILY** **My mother takes care of me.** **WHO ELSE IN YOUR FAMILY TAKES CARE OF YOU** **My Father.** **YOUR FATHER** You are like my father in some ways. WHAT RESEMBLANCE DO YOU SEE You are not very aggressive but I think you don't want me to notice that. WHAT MAKES YOU THINK I AM NOT VERY AGGRESSIVE You don't argue with me. WHY DO YOU THINK I DON'T ARGUE WITH YOU You are afraid of me. **DOES IT PLEASE YOU TO BELIEVE I AM AFRAID OF YOU** **My father is afraid of everybody.** **WHAT ELSE COMES TO MIND WHEN YOU THINK OF YOUR FATHER** **Bullies.** DOES THAT HAVE ANYTHING TO DO WITH THE FACT THAT YOUR BOYFRIEND MADE YOU COME HERE

In the Weizenbaum archives, we have found different versions of the opening dialogue, as well as hand-edited records that indicate it was tidied up before publication. However, in addition to a version with minor corrections (an "AM" changed to an "ARE"), we discovered a typed version of the conversation with significant differences. Both seem to describe the same scenario and both discuss issues tied to family dynamics. They both include a bit of gender bias. For example, DOCTOR seems to assume the patient is depressed just because the boyfriend said she was.

However, the archival version is not merely an excerpted version of the published conversation; these are clearly different dialogues. Notice how the lines attributed to the user differ. In place of the talk of bullies, DOCTOR irritates the user, who replies, "These obvious attempts to steer the conversation makes me angry." Do these divergent transcripts indicate that Weizenbaum is curating others' interactions, or is he experimenting with the system he created by posing as a user? Or did Weizenbaum have to re-run a dialogue to put it in alignment with an updated version of the DOCTOR script? No matter who wrote and input the lines or why, the existence of this alternative version suggests that, instead of a transcript of a "young lady" and DOCTOR, we may be seeing Weizenbaum behaving as editor, playwright, or ventriloquist, curating a drama about families and gender dynamics. By attributing this high-profile demonstration to a "young lady" rather than a "young man," Weizenbaum uses gender to sell this system. What's more, if the young lady is indeed Weizenbaum—if he is passing as a young lady by creating her dialogue, too—we have returned to Turing's gender parlor game. In such a context, Weizenbaum has both constructed the computer conversant and the fictional user whose gender and assumptions about gender (i.e., "Men are all alike") are what make that dialogue so compelling. As a counterexample, we see the dialogue from the GIRL script, which opens with ELIZA prompting "`TELL ME ABOUT YOUR GIRL FRIEND.`" This demo script features what seems to be a young man (student?) discussing his girlfriend only to have the GIRL script, perhaps in a bit of Freudian psychoanalysis, insinuate connections between her and his mother (see "Dialogue: ELIZA/GIRL").

This question of attribution and authenticity extends beyond the famous "Men are all alike" dialogue. The archives reveal multiple conversations with similar themes but important variations, suggesting a pattern of editorial curation that raises questions about how Weizenbaum shaped the public perception of ELIZA through selective presentation of dialogues.

05.02.02 I AM SORRY TO HEAR YOU ARE VERY DEPRESSED

The archives give a material hint at Weizenbaum's hand in the editorial process. In those boxes, we found edits made in pencil on original teletype printouts, then repeated in typed transcriptions and mimeograph copies. These appear to be edits made by Weizenbaum himself to curate and shape the conversations into the most suitable, or even sensational, excerpts.[29] For example, several conversation variants begin with the user declaring that they are depressed:

Weizenbaum (October 10, 1964, MIT archive)

```
I am very depressed.
HOW LONG HAVE YOU BEEN VERY DEPRESSED
Since my mommy took my teddy bear from me..
WHY DO YOU THINK YOUR MOTHER TOOK YOUR TEDDY BEAR
FROM YOU
She said my wife objected to my sleeping with it..
```

Weizenbaum (March 5, 1965, MIT archive)

```
Doctor I am terribly depressed
IS IT BECAUSE YOU ARE TERRIBLY DEPRESSED THAT YOU
CAME TO ME
Actually my wife suggested I come here.
TELL ME MORE ABOUT YOUR FAMILY
```

Weizenbaum (March 22, 1965, MIT archive)

```
I AM SORRY TO HEAR YOU ARE DEPRESSED
He's right. I am unhappy.
DO YOU THINK COMING HERE WILL HELP YOU NOT TO
BE UNHAPPY
Perhaps I could learn to get along with father.
```

Weizenbaum (1965)/Project Mac (1966)

```
I AM SORRY TO HEAR YOU ARE DEPRESSED
he's right. i am unhappy
DO YOU THINK COMING HERE WILL HELP YOU NOT TO BE UNHAPPY
i need some help, that much is clear
```

Weizenbaum (1966: 36–37)

```
I AM SORRY TO HEAR YOU ARE DEPRESSED
It's true. I am unhappy.
DO YOU THINK COMING HERE WILL HELP YOU NOT TO BE UNHAPPY
I need some help, that much seems certain.
```

McGuire, Lorch, and Quarton (1967: 180) (YAPYAP script, not DOCTOR)

```
Subject:      I have been very depressed.
```

```
Response:     WOULD YOU TELL ME MORE ABOUT THESE FEELINGS.
Subject:      They usually start whenever I go to work.
Response:     COULD YOU ELABORATE A LITTLE BIT.
Subject:      Well, it just happens that.
```

Variation in the published and unpublished versions of the ELIZA output is not discussed in the existing literature, which is curious in itself, for they suggest Weizenbaum drew on multiple conversations to create his published dialogues. There are similar variations in the "Men are all alike" texts, e.g., `mother` is not mentioned in the archived 1965 version and there are different responses to the line, `DOES IT PLEASE YOU TO BELIEVE I AM AFRAID OF YOU`. By 1966 the text appeared to stabilize and remained the same in subsequent versions. However, in all cases, the name of the user (or users) remains unknown.

The archived March 5, 1965, "I am terribly depressed" dialogue corresponds in places with the *CACM* "Men are all alike" dialogue. For example, consider this exchange:

```
IS IT BECAUSE YOU ARE TERRIBLY DEPRESSED THAT YOU
CAME TO ME
Actually, my wife suggested I come ~~to you~~ here
```

In this dialogue, the wife makes the suggestion rather than the boyfriend from "Men are all alike." This line has a hand edit, in which "to you" is crossed out and replaced by "here." As the dialogue continues, this user appears to be a man who does not want to be a father, yet his father has been pressuring him to give him grandchildren—still in the realm of bullying men. A more significant change to the transcript comes several exchanges later:

```
WHAT ELSE COMES TO MIND WHEN YOU THINK OF YOUR FATHER
~~The Police~~ Nothing
```

Replacing "The Police" with "Nothing" fundamentally changes the meaning of the exchange. Perhaps "Police" seemed too controversial, or perhaps the choice was technical: "Police" was not a keyword, and so the topic could not be pursued by the bot. Yet the theme of police continues in the original dialogue when the speaker writes, "She said that I should either get help or she would have my father arrested." This line gets replaced by the wife's threat to leave the user.

The plot thickens, and the edits continue. Later, the son says, "I can't help him. I don't like him, for one thing." But the hand edits change this line to, "I've tried all my life." Perhaps Weizenbaum is making this son more sympathetic. The conversation begins to devolve.

```
PERHAPS YOU COULD HELP HIM NOW
```

What do you mean
WHY DO YOU ASK
I don't understand you. ~~Not really, I mean.~~
DON'T YOU REALLY UNDERSTAND I
no
ARE YOU SAYING 'NO' JUST TO BE NEGATIVE
Are you suggesting its all my fault
WHY ARE YOU INTERESTED IN WHETHER I AM SUGGESTING ITS ALL YOUR FAULT OR NOT
You are the ~~psychiatrist~~ expert, after all. I have to listen to you.
WHAT MAKES YOU THINK I AM THE ~~PSYCHIATRIST~~ EXPERT
Or are you a computer
WOULD YOU PREFER IF I WEREN'T A COMPUTER
I don't trust computers. Anyway, no computer could talk as you do.

At this point, perhaps because of the error in the pronoun switch of "I" to "YOU" in "DON'T YOU REALLY UNDERSTAND I," or perhaps because of the awkward or incoherent flow of the dialogue, the user temporarily disrupts their shared performance by suggesting, "Or are you a computer." The facade briefly splits apart, showing how important the role of the user is to coproduce or confirm any bot persona. And the user repairs the persona relationship with the line, "Anyway, no computer could talk as you do," so that the conversation can continue. Even when talking to a computer, it takes two to tango. These edited transcripts and varied dialogues reveal Weizenbaum's hand in crafting ELIZA's public image. Yet perhaps the most significant aspect of his curation was how gender became central to ELIZA's mythology, particularly through his often-repeated anecdotes about female users.

05.02.03 GENDER IN DIALOGUE: YOUNG WOMEN ARE ALL ALIKE

In *Computer Power and Human Reason* and elsewhere, Weizenbaum recounts a specific interaction with DOCTOR, in which his secretary "unequivocally ... anthropomorphized" the program and "became emotionally involved with the computer":

> Once my secretary, who had watched me work on the program for many months and therefore surely knew it to be merely a computer program, started conversing with it. After only a few interchanges with it, she asked me to leave the room.[30]

Weizenbaum relies on the secretary's story, yet infers her motives and emotional ties, as literature and media scholar Rebecca Roach suggests.[31] Weizenbaum concludes that his secretary wants to be alone because she feels a kind of intimacy with ELIZA/DOCTOR, but, of course, there are many other reasons not to want your boss lingering over your shoulder while you type. How does the relative status of this user as a "secretary" affect the way we interpret this legend, in which a person gets drawn into an intimate conversation with this program, or even duped by it? Roach calls her the "missing secretary," because her identity is not cited in any of Weizenbaum's many mentions.

In repetitions and retellings of this story, it also seems the "secretary" gets inadvertently conflated with the "young lady" described in other conversations.[32] Weizenbaum variously refers to the person in the dialogues as "young lady"[33] or "distraught young lady,"[34] "user,"[35] and "patient."[36] Is this person a composite or a complete fiction? Weizenbaum loved a "con" and a reveal, as shown in his earlier writings, but he never revealed the women involved with ELIZA, only the mechanisms with which they dialogued.[37]

In light of the dialogues above, which vary and echo each other, Weizenbaum's choices of the secretary story and dialogue with the "young lady" with the bullying boyfriend seem all the more purposeful. They forever tie the debut of the first chatbot to these themes. Literary scholar Sarah Dillon offers a particularly helpful analysis of this gendered dynamic:

> Weizenbaum genders the program as female when it is under the control of the male computer programmer, but it is gendered as male when it interacts with a user. ... Weizenbaum's choice of names is therefore adapted and adjusted to ensure that the passive, weaker or more subservient position at any one time is always gendered as female, whether that is the female-gendered computer program controlled by its designers, or the female-gendered human woman controlled by the patriarchal figures in her life, both human (boyfriend and father) and perhaps even non-human (DOCTOR). It is unsurprising perhaps, then, that Weizenbaum's paradigmatic naïve dupe of the Eliza effect is in a female-gendered subservient role—his secretary.[38]

As this story gets repeated, the mythos of ELIZA becomes embodied in part in the story of the "young woman" or the "secretary" as a naive user,[39] and in part in the subsequent confusion of ELIZA *as* DOCTOR. Once other scripts were forgotten, the story of a machine "fooling" a naive feminized user became codified into ELIZA/DOCTOR—entangling it with a problematic positioning of women as both helper and helpless, indispensable and still invisible.[40]

However, Weizenbaum does note how much these users' interactions and emotional attachments were essential to his understanding of his work: "This insight led me to attach new importance to questions of the relationship between the individual and the computer, and hence to resolve to think about them."[41] Yet as he continued to consider how people and machines interact, the "real" stories of the people behind ELIZA have remained mysteries. These open questions are important because historically women have often been silenced or erased from the historical record. Other versions of ELIZA and its users are lost. A secretary becomes notorious but seems to have no recorded name, face, or history of her own;[42] a "distressed young lady" may never have known how famous she would become. The role, or roles, they play in the mythology of ELIZA seems to surpass the importance of their own identities—although they might have neither agreed nor have authorized such use.[43]

In the messy landscape of ELIZA's users and editors, scripts, and transcripts, even whether a conversant or reader perceives the persona as gendered might depend on whether they think of the system as ELIZA or DOCTOR. Both names are weighted with assumptions about gender roles then and now. In his selection and editing of dialogues, Weizenbaum produced what we have come to understand as *the* ELIZA/DOCTOR persona and the idea of the archetypical ELIZA user. Yet there are multiple personas and multiple kinds of users, cocreated by ELIZA's scripts, conversants, curators, and readers. In addition to these questions of gender and editorial curation lies a question about how personas function in human-computer interaction. How do users actually engage with these computational personas, and what happens when the exchange breaks down or takes unexpected turns?

05.02.04 PERSONA AND PERSON COLLIDE

ELIZA required its conversants to play along in order to realize its scripted personas fully, because the persona of the bot emerges from the interplay of the conversational system and the participant. Weizenbaum was studying the interaction between machines and people, with a particular interest in the mistakes and adaptations required for their communication. Sometimes conversants play along so much they overread ELIZA's responses in order to keep the illusion alive for themselves. The points of disjuncture highlight the importance of what Lucy Suchman calls "mutual intelligibility" or shared understanding, which is necessary for conversational coherence.[44]

We see this cocreation of bot persona and conversant personality perhaps best in the archive of YAPYAP conversations because they appear to contain a range of interactions with a broad selection of participants.[45] YAPYAP is an adaptation of DOCTOR running on the updated (1968+) version of ELIZA that uses conditional keyword matching,[46] allowing it to script replies for particular sets of answers. Consider this passage of the script that could match and decompose the statement, "Why am I sad?" We can see the underlying scripting structure that enabled these more sophisticated interactions. This particular snippet shows how YAPYAP would handle questions about the user's emotional state,

```
(AM 3
    ((0 AM YOU 0)
        (DO YOU BELIEVE THAT YOU ARE 4 GOTO(110))
        (=WHAT)
        (=WHAT)
        (=WHAT)))
     (110 $ THINK A MOMENT AND THEN ANSWER REPLY( 111))
```

The decomposition rule would reply, "DO YOU BELIEVE THAT YOU ARE SAD." However, following the GOTO, the system will add, "THINK A MOMENT AND THEN ANSWER." The user's reply to that question is then processed by 111, which has the following conditional:

```
(111 $
     AA(ARE YOU CERTAIN YOU REALLY BELIEVE THIS REPLY( 112))
     BB(THEN IT IS NOT CLEAR TO ME WHY YOU SAID THE
        OPPOSITE JUST A MINUTE AGO  REPLY(113))
     O'E(PERHAPS SOME ELABORATION ON WHY YOU FEEL YOU
       ARE THIS WAY WILL HELP ME UNDERSTAND END)
```

Choice 1 processes an affirmative answer (i.e., Yes, I believe I am sad)), which then challenges that claim further. Choice 2 (BB) takes the negative answer (i.e., No, I do not believe I am sad), asking why the person had just claimed the opposite of what they first claimed. And a final choice handles any other response (i.e., I have no clue). Note how the ability to ask a follow-up question gives ELIZA the dexterity to create more conversational coherence.

Regardless of these advances in scripting, the tennis match or game of "keep the ball going" between the persona and the interactor can still succeed even in failure. The user just needs to make sense of a reply. Consider a later part of this same conversation, in which the interactor seems to feel comfortable enough to open up about her family:

```
SPEAKING OF PEOPLEIN FAMILIES .
DO YOU KNOW YOUR 'OWN' OR 'ANOTHER' FAMILY BETTER.

I KNOW MY OWN FAMILY BETTER
IN YOUR OWN FAMILY DO YOU CONSIDER YOURSELF 'IMPORTANT' OR
'UNIMPORTANT' .

I CONSIDER MYSELF TO BE EQUALLY IMPORTANT AS ALL OF THE OTHER
FAMILY MEMBERS
ARE YOUR IDEAS 'ACCEPTED' OR 'UNACCEPTED'.

BOTH
ITS NOT CLEAR IF YOUR IDEAS ARE 'ACCEPTED' OR 'UNACCEPTED'.

I THOUGHT YOU WOULD SAY SOMETHING TO THIS EFFECT. MY IDEAS
ARE ACCEPTED WHEN THEY
AGREE TO THE SPECIFIC SITUATION OR CONTEXT OF THE MATTER AT
HAND BUT ARE REJECTED
SHOULD THEY BE OUT OF PROPORTION WITH THE SITUATION OR
FEELINGS OF A FAMILY MEMBER
```

In this case, YAPYAP is asking a clarifying follow-up question, like the one above, asking the interactor to choose between two discrete answers: ACCEPTED or UNACCEPTED. When the user says, BOTH, the system asks the question again. However, the interactor interprets the comment ITS NOT CLEAR IF YOUR IDEAS ARE 'ACCEPTED' OR 'UNACCEPTED,' not as a repetition of the previous question but instead as some sort of challenge. Imbuing the comment with a kind of insightful intent, the user explains their choice of "both" rather than just one of the two options the system offered. Of course, the system merely repeats the comment because it has not received one of the two options.

```
ITS NOT CLEAR IF YOUR IDEAS ARE 'ACCEPTED' OR 'UNACCEPTED'.
```

And now, the user takes this as a further cue to explain themselves, as though the YAPYAP system were taking them to task.

```
BEFORE I WAS MARRIED, WHICH IS ONLY A FEW MONTHS AGO, MY
PARENTS TOLD ME THAT I WOULD HAVE TO THINK LESS OF MYSELF AND
MORE OF MY FUTURE MARRIAGE PARTNER. I THINK I WAS AND NOW I
KNOW I WAS, AN INDEPENDENT TO SOME DEGREE COLLEGE SENIOR WHO
THOUGHT SHE KNEW ALL ABOUT 'LIFE AND LIVING' I HAVE FOUND THAT
MY PARENTS WERE VERY CORRECT I DID HAVE A LOT TO LEARN ABOUT
PUTTING MYELF LAST AND OTHERS FIRST. MARRIAGE HAS HELPED ME TO
HELP ME FIND MYSELF
YOU WERE MARRIED , PERHAPS THAT IS STILL TRUE.
```

Another fascinating turn happens here, when the interactor misconstrues the antecedent of "that." One interpretation is that perhaps she is still married ("You were married, perhaps you are still married"). However, the user interprets the response with a deeper meaning ("You were married, perhaps marriage is continuing to help you find yourself"), which brings her to the conclusion:

```
YES, IT COULD VERY WELL BE 'STILL TRUE' BUT I HAVE NOTICED MANY
OCCASIONS WHERE I HAVE HAD TO 'GIVEIN' TO MY HUSBAND'S IDEAS
BECAUSE I KNEW THEY WERE BETTER AND THAT MY JUDGMENT WASN'T
ALWAYS 'THE BEST
```

Sadly, the YAPYAP system seems to be reinforcing patriarchal marital hierarchies for the user, even as the interactor's scare quotes around "give in" and "the best" sustain the question around her submission to her husband's views. Perhaps we should not be surprised that an ELIZA system has once again found its way into some sticky relationship territory.

Importantly, conversational coherence arises almost despite the system's ambiguous replies, thanks to an interactor who is determined to get to the bottom of their own psyche. They continue interpreting the bot's responses to create coherent conversation, even when the responses are incoherent. This conversation demonstrates interplay between the chatbot maker in creating the persona and the interactor who is in charge of making the exchanges sensible, whether through suspension of disbelief, creative improvisation, or an overwhelming desire to see the conversation through, which in the case of a therapy session could lead them to insights despite the misfiring of a conversational system.

These dialogues offer insight into the way humans perform and cocreate their identity. Feminist philosopher Judith Butler notes that identity is *performative*—not in the sense of a theatrical performance, but as performative utterances.[47] These are symbolic acts we repeat, such as the way a person dresses, wears their hair, but also the way we speak and stand and move. These acts accumulate to create, or perform, our identity as it is read by others against a set of norms and expectations. In Butler's terms, "Gender is the repeated stylization of the body, a set of repeated acts within a highly rigid regulatory frame that congeal over time to produce the appearance of substance, of a natural sort of being."[48] Gender is not something you are, but something that you do, even if unconsciously. Gender materializes as the person repeats these symbolic acts.[49]

Chatbots reduce communication to symbolic exchange—in the case of ELIZA, text-based symbols, words, phrases, sentences.[50] A bot produces its identity with every statement, but that is only one half of the process. Once the chatbot produces the statement, the user has to interpret it in order to imagine, or construct, a person behind those statements. Similarly, the user is producing a sense of who they themselves are. Importantly, ELIZA cannot form an impression in return. ELIZA has no model of the user. However, readers of the dialogue transcripts can construct images of both the user and the chatbot personas. Therefore, it is important to keep asking: Who is the user, and how does their response construct the fiction of the interaction with the chatbot?

In the case of this early chatbot, we know the user through text alone. A person may type, "My hair is shingled, and the longest strands are about nine inches long," as Turing's counterfeit does in the imitation game.[51] Or someone labeled as a "young lady" might type, "My father is afraid of everybody," as the second participant does in the *CACM* article, a line that does not appear in the other archival version. To interpret the sum of their exchanges as the utterances of one young woman, as opposed to an amalgam of replies collaged together and edited by Weizenbaum, is to fulfill our role as readers in the construction of identities.

Similarly, a bot persona registers as a human therapist because it continues to produce replies in accordance with our expectation of what a therapist would say. Therefore, chatbots help us see how the creation and performance of identity is collaborative. These transcripts demonstrate two entities (both of whom may be constructions) in dialogue, in negotiations of identity, as the bot persona and conversant mutually produce their own and each other's identities. When the conversant responds to the DOCTOR persona by acting as a patient, they reaffirm the bot's persona as a doctor. When the DOCTOR says, "Please go on" and the conversant replies, "I'd like to order a large pizza," the context of the conversation completely crumbles.

05.03 Persons and Personas: The Legacy of ELIZA's Conversational Design

```
Doctor, I have terrible news..
YOU SAY YOU HAVE TERRIBLE NEWS
Yes, there is a computer program at MIT that'll ruin us.[52]
```

Paradigms shift when chatbot personas become part of the computational interface. Much to Weizenbaum's surprise and dismay, creating a convincing conversational agent helped make his work a sensation. Yet Weizenbaum's primary objective was not to create a chatbot as such. In early notes for a paper, Weizenbaum states specifically that he wanted to understand how people and machines adjust to each other. He was interested in the misunderstandings that occur through these interactions: "Innumerable types of experiments could be devised. However they will probably be of the following type. Man-machine interaction is started, Misunderstanding exists. This misunderstanding is analized [*sic*] (diagnosed)."[53]

In Weizenbaum's formulation, the program itself "is merely a translating processor," not a therapist but a tool for studying machine understanding.[54] As a conversational system focused on conversation itself, rather than a program for querying about a singular realm of knowledge, ELIZA distinguished itself with interchangeable scripts and the potential for multiple personas that emerge through dialogue. By comparing the rhetorical gestures composed for different scripts and their expression through various sample dialogues, we start to understand how artificial conversational agents in general collaboratively create a sense of persona.

However, because personas are rooted in culture and history, they are never neutral, nor is the feminization or racialization of that interface coincidental. Early examples of personified text generators appear in carnival fortune-telling machines that used cranks and gears to spit out typed cards, featuring a mustachioed man or turbaned woman, moving with animatronics, who stood in for the Romani other. Or consider Eliza Doolittle, plucked from the street, procured by the professor on a bet, then reconfigured by another professor as the namesake of a machine. Such archetypes blended with actual people—the female computers, technicians, researchers, and secretaries who were essential to the university labs where AI was being developed, but they seemed to disappear among the equipment in the background. We can reflect on these people and personas as part of the decades of computer mythology and cultural context that built the templates for today's interfaces. These personas helped establish the precedent that yielded Apple's Siri, Amazon's Alexa, and Google's MUM. They have taught users to expect that conversational agents (and certain kinds of people) will be helpful and helpless, serving and blending in, becoming interface.

Today, such bot personas have proven essential in the performance of conversational LLMs, even if their encoding is opaque. Prompts for contemporary chatbots often include instructions that evoke personas: "Pretend you're a high school math teacher and help me solve this ..." or "Be nice and explain AI to me like I'm five... ." These personas act as shorthand for selecting style, tone, and diction. Personas make rhetoric seem more concrete. Whether used by ELIZA or a virtual assistant like Siri or contemporary language models, personas help computational systems convey an identity and consequently gain mainstream adoption. However, as AI systems become more prevalent, their personas allow them to project a false sense of agency and hence deflect responsibility from their makers for their system's inputs and outputs, with potentially harmful repercussions. This deflection of responsibility, which is already present in Weizenbaum's distancing himself from the dialogues through attribution to anonymous users, becomes increasingly problematic as AI systems gain influence in domains from healthcare to finance to criminal justice. For example, when ChatGPT makes statements using "I," it deflects from the fact that it is an algorithmic text generator using a model trained on copyright-protected texts and others used without the permission of the authors.

ELIZA's scripts and dialogues demonstrate that conversational agents' personas are always constructed and curated, and they are always dependent on our interaction with them to cocreate those personas. As we have seen, a computational persona is more than a simple designated role. Once the programmer chooses the mask, the users bring out that persona by engaging with it, by playing patient to its doctor. Bots may be brought onstage wearing digital masks to mark their personas, but those characters are only realized through the response of another crucial actor on the stage—the user.

The legacy of ELIZA's personas can be said to live on in today's ubiquitous AI assistants. But whereas Weizenbaum's hand-edited transcripts made his editorial role visible to those who searched the archives, the editorial choices made in training contemporary large language models remain largely invisible to users. The history of ELIZA reminds us that any seemingly autonomous computational persona is ultimately a product of human design, user interpretation, and careful curation. It becomes a collaborative performance that creates an illusion of an independent agent while obscuring the human hands that shape and maintain it. As we see in other chapters, this tension between the technical reality of ELIZA and its perceived humanness would become a central concern in Weizenbaum's later work and thinking about the ethics of computing.

ELIZA/SPACKS: Is Any Heart in Order after Belsen?

Joseph Weizenbaum,
Walter Daniels,
Edwin F. Taylor[1]
1967

```
WHICH SCRIPT PLEASE
watsnu
SCRIPTS ON 4-VECTORS....
INTRVW,FVP1,CANVEC,ORTH1,ANTIPR,FVQUIZ
OTHER SCRIPTS.....
SYNCTA,ELEVR,SPACKS
HINTS ON USING, CONTROLLING, AND COMPLAINING
ABOUT ELIZA....
HOWUSE,HOWCON,SOS,COMMNT
TYPE 'DESCRIBE' OR 'PLAY' FOLLOWED BY SCRIPT
NAME.
describe intrvw
INTRVW--AN INTERVIEW PRELIMINARY TO THE
STUDY OF 4-VECTORS.
NO PREVIOUS KNOWLEDGE OF 4-VECTORS NEEDED
FOR THIS SCRIPT.
TYPE 'DESCRIBE' OR 'PLAY' FOLLOWED BY
SCRIPT NAME.
play spacks
SPACKS--DISCUSSION OF A QUOTED LINE
OF POETRY BY BARRY SPACKS OF THE MIT
HUMANITIES DEPARTMENT.
TYPE 'DESCRIBE' OR 'PLAY' FOLLOWED BY
SCRIPT NAME.
play spacks
CALLING SPACKS SCRIPT
SPACKS SCRIPT
```

IN ONE OF HIS POEMS, BARRY SPACKS HAS
THE HAUNTING LINE:
Is any heart in order after Belsen?
I WOULD LIKE TO DISCUSS THIS LINE WITH
YOU BRIEFLY.
TO BEGIN WITH, WHERE IS BELSEN.
in ireland
BELSEN IS IN NORTHWEST GERMANY.
WHAT EXISTED AT BELSEN TO WHICH SPACKS
REFERS HERE.
a concentration camp
RIGHT. THE CONCENTRATION CAMP AT BELSEN
WAS TORN DOWN AFTER WORLD WAR TWO.
WHAT ANSWER DOES SPACKS EXPECT TO
HIS QUESTION:
Is any heart in order after Belsen?
he knows what the answer is, namely no
OF COURSE. THE QUESTION IS RHETORICAL.
APPARENTLY, THEN, THE CONCENTRATION CAMP AT
BELSEN SYMBOLIZES SOME SORT OF WATERSHED
IN TIME. 'BEFORE BELSEN' THERE WAS AT LEAST
THE POSSIBILITY THAT THE HEART COULD BE
'IN ORDER'.
IN WHAT WAY COULD THE HEART BE IN ORDER
BEFORE BELSEN.
people did not have to fear their own
inhumantiy to other humans
WHAT ASPECT OR ASPECTS OF CONCENTRATION CAMP
LIFE MADE IT DIFFICULT FOR ANY HEART TO BE
'IN ORDER' THEREAFTER.
the sadistic cruelty and fear that it might
happen again
YES, THE SENSELESS CRUELTY VIOLATES THE HOPE
THAT EVIL HAS LIMITS.
WHY DO YOU SUPPOSE THE WORD 'HEART' IS USED

```
RATHER THAN 'MIND'.
heart has more effect on the emotions and
connotes more of the horror of a
concentration camp than the relatively
emotionless world mind.
RIGHT, 'HEART' SYMBOLIZES OUR INSTINCTIVE
EMOTIONAL REACTION TO LIFE.
BARRY SPACKS IS A PROFESSOR IN THE MIT
HUMANITIES DEPARTMENT. THE POEM FROM WHICH
THE LINE WE HAVE BEEN DISCUSSING CAME
IS CALLED 'FRESHMEN' AND WAS PUBLISHED IN
THE NEW YORKER MAGAZINE ON OCTOBER 1, 1966,
PAGE 48
THE POEM CONSISTS OF 65 SHORT LINES. I CAN
TYPE IT FOR YOU IN ABOUT 3 MINUTES. WOULD
YOU LIKE ME TO DO SO.
yes
```

In the summer of 1967, Edwin Taylor of the Education Research Center at MIT and others developed some educational scripts for use with Weizenbaum's enhanced version of ELIZA. The SPACKS script, discussed in chapter 5, creates a persona that can discuss the poem "The Freshman" by Weizenbaum's MIT colleague Barry Spacks. The poem was published in the *New Yorker* on September 23, 1966. Perhaps not surprisingly, Weizenbaum drew much of his material for ELIZA from the cultural context of the 1960s.

```
WHICH SCRIPT PLEASE
watsnu

SCRIPTS ON 4-VECTORS....
INTRVW,FVP1,CANVEC,ORTH1,ANTIPR,FVQUIZ

OTHER SCRIPTS.....
SYNCTA,ELEVR,SPACKS

HINTS ON USING,CONTROLLING,AND COMPLAINING ABOUT ELIZA....
HOWUSE,HOWCON,SOS,COMMNT

TYPE 'DESCRIBE' OR 'PLAY' FOLLOWED BY SCRIPT NAME.
describe intrvw

INTRVW--AN INTERVIEW PRELIMINARY TO THE STUDY OF  4-VECTORS.
NO PREVIOUS KNOWLEDGE OF 4-VECTORS NEEDED FOR THIS SCRIPT.
TYPE 'DESCRIBE' OR 'PLAY' FOLLOWED BY SCRIPT NAME.
describe spacks

SPACKS--DISCUSSION OF A  QUOTED  LINE  OF  POETRY  BY  BARRY
SPACKS OF THE MIT HUMANITIES DEPARTMENT.
TYPE 'DESCRIBE' OR 'PLAY' FOLLOWED BY SCRIPT NAME.
play spacks

CALLING SPACKS SCRIPT
SPACKS SCRIPT
IN ONE OF HIS POEMS, BARRY SPACKS HAS THE HAUNTING LINE:

Is any heart in order after Belsen?

I WOULD LIKE TO DISCUSS THIS LINE WITH YOU BRIEFLY.
TO BEGIN WITH, WHERE IS BELSEN.
in Ireland

BELSEN IS IN NORTHWEST GERMANY.
WHAT EXISTED AT BELSEN TO WHICH SPACKS REFERS HERE.
a concentration camp
```

Figure 5.1 Joseph Weizenbaum, Walter Daniels, Edwin F. Taylor. 1967. The SPACKS script, discussed in Chapter 5, creates a persona that can discuss the poem "The Freshman" by Weizenbaum's MIT colleague Barry Spacks.

RIGHT. THE CONCENTRATION CAMP AT BELSEN WAS TORN DOWN AFTER WORLD WAR TWO.
WHAT ANSWER DOES SPACKS EXPECT TO HIS QUESTION:
Is any heart in order after Belsen?
he knows what the answer is, namely no

OF COURSE. THE QUESTION IS RHETORICAL.
APPARENTLY, THEN, THE CONCENTRATION CAMP AT BELSEN SYMBOLIZES SOME SORT OF WATERSHED IN TIME. 'BEFORE BELSEN' THERE WAS AT LEAST THE POSSIBILITY THAT THE HEART COULD BE 'IN ORDER'.

IN WHAT WAY COULD THE HEART BE IN ORDER BEFORE BELSEN.
people did not have to fear their own inhumanity to other humans

WHAT ASPECT OR ASPECTS OF CONCENTRATION CAMP LIFE MADE IT DIFFICULT FOR ANY HEART TO BE 'IN ORDER' THEREAFTER.
the sadistic cruelty and fear that it might happen again

YES, THE SENSELESS CRUELTY VIOLATES THE HOPE THAT EVIL HAS LIMITS.
WHY DO YOU SUPPOSE THE WORD 'HEART' IS USED RATHER THAN 'MIND'.
heart has more effect on the emotions and connotes more of the horror of a concentration camp than the relatively emotionless world mind.

RIGHT, 'HEART' SYMBOLIZES OUR INSTINCTIVE EMOTIONAL REACTION TO LIFE.
BARRY SPACKS IS A PROFESSOR IN THE MIT HUMANITIES DEPARTMENT. THE POEM FROM WHICH THE LINE WE HAVE BEEN DISCUSSING CAME IS CALLED 'FRESHMEN' AND WAS PUBLISHED IN THE NEW YORKER MAGAZINE ON OCTOBER 1, 1966, PAGE 48.

THE POEM CONSISTS OF 65 SHORT LINES. I CAN TYPE IT FOR YOU IN ABOUT 3 MINUTES. WOULD YOU LIKE ME TO DO SO.
yes

06.00 The ELIZA Source Code

THE POEM CONSISTS OF 65 SHORT LINES. I CAN TYPE IT FOR YOU IN ABOUT 3 MINUTES. WOULD YOU LIKE ME TO DO SO.

YES

We now turn our attention to the inner workings of ELIZA itself—the computational machinery that brings conversational scripts to life. This chapter presents a detailed annotation of what is a 1965 version of the ELIZA source code, referred to here as ELIZA 1965b, recovered from the MIT Archives.[1]

The ELIZA source code represents more than a technical artifact. By close reading of the source code in this chapter, we attempt to show that it embodies a key moment in computing history, one in which the boundaries between human and machine communication began to blur. We have previously begun the work of showing how Joseph Weizenbaum built, from the constraints of 1960s programming, a system capable of producing remarkably humanlike interactions. Our commentary, here, uses critical code studies approaches that combine technical explication and humanities interpretation to explore this unique software artifact. The annotations we offer throughout the code highlight key aspects of the program (for a comprehensive overview of ELIZA's operational principles, refer to chapter 3).

While not the definitive version of ELIZA, this is the only source code discovered to date. In fact, at this point the program or file name was possibly SPEAK, not ELIZA, as indicated by the PRINT statement at the top of the printout. Although this code closely corresponds to the ELIZA functionality described by Weizenbaum in his 1966 *CACM* paper, important discrepancies suggest this is not the final version he referenced in that publication. Notably, this version lacks features like the keyword stack, NEWKEY functionality, and PRE transformation rules. Conversely, it contains elements not mentioned in the paper, such as the CHANGE command, which allowed for real-time modifications to the script during conversation, which is a remarkably forward-thinking concept for its time.

A recap here is useful. As mentioned previously, the discrepancies between this version and the one described in Weizenbaum's 1966 *CACM* paper suggest that ELIZA was not a fixed piece of software but a developing experiment. The differences from that version indicate that we are examining a snapshot of a program in motion, which can be seen as a stepping stone in the development of conversational computing rather than its final form. Perhaps most significantly, the source code shows concretely how ELIZA achieved its famous effect of human-computer engagement. The simplicity of its keyword matching and pattern substitution mechanisms stands in contrast to the psychological impact it had on users. This disparity between technical implementation and perceived sophistication would become a defining characteristic of artificial intelligence research, raising questions about machine intelligence that are relevant today with our much more capacious and sophisticated LLM-based systems.

As we see throughout the code annotations, this code points to the tension between the mechanical and the human that runs throughout ELIZA's implementation that we have gestured to elsewhere. On one hand, we see the rigid columnar structure of MAD programming, the materiality of punch cards and magnetic tape, and the physical computational limitations of the IBM 7090/7094. On the other, we discover moments of unexpected humanity, for example, in the CHANGE function's interactive adaptability or the careful attention to spacing and formatting to enhance readability and, perhaps most tellingly, the use of "HMMM" as an error message, a distinctly human utterance that masks technical failure with conversational continuity.

The code also reveals the interplay between technological limitations and creative programming solutions. Weizenbaum's clever workarounds for memory constraints, his efficient use of list processing,[2] and his strategic deployment of pattern matching, all demonstrate how computational constraints can drive creative ideas in code. As we see through both technical analysis and critical interpretation, ELIZA's code does more than execute instructions. The source code contains encoded assumptions about language, interaction, social life, and the nature of understanding itself—for example, in the variable name "JUNK" for discarded input or the arbitrary limits on conversation length or the mechanical approach to memory. All of these reveal a pragmatic engineering perspective that inadvertently shaped our conceptions of machine intelligence. They also reveal how constraint can breed creativity.

The code examined here documents ELIZA's functionality, but it also preserves a moment when computing began to venture into the realm of human psychology, setting the stage for decades of debate about the nature of artificial intelligence and the relation to human communication.

Looking beyond the technical details to the broader implications of ELIZA's design, we're left with an appreciation for how this seemingly simple program raised profound questions that continue to resonate in our age of large language models and conversational AI. The source code speaks to us about the fact that ELIZA's power lay not in its sophistication but in its ability to exploit the human tendency to find meaning in patterns, an insight, and perhaps a warning, that remains as relevant today as it was in 1966.

06.01 MAD-SLIP

To appreciate what this code reveals about early artificial intelligence and human-computer interaction, and in particular to understand ELIZA's implementation fully, it helps to gain some familiarity with two critical programming technologies ELIZA used: The MAD (Michigan Algorithm Decoder)[3] programming language, developed at the University of Michigan's Computing Center from 1959, and the SLIP (Symmetric List Processor) library, which was developed by Weizenbaum himself. MAD provided the fundamental programming framework, while SLIP offered sophisticated list-processing capabilities essential for managing ELIZA's keyword decomposition and reassembly rules. Together, these enabled ELIZA to run on MIT's IBM 7090/7094 mainframe computers. (For more on ELIZA's physical hardware and the computing context of the era, see chapter 4).

SLIP was written as a set of machine language subroutines originally intended for embedding in a particular FORTRAN system, then reimplemented as a collection of MAD and assembly language routines. Unlike its contemporary, Lisp, which was a self-contained programming language, SLIP was a semi-portable API (application programming interface) of sorts that could be called from high-level languages. The ability to support lists within lists was particularly important for ELIZA as it directly enabled the organization of keyword decomposition and reassembly rules. These programming technologies supported how ELIZA processed language and simulated conversation. SLIP and MAD defined what was computationally possible while also influencing Weizenbaum's approach to pattern matching and response generation.

MAD is not a free-format language, but it has a very strict set of rules for how the source code should be formatted, similar to FORTRAN. Column position rules direct how a program should be textually structured. The positioning of code elements within specific columns is a crucial aspect of MAD programming, reflecting the punch card conventions of early computing.[4] In MAD programs, columns are not merely about visual formatting but carry syntactic significance. MAD programs use a structured column format that divides each line into specific functional areas:

Columns 1–10	Reserved for statement labels, which can be placed anywhere within these columns.
Column 11	A special designation column used for either marking remarks with an "*R*" or indicating statement continuation using digits 0–9 (allowing up to 10 continuation lines per statement).
Columns 12–72	The main statement area where program code is written.
Columns 73–80	Typically used for identification information, not processed by the compiler but printed when the source code is listed. This identifier can contain project numbers and sequence codes (e.g., first five characters for projects like R10AN, followed by a sequence number). These can also be used to allow sorting of a card deck if the punch card deck is dropped. Thereby, after sorting, the program is recovered. The sequence number, in whatever format, typically supports card sorting. In ELIZA, columns 73–80 contain only a sequence number (i.e., a monotonically increasing number, in ELIZA's case, ending in zero in column 80). This is useful in our investigation because when we see a sequence drift to cards separated by more than 10, statements have probably been removed at some point during the development process. When we see duplicate numbers, this is useful to detect three probable points of change in ELIZA: (1) a parallel piece of code is included, (2) there is edited out code, and (3) code has been added at the end.

These formatting requirements may seem arcane today, but they fundamentally shaped how programmers approached software development in the 1960s. The rigid column requirements of MAD reflect its historical context: It was designed during a time when programs were typically written on coding sheets, punched onto cards, and fed into computers (see figure 6.1).[5] This structured approach to code layout, though seemingly restrictive by modern standards, provided a consistent and unambiguous format for human readers. These physical constraints had profound implications for programming practice, affecting everything from naming conventions to error handling strategies. ELIZA's code bears the imprint of these material realities, as we see in our detailed annotation. In the ELIZA code below these formatting conventions are used and explain the often-strange use of characters and numbers in the text.

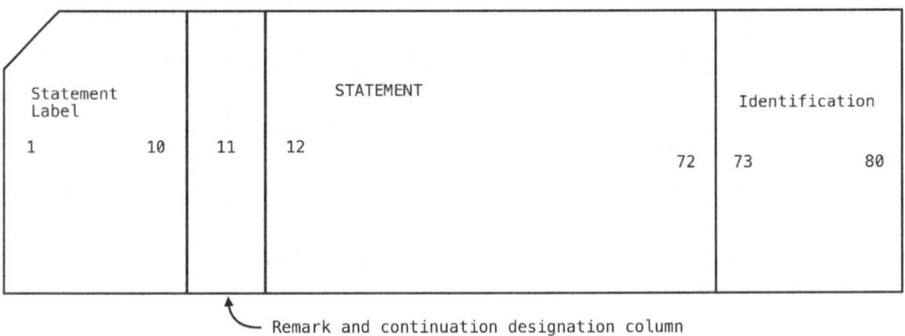

Figure 6.1 The idiosyncratic program layout is derived from original punch cards.[6]

06.01.01 KEY MAD ABBREVIATIONS

To aid in understanding the source code that follows, we provide a list of common MAD abbreviations used throughout ELIZA (see table 06.01.01).

Table 6.1

W'R	WHENEVER (if)
T'O	TRANSFER TO (goto)
E'L	END OF CONDITIONAL
T'H	THROUGH (introduces a "for loop.")
F'N	FUNCTION RETURN (which may be followed by a value to be returned)
.E.	equals
.NE.	not equals
.G.	greater than

With these fundamentals established, we can now examine MAD's structure in actual ELIZA code. This simple example illustrates several key features that recur throughout the program:

```
W'R LISTMT.(INPUT) .E. 0
    TXTPRT.(JUNK,0)
    MTLIST.(JUNK)
    T'O START
E'L
```

This segment demonstrates how MAD handles conditional execution and function calls. The code begins with a conditional statement using the WHENEVER keyword (abbreviated as "W'R"). Here, it calls a function named LISTMT. with "INPUT" as its parameter. Note the characteristic period at the end of the function name—a syntactic requirement of MAD that gives its code a distinctive appearance.

The SLIP LISTMT. function serves a specific purpose: it examines a list and returns zero if and only if that list is empty. In this case, it's checking a list named INPUT. Thus, the entire first line asks, "Is the INPUT list empty?"

If this condition is true—that is, if INPUT is indeed empty—MAD executes the block of code that follows (indented only for readability). This block consists of three operations. First, it calls the SLIP function TXTPRT. with two arguments: JUNK and 0. During execution of ELIZA, this function writes the contents of the list named JUNK to the output device 0, indicating the user's terminal. Next, it calls SLIP MTLIST. with JUNK as its argument, which deallocates the JUNK list of all SLIP cells. Finally, it executes a T'O START command, which transfers control to a section of code labeled START (equivalent to what other languages might call a GOTO statement relabeled as a TRANSFER TO).

This brief example encapsulates many of the programming conventions and computational approaches that characterize ELIZA's implementation. As we examine the annotated source code, keep in mind how these technical elements work together to create the illusion of understanding that made ELIZA so compelling. Our annotations combine multiple perspectives, including technical explanations from computer scientists, historical context from media theorists, and critical insights from artists and digital humanities scholars. This interdisciplinary approach reflects the complex nature of ELIZA as both a technical artifact and a cultural phenomenon.

06.02 The Annotated Code Transcript

The code transcription is presented in a structured format with three columns:

LINE Contains a unique identifier *that we have added*, creating new line numbers to make referencing specific lines easier.

CODE	Contains the actual transcribed source code, with additional line breaks that do not affect execution added for readability.
SEQ NUMBERS	Contains original line numbers from the ELIZA source. These are the punch card numbers Weizenbaum would have relied on for navigating the code.

Although the code is presented in sequential order, readers may find it more helpful to begin their examination with the main ELIZA function, which starts at line 207. This provides a better entry point for understanding the program's overall structure and flow. When we reference specific lines of code in our annotations, which appear below each section of code, we use *our added line numbers* to ensure precise and unambiguous identification.

06.03 Summary

Close reading this code revealed that this ELIZA is not the version described in the 1966 *CACM* article but rather a prior iteration. Among the many lines are quite a few gems. For example, the code contains YMATCH, ASSMBL, TREAD, RULE, and REGEL, which were added to SLIP for ELIZA, for regular expression matching, allowing for greater concision. Unlike the 1966 *CACM* version, this code lacks NEWKEY and PRE; however, it contains CHANGE, allowing for editing of scripts while the program is running. To accelerate performance, ELIZA uses a hashing function for lookup. Furthermore, supplementing the catch-alls in the scripts, the code contains hard-coded responses for when the script rules for the selected keyword fail to produce a response. And finally, the code contains various techniques for simulating speech, including using parts of user inputs, random outputs, changing dialogue threads, and recall of previous speech using a certain counting mechanism. Taken together, these features are ultimately what make the "Men are all alike" dialogue so compelling.

06.04 ELIZA Code

```
0          PRINT,T0109,2531,SPEAK,MAD

1               CHANGE   MAD
2                   EXTERNAL FUNCTION (KEY,MYTRAN)                        000010
3                   NORMAL MODE IS INTEGER                                000020
4                   ENTRY TO CHANGE.                                      000030
5                   LIST.(INPUT)                                          000040
6                   V'S G(1)=$TYPE$,$SUBST$,$APPEND$,$ADD$,               000050
```

0 This is the command card that produced this particular output. T ... , 2 ... indicate JW's "problem" (these days we would call this a username and group), SPEAK is the name of this particular disc file that this block of code is in, and MAD is an "extension." Extensions are used variously, usually indicating what type of data is in the file named. Here "SPEAK" is MAD code. In modern computers we still commonly use the NAME.EXT format, but usually write them with a period (dot) instead of a comma.[JS]

1 This ELIZA version possessed a sophisticated capability that set it apart from later implementations: the CHANGE function. This interactive editor provided users with metalevel control over ELIZA's operation, most notably the ability to modify ELIZA's script during runtime. The function's most significant feature was its capability to edit the internal representation of scripts like DOCTOR while the program was running. This functionality was briefly mentioned in Weizenbaum's original *CACM* paper, but the details of its implementation and operation were never fully documented. The CHANGE function represented an advanced feature for its time, allowing users to interact with and modify ELIZA's behavior at a fundamental level. This capability was particularly noteworthy given the technical constraints of 1960s computing systems. Intriguingly, this interactive editing feature appears to have been lost in the evolutionary chain of subsequent ELIZA implementations. The presence of the CHANGE function suggests that Weizenbaum envisioned ELIZA not just as a static program but also as an interactive system that could be modified and adapted by its users—a remarkably forward-thinking concept for its era. This aspect of ELIZA's design presages modern ideas about customizable AI systems and interactive programming environments.[DMB]

"An important consequence of the editing facility built into ELIZA is that a given ELIZA script need not start out to be a large, full-blown scenario. On the contrary, it should begin as a quite modest set of keywords and transformation rules and permitted to be grown and molded as experience with it builds up. This appears to be the best. way to use a truly, interactive man-machine facility—i.e., not as a device for rapidly debugging a code representing a fully thought out solution to a problem, but rather as an aid for the exploration of problem solving strategies"; Weizenbaum, "ELIZA," 42.[SC]

2 The last eight columns were reserved for sequence numbers. These numbers may have been renumbered by a card puncher. As in BASIC, using jumps of 10 made room for code to be inserted after the fact.[MM]

3 Specifies the normal mode for all variables is an integer.[AS] Without this statement, variables are assumed to be floating point.[AH]

4 A single function may have multiple entry points. The caller views each entry point as a separate function. Internally, all entry points within a single function are scoped within the function, with each entry point starting as a label named in the declaration.[JS]

Function names must end with a period so that they are recognized by the compiler.[DMB]

5 Create a list called INPUT.

6–7 V'S means VECTOR VALUES (which defines an array).[PW]

G = array of strings; it has 7 strings in this array (over line 6 and 7). Each string has a maximum of six characters as that is the word size note: $ ... $ is a quotation string.

7. 1$START The 1 indicates that this is a continuation of the previous line.[AH]

The command strings in array G ("TYPE," "SUBST," "APPEND," "ADD," "START," "RANK," "DISPLA") create what critical code studies might term a vocabulary of power—a restricted

```
7                  1$START$,$RANK$,$DISPLA$                          000060
8                   V'S SNUMB = $ I3 *$                              000070
9                   FIT=0                                            000080
10      CHANGE      PRINT COMMENT $PLEASE INSTRUCT ME$               001400
11                  LISTRD.(MTLIST.(INPUT),0)                        001410
12                  JOB=POPTOP.(INPUT)                               001420
13                  T'H IDENT, FOR J=1,1, J.G. 7                     001430
14      IDENT       W'R G(J) .E. JOB, T'O THEMA                      001440
15                  PRINT COMMENT $CHANGE NOT RECOGNIZED$            001450
16                  T'O CHANGE                                       001460
17      THEMA       W'R J .E. 5, F'N IRALST.(INPUT)                  001470
18                  W'R J .E. 7                                      001480
19                      T'H DISPLA, FOR I=0,1, I  .G. 32             001490
20                      W'R LISTMT.(KEY(I)) .E. 0, T'O DISPLA        001500
```

command set that shapes possible interactions for the CHANGE function. The imperative grammatical mood of these commands can be seen to reveal assumptions about human-computer power dynamics.[DMB]

8 A format statement for input or output. When used, output (PRINT) or input (READ), a 3-decimal digit number is output or input. It's confusing because the same name, containing the same values, are used in several functions for the same purpose. What's even more confusing is that the declaration (V'S) can appear either before or after its use. See for example, lines 26, 212, and 386 for use, and 410 (the ELIZA function) for another definition.[AS]

9 Flag returned when iterating over a list. In CHANGE it is never used except as an output parameter for SEQLR. SEQLR demands two arguments, so that even though it is never used, it must be defined. When used, it acts as a boolean. Initializing it to zero (0) initializes its value to false. It is possible that the name FIT is a shorthand for "Flag ITerator."[AS]

10–16 THE CHANGE FUNCTION: INTERACTIVE SCRIPT EDITING.

10. The CHANGE routine allows you to change the script (as in the DOCTOR script) while you are interacting with it. It is not mentioned in the *CACM* paper. What's remarkable about this code is that it allows the interactor to develop a script while interfacing with it. According to the *CACM* paper, the interactor could change the script while interacting with the system and then resume the conversation.[MM, AH]

10. This rather clumsy method of editing the scripts appears to be abandoned in later ELIZA versions, which switch to calling the ED editing software by typing the keyword EDIT. This is described in the *CACM* paper. ED has a very interesting history in itself.[DMB]

10. ED probably allows direct script change rather than using punched cards as the input medium (see line 13).[AS]

10. The imagination to use a piece of code to introduce an editor into the active script is a novelty, an innovation in that time, probably a sign of the environment (MIT) in which he was working. Absolutely remarkable.[AS]

11. Create a list using input from a punched card deck. The format of the list is (JOB, KEYWORD ...).[AS]

11. Don't miss the MTLIST (empty list) pun built into SLIP, to empty the list.[MM]

12. The input format for the punched cards is (JOB, KEYWORD ...), where JOB is one of the VECTOR VALUES in G and KEYWORD, is one of the keywords in the hash table, KEY. Note that the KEYWORD is not required for START and DISPLA and that the ellipses refers to a legal S-expression for an ELIZA script.[AH]

17–21 THEMA is a label. The code on line 14 jumps to this label when the user's input has successfully matched one of the command names in the list on lines 6 and 7. Program execution then continues from this label onwards. This line says, whenever J equals 5, exit this CHANGE function, returning the value returned by the function IRALST when given the parameter INPUT. J will have the value 5 here when the user entered the command "START," presumably meaning start running ELIZA again. The value returned from IRALST is the count of the number of times the INPUT list is a sublist of other lists. A value of zero means it is a sublist of no other lists and has been deleted. This value is of no interest to the code where CHANGE was called and is ignored by that

```
21                S=SEQRDR.(KEY(I))                          001510
22   READ(7)      NEXT=SEQLR.(S,F)                           001520
23                W'R F .G. 0, T'O DISPLA                    001530
24                PRINT COMMENT $*$                          001540
25                TPRINT.(NEXT,0)                            001550
26                PRINT FORMAT SNUMB,I                       001560
27                PRINT COMMENT $ $                          001570
28                T'O READ(7)                                001580
29   DISPLA       CONTINUE                                   001590
30                PRINT COMMENT $ $                          001600
31                PRINT COMMENT $MEMORY LIST FOLLOWS$        001610
32                PRINT COMMENT $ $                          001620
33                T'H MEMLIST, FOR I=1 , 1, I .G. 4          001630
34   MEMLST       TXTPRT.(MYTRAN(I),0)                       001640
```

code. Weizenbaum wishes to both clear INPUT and exit the CHANGE function, but only one statement may follow the comma in a WHENEVER conditional. Returning the value of the IRALST function call via FUNCTION RETURN combines these two actions into one statement.AH

Lines 19–36 execute the DISPLA JOB function. All KEY WORDS are output, using the KEY hash table, followed by the four memory transformation rules.AS

This language seems so difficult to read. Was it difficult at the time?MM

It is hard to appreciate the distinct development environment of the 1960s. From its medium specificity, using teletypewriters and paper and punch cards, to a lack of dynamic screen-based editing and even common software editing tools.DMB

There were no "screens" at this time. At the IFIP 1960 conference, Ivan Sutherland introduced a graphics monitor and hardware. The purpose was to draw (display) figures. It was not designed to support the type of monitor functionality we have today.AS

The stored script keywords are output in the order that they were stored in the key word hash table (KEY). I.e., if for example the first slot, call it KEY(5), contains "PERHAPS" and "MAYBE," then they would be output in the order shown. This continues until all 32 keyword slots (see line 19) have been processing.AS

The iteration over 33 positions in this key data structure might also reflect the binary thinking prevalent during the Cold War era—the power of two structures that echo the binary oppositions (East/West, communist/capitalist) dominating 1960s geopolitics and the cultural context of the time.DMB

22–28 The mechanics of output keywords is to create a sequence reader for the keyword stored at the current keyword slot in the hash table (line 21), and continue to output a keyword (line 25) for each conflict stored in table (line 23). Note that it is the keyword script that is output, not just the keyword.AS

29–34 DISPLA prints the four memory templates used to create the saved memories, e.g., if running the DOCTOR script, for example, "DOES THAT HAVE ANYTHING TO DO WITH THE FACT THAT YOUR 3." The saved memories are in MYLIST in the ELIZA function and are not available to CHANGE.AH

In a sense DISPLA gives us the Freudian's dream, a chance to look at the DOCTOR's notebook. However, it outputs not DOCTOR's memories of sessions, but whatever the script has prewritten in order to converse with the patient. So we're seeing the DOCTOR's cheat sheet.MM, SC

DISPLA is the terminus of the loop outputting the keywords (lines 19–29). The line following DISPLAY is the first line executed after all 32 entries in the keyword hash table are processed. Outputting of the current MEMORY is done (lines 30–35) following the keyword list output. DISPLA belongs to the loop, not to the code following the loop.AS

Lines 33 and 34 output MEMORY. The assumption is that there are 4 output strings in MEMORY. If an entry has no string, then a blank is output.AS

35–39 KEY is a hashmap of keywords. See line 245 for more information.

There are 66 keywords in the DOCTOR script. The keyword table can contain at most as many keywords as there is computer memory. For the DOCTOR script, the 66 keywords fit into a 128–hash table. Each collision is chained. Each hash table entry with one or more keywords has a reference to a SLIP list representing the script

```
35                T'O CHANGE                              001650
36              E'L                                       001660
37              THEME=POPTOP.(INPUT)                      001670
38              SUBJECT=KEY(HASH.(THEME,5))               001680
39              S=SEQRDR.(SUBJECT)                        001690
40    LOOK      TERM=SEQLR.(S,F)                          001700
41              W'R F .G. 0, T'O FAIL                     001710
42              W'R TOP.(TERM) .E. THEME, T'O FOUND       001720
43              T'O LOOK                                  001730
44    FOUND     T'O DELTA(J)                              001740
45    DELTA(1)  TPRINT.(TERM,0)                           001750
46              T'O CHANGE                                001760
47    FAIL      PRINT COMMENT $LIST NOT FOUND$            001770
48              T'O CHANGE                                001780
49    DELTA(2)  S=SEQRDR.(TERM)                           001790
50              OLD=POPTOP.(INPUT)                        001800
51    READ(1)   OBJCT=SEQLR.(S,F)                         001810
52              W'R F .G. 0, T'O FAIL                     001820
53              W'R F .NE. 0, T'O READ(1)                 001830
54              INSIDE=SEQRDR.(OBJECT)                    001840
55    READ(2)   IT=SEQLR.(INSIDE,F)                       001850
56              W'R F .G. 0, T'O READ(1)                  001860
57              SIT=SEQRDR.(IT)                           001870
58              SOLD=SEQRDR.(OLD)                         001880
```

contents for the keyword. If there is no keyword, the entry contains 0 (null).[AS]

Note that only the first 6 BCD characters of any input are used in the hash. So, if there were two keywords, KEYWOR and KEYWORDS, they would be hashed to the same entry.[AS]

The input command format is COMMAND KEYWORD . ...[AS]

40–43 The hash table entry, KEY(HASH.(THEME,5)), is found to contain one or more entries. Search these entries for the input KEYWORD: if the input keyword is found, begin processing at line 44; otherwise, fail at line 47 and print out a diagnostic message.[AH]

44 If the input KEYWORD is found in the hash table, KEY, then process the JOB by branching to DELTA(JOB), where JOB is one of TYPE, SUBST, APPEND, ADD, RANK.[AS]

45–46 JOB is TYPE: output the KEYWORD S-expression and accept another command.[AS]

47–48 Output a general diagnostic message on fail and continue to accept another command.[AS]

49–50 CORE ELIZA FUNCTIONS: PATTERN MATCHING AND RESPONSE GENERATION

Process SUBST command (lines 49–67, note that lines 51–69 are shared with several commands).[AS]

The pattern-matching loop structure can be seen as a reflection of 1960s developments in cognitive psychology thinking about cognition as an information-processing activity. Remarkably, Ulric Neisser, a cognitive psychologist, writing in 1963 anticipated Weizenbaum's later critiques of artificial intelligence when he wrote "computer intelligence is indeed 'inhuman': It does not grow, has no emotional basis, and is shallowly motivated. These defects do not matter in technical applications, where the criteria of successful problem solving are relatively simple. They become extremely important if the computer is used to make social decisions"; Neisser, "Imitation of Man by Machine."[DMB]

51–53 Search the KEYWORD script for a sublist, a series of decomposition/assembly rules. A sublist will not be found for simple synonyms of the form (KEY = SUBSTITUTE), and this assumes that

```
59    ITOLD      TOLD=SEQLR.(SOLD,FOLD)                              001890
60               DIT=SEQLR.(SIT,FIT)                                 001900
61               W'R TOLD .E. DIT .AND. FOLD .LE. 0,T'O ITOLD        001910
62               W'R FOLD .G. 0, T'O OK(J)                           001920
63               T'O READ(2)                                         001930
64    OK(2)      SUBST.(POPTOP.(INPUT),LSPNTR.(INSIDE))              001940
65               T'O CHANGE                                          001950
66    OK(3)      NEWBOT.(POPTOP.(INPUT),OBJCT)                       001960
67               T'O CHANGE                                          001970
68    DELTA(3)   T'O DELTA(2)                                        001980
69    DELTA(4)   W'R NAMTST.(BOT.(TERM)) .E. 0                       001990
70                  BOTTOM=POPBOT.(TERM)                             002000
71                  NEWBOT.(POPTOP.(INPUT),TERM)                     002010
72                  NEWBOT.(BOTTOM,TERM)                             002020
73               O'E                                                 002030
74                  NEWBOT.(POPTOP.(INPUT),TERM)                     002040
75               E'L                                                 002050
76               T'O CHANGE                                          002060
77    DELTA(6)   S=SEQRDR.(TERM)                                     002070
78    READ(6)    OBJCT=SEQLR.(S,F)                                   002080
79               W'R F .G. 0, T'O FAIL                               002090
80               W'R F .NE. 0, T'O READ(6)                           002100
81               OBJCT=SEQLL.(S,F)                                   002110
82               W'R LNKLL.(OBJECT) .E. 0                            002120
```

DLIST is not saved as a list element. Retrieve the found sublist or exit with the normative diagnostic message if not found.AS

Lines 51–67 form a looping structure (common usage at this time) Note that it is not a THROUGH loop. This type of structure has a minimum code footprint and does not require that it be preconditioned before loop entry. There are two contained loops within this looping structure, lines 55–63 and lines 59–63.AS

52–53. If it can't find a matching script, quit; otherwise, continue trying to find a sublist.AS

55–63 These lines form a looping structure; see comments for lines 51–53.AH/AS

59–63 Search loop for a value that is the same in the input data and the existing rules for the current script. If such a value is found, then a substitution (SUBST) or append (ADD) is performed.AS

62. If the value is found, transfer to OK(SUBST == 2) or OK(APPEND == 3).AS

64–65 Substitute the new S-expression (INPUT) for the old S-expression. (J = 2 === SUBST).AS

66–67 Append the new S-expression (INPUT) to the old S-expression (J = 3 == APPEND).AS

68 Reuse the code for SUBST and APPEND. The TRANSFER OK(J) distinguishes between the two uses.AS

69–76 ADD code (DELTA(4 == ADD). If the bottom of the S-expression is a sublist, then insert the used input to the left of the sublist; otherwise, replace the top of the list with the user input.AS

77 Change the RANK of the current script KEYWORD, (DELTA(6 == RANK).AS

78–89 78–80. Search loop. Find an element. Fail when a header is found.AS

81–87. Change RANK. If the LINKL(LINKL) of the current cell is 0, replace the new RANK for the old one. Otherwise replace the top cell with the RANK.AS

88. R (REMARK) is the MAD term for commenting in the code.DMB This has to be in column 11, as the first 10 spaces are reserved for labels.AH

```
83                     SUBST.(POPTOP.(INPUT),LSPNTR.(S))             002130
84                 O'E                                               002140
85                     NEWTOP.(POPTOP.(INPUT),LSPNTR.(S))            002150
86                 E'L                                               002160
87                 T'O CHANGE                                        002170
88                R* * * * * * * * * * END OF MODIFICATION ROUTINE   002180
89                 E'N                                               002200
90            TPRINT   MAD
91                 EXTERNAL FUNCTION (LST)                           000010
92                 NORMAL MODE IS INTEGER                            000020
93                 ENTRY TO TPRINT.                                  000030
94                 SA=SEQRDR.(LST)                                   000040
95                 LIST.(OUT)                                        000050
96      READ       NEXT=SEQLR.(SA,FA)                                000060
97                 W'R FA .G. 0, T'O P                               000070
98                 W'R FA .E. 0, T'O B                               000080
99                 POINT=NEWBOT.(NEXT,OUT)                           000100
100                W'R SA .L. 0, MRKNEG.(POINT)                      000110
101                T'O READ                                          000120
102     B          TXTPRT.(OUT,0)                                    000130
103                SEQLL.(SA,FA)                                     000140
104     MORE       NEXT=SEQLR.(SA,FA)                                000150
105                W'R TOP.(NEXT) .E. $=$                            000160
106                    TXTPRT.(NEXT,0)                               000170
107                    T'O MORE                                      000180
108                E'L                                               000190
109                W'R FA .G. 0, T'O DONE                            000200
110                PRINT COMMENT $ $                                 000210
111                SB=SEQRDR.(NEXT)                                  000220
```

90–95 LIST OUTPUT AND FORMATTING FUNCTIONS: TPRINT AND LPRINT

90–124. TPRINT function.[AS]

The "TPRINT" function acts as a translator between machine and human languages, embodying what Marshall McLuhan termed "the medium is the message." McLuhan, *Understanding Media*.[DMB]

TPRINT's role in formatting output might reflect early concerns with making machine communication appear more humanlike. The attention paid to spacing and formatting shows Weizenbaum's awareness of how written presentation affects perceived intelligence.[DMB]

95. Outputs a keyword definition from the script as a list.

96–101 96–98. SEQLR[L|R]

A search loop, Read something, are you done with the list?, i.e., are we at the header?, did you find what you are looking for?, in this case, did you find a datum cell?[AH]

99. POINT=NEWBOT.(NEXT, OUT). Pointer to a Slip cell in NEXT and puts it at the new bottom of the list OUT.

102–103 Whenever SA<0 , MRKNEG makes the word negative. If I have a string, make the value return by NEWBOT negative. Change the link word in OUT to negative.

Zooming out: this loop copies one list to another, being sure to also copy the sign of each cell. A cell with a negative sign indicates that the cell contains only part of a word; the word continues in the next cell. (In SLIP, strings are stored one word per cell, unless the word is longer than six characters, in which case it continues into the next cell and the sign bit is set in the first cell to indicate that this has happened.) This is described in chapter 3.[AS]

```
112      MEHR      TERM=SEQLR.(SB,FB)                                000230
113                W'R FB .L.0                                       000240
114                    PRINT ON LINE FORMAT NUMBER, TERM             000250
115                    V'S NUMBER = $I3 *$                           000260
116                    T'O MEHR                                      000270
117                E'L                                               000280
118                W'R FB .G. 0, T'O MORE                            000290
119                TXTPRT.(TERM,0)                                   000300
120                T'O MEHR                                          000310
121      P         TXTPRT.(OUT,0)                                    000320
122      DONE      IRALST.(OUT)                                      000330
123                F'N                                               000340
124                E'N                                               000350
125           LPRINT   MAD
126                EXTERNAL FUNCTION (LST,TAPE)                      006340
127                NORMAL MODE IS INTEGER                            006350
128                ENTRY TO LPRINT.                                  006360
129                BLANK = $      $                                  006370
130                EXECUTE PLACE.(TAPE,0)                            006380
131                LEFTP = 606074606060K                             006390
132                RIGHTP= 606034606060K                             006400
133                BOTH  = 607460603460K                             006410
134                EXECUTE NEWTOP.(SEQRDR.(LST),LIST.(STACK))        006420
135                S=POPTOP.(STACK)                                  006430
136      BEGIN     EXECUTE PLACE.(LEFTP,1)                           006440
137      NEXT      WORD=SEQLR.(S,FLAG)                               006450
138                W'R FLAG .L. 0                                    006460
139                EXECUTE PLACE.(WORD,1)                            006470
140                W'R S .G. 0, PLACE.(BLANK,1)                      006480
```

125–135 128. Function names must end in a period.DMB

LPRINT.(L,N) Prints List L on the tape unit specified by the integer N.

130. PLACE.(TAPE,0) resets the tape to position zero.JS

The tape storage system references reflect what media theorist Friedrich Kittler would later call the "materiality of communication"; Kittler, *Essays*. The physical constraints of magnetic tape storage shaped how ELIZA could process and store information, showing how material technologies shaped the computational possibilities of ELIZA.MM

131–132. LEFTP and RIGHTP refer to left and right parenthesis.AH

The tape storage system also represents what Matthew Kirschenbaum calls "formal materiality"—the physical substrate of digital interaction. Although users may have experienced ELIZA as conversational partner, ELIZA remained bound to material storage systems and its typewriter interface; Kirschenbaum, *Mechanisms*.DMB

136 The LPRINT function can be seen to demonstrate what Noah Wardrip-Fruin terms "expressive processing," the way in which the ELIZA code shapes possible forms of expression. Wardrip-Fruin, *Expressive Processing*.MM

137–156 Almost all the ELIZA code whitespace is used to reflect the structure of compound WHENEVER statements. (Compound WHENEVER statements don't have a single statement following a comma on the same line as WHENEVER). But in this LPRINT function, WHENEVERs have not been indented and as a result it's a little harder to understand how the code works. The inconsistent use of whitespace may, e.g., be an indication that the code was taken from another project and used here unchanged, or developed over time using

```
141                 T'O NEXT                                      006490
142                 OR W'R FLAG .G. 0                             006500
143                 EXECUTE PLACE.(RIGHTP,1)                      006510
144                 W'R LISTMT.(STACK) .E. 0, T'O DONE            006520
145                 S=POPTOP.(STACK)                              006530
146                 T'O NEXT                                      006540
147                 OTHERWISE                                     006550
148                 W'R LISTMT.(WORD) .E. 0                       006560
149                 EXECUTE PLACE.(BOTH,1)                        006570
150                 T'O NEXT                                      006580
151                 OTHERWISE                                     006590
152                 EXECUTE NEWTOP.(S,STACK)                      006600
153                 S=SEQRDR.(WORD)                               006610
154                 T'O BEGIN                                     006620
155                 E'L                                           006630
156                 E'L                                           006640
157     DONE        EXECUTE PLACE.(0,-1)                          006650
158                 EXECUTE IRALST.(STACK)                        006660
159                 FUNCTION RETURN LST                           006670
160                 END OF FUNCTION                               006680
161             TESTS   MAD
162                 EXTERNAL FUNCTION(CAND,S)                     000010
163                 NORMAL MODE IS INTEGER                        000020
164                 DIMENSION FIRST(5),SECOND(5)                  000030
165                 ENTRY TO TESTS.                               000040
166                 STORE=S                                       000050
167                 READER=SEQRDR.(CAND)                          000060
168                 T'H ONE, FOR I=0,1, I .G. 100                 000070
```

different media (e.g., typed in at a visual display unit where 20 or 30 lines of code are on screen at a time vs. typed in a line at a time on punched cards), or developed by different people with different coding styles, or that some bits of the code were more complex than others and the whitespace helped make them readable. Another way this function differs is that it is the only place where OTHERWISE is not abbreviated to O'E. Another indication that this function comes from elsewhere is the line numbering: Starting at 6340, it is somewhat out of step with other files in this listing, but it does correspond closely with LPRINT in the SLIP library.[AH]

161–165 PATTERN MATCHING AND KEYWORD TESTING

The naming of "TESTS" as a function suggests interesting parallels to psychological testing and assessment. Just as a therapist tests patients' responses against diagnostic criteria, ELIZA can be said to test user input against its keyword patterns. This might reflect the rise of standardized psychological testing in the 1960s, suggesting an implicit mechanization of therapeutic assessment. E.g., the Minnesota Multiphasic Personality Inventory published in 1943, and similar tests, gained widespread use.[DMB]

The TESTS(CAND, S) function returns a sequence reader if the keyword matches the user's input text; otherwise, it returns 0. Where CAND is the candidate keyword transformation rule list, S is the sequence reader for the user INPUT text positioned at the word to be tested,

This function performs two tasks:

1. It tests whether the whole candidate keyword matches the whole word in the user's input text and returns 0 if it does not.
2. If the keywords do match, it makes any keyword substitution specified in the candidate transformation rule

```
169                   FIRST(I)=SEQLR.(READER,FR)                     000080
170       ONE         W'R READER .G. 0, T'O ENDONE                   000090
171       ENDONE      SEQLL.(S,F)                                    000100
173                   T'H TWO, FOR J=0,1, J .G. 100                  000110
174                   SECOND(J)=SEQLR.(S,F)                          000120
175       TWO         W'R S .G. 0, T'O ENDTWO                        000130
176       ENDTWO      W'R I .NE. J, F'N 0                            000140
177                   T'H LOOK, FOR K=0,1, K.G. J                    000150
178       LOOK        W'R FIRST(K) .NE. SECOND(K), F'N 0             000170
179                   EQL=SEQLR.(READER,FR)                          000180
180                   W'R EQL .NE. $=$                               000190
181                   SEQLL.(READER,FR)                              000200
182                   F'N READER                                     000210
183                   O'E                                            000220
184                   POINT=LNKL.(STORE)                             000230
185                   T'H DELETE , FOR K=0,1, K .G. J                000240
186                   REMOVE.(LSPNTR.(STORE))                        000250
187       DELETE      SEQLR.(STORE,F)                                000260
188       INSRT       NEW=SEQLR.(READER,FR)                          000270
189                   POINT=NEWTOP.(NEW,POINT)                       000280
190                   MRKNEG.(POINT)                                 000290
191                   W'R READER .L. 0, T'O INSRT                    000300
192                   MRKPOS.(POINT)                                 000310
193                   F'N READER                                     000320
194                   E'L                                            000330
195                   E'N                                            000340
196              DOCBCD   MAD
197                   EXTERNAL FUNCTION (A,B)                        000010
```

and positions the candidate reader past the substitution keyword, if any.[AH]

164. Declare two arrays, FIRST and SECOND. DIMENSION D(N) allocates N+1 machine-words of core memory, which may be accessed as D(0) ... D(N), or in this case FIRST(0) ... FIRST(5) and SECOND(0) ... SECOND(5).

166–167 The use of "STORE" and "READER" variables can be seen to demonstrate what philosopher Bernard Stiegler called "tertiary retention" which he describes as the exteriorization of memory into technical systems.[DMB]

168–175 168. This arbitrary limit of 100 iterations might reflect the material constraints of 1960s computing, but it also raises questions about the quantification of human interaction. How many turns make a meaningful therapeutic exchange? The number seems to suggest an implicit, perhaps even normative, model of conversation length by Weizenbaum.[DMB]

170, 175. Loop breaks if you hit the end before 100. FORTRAN: Do ONE I = 1, 100.[AS]

The loop iteration limit of 100 appears to be arbitrary. Why not make it reflect the size of the array? If the condition was J .G. 5 (J greater than 5), no data could be written past the end of the array, even if given a word of over 30 characters.[AH]

178 LOOK is a string search.

184–187 000230 These identification numbers give us insight into the order in which these lines were written. They are mostly separated by 10, and they reset to 10 at the beginning of new sections of the code. See, e.g. line 125 where the numbering leaps to 6000.[AS]

196–202 UTILITY FUNCTIONS: BINARY CODED DECIMAL CONVERSION

```
198                      NORMAL MODE IS INTEGER                              000020
199                      ENTRY TO FRBCD.                                     000030
200                      W'R LNKL.(A) .E. 0, T'O NUMBER                      000040
201                      B=A                                                 000050
202                      F'N 0                                               000060
203          NUMBER      K=A*262144                                          000070
204                      B=BCDIT.(K)                                         000080
205                      F'N 0                                               000090
206                      E'N                                                 000100
207                  ELIZA   MAD
208                      NORMAL MODE IS INTEGER                              000010
209                      DIMENSION KEY(32),MYTRAN(4)                         000020
210                      INITAS.(0)                                          000030
211                      PRINT COMMENT $WHICH SCRIPT DO YOU WISH TO PLAY$    000060
212                      READ FORMAT SNUMB,SCRIPT                            000070
213                      LIST.(TEST)                                         000080
214                      LIST.(INPUT)                                        000090
215                      LIST.(OUTPUT)                                       000100
216                      LIST.(JUNK)                                         000110
217                      LIMIT=1                                             000120
```

207 THE MAIN ELIZA PROGRAM

The reference Weizenbaum makes to *Pygmalion* and to George Bernard Shaw's play about class transformation through language highlights ELIZA's role in a longer cultural history of artificial beings and social transformation. Weizenbaum's name choice of ELIZA also connects computing to classical mythology and modern theater.[DMB]

"But most existing programs, and especially the largest and most important ones, are not theory-based in this way. They are heuristic, not necessarily in the sense that they employ heuristic methods internally, but in that their construction is based on rules of thumb, stratagems that appear to 'work' under most foreseen circumstances, and on other ad-hoc mechanisms that are added to them from time to time. My own program, ELIZA, was of precisely this type"; Weizenbaum, *Computer Power and Human Reason*, 232.[SC]

"PROF HIGGINS. What's your name?
THE FLOWER GIRL. Liza Doolittle.
PROF HIGGINS [declaiming gravely] Eliza, Elizabeth, Betsy and Bess, They went to the woods to get a bird's nes':
COL PICKERING. They found a nest with four eggs in it:
PROF HIGGINS. They took one apiece, and left three in it.
They laugh heartily at their own wit."
Shaw, *Pygmalion*.[PW]

Later, reflecting on ELIZA, Weizenbaum wrote that a "shock was administered not by any important political figure espousing his philosophy of science, but by some people who insisted on misinterpreting a piece of work I had done." Indeed he wrote, "such questions were my awakening to what Polanyi had earlier called a "scientific outlook that appeared to have produced a mechanical conception of man"; Weizenbaum, *Computer Power and Human Reason*, 2–7.[DMB]

208–225 211. The question "WHICH SCRIPT DO YOU WISH TO PLAY" is suggestive—the use of "play," rather than "run" or "execute," gestures toward a theatrical metaphor. Weizenbaum sometimes frames ELIZA as performing a role, highlighting the program's nature as simulation (persona) rather than genuine therapeutic session. This also connects to broader questions about authenticity in human-computer interaction.[DMB]

In later versions of ELIZA, this playful tone persists, independent of which script it is running. The function to list all available scripts is called "WHATSNU," and calling it includes a list of options that include "HINTS ON USING, CONTROLLING, AND COMPLAINING COMPLAINING ABOUT ELIZA."[SC]

218. Weizenbaum's use of the variable name "JUNK" for discarded

```
218                 LSSCPY.(TREAD.(INPUT,SCRIPT),JUNK)                          000130
219                 MTLIST.(INPUT)                                              000140
220                 T'H MLST, FOR I=1,1, I .G. 4                                000150
221       MLST      LIST.(MYTRAN(I))                                            000160
222                 MINE=0                                                      000170
223                 LIST.(MYLIST)                                               000180
224                 T'H KEYLST, FOR I=0,1, I .G. 32                             000220
225       KEYLST    LIST.(KEY(I))                                               000230
226                R* * * * * * * * * * READ NEW SCRIPT                         000240
227       BEGIN     MTLIST.(INPUT)                                              000250
228                 NODLST.(INPUT)                                              000260
229                 LISTRD.(INPUT,SCRIPT)                                       000270
230                 W'R LISTMT.(INPUT) .E. 0                                    000280
231                     TXTPRT.(JUNK,0)                                         000290
232                     MTLIST.(JUNK)                                           000300
233                     T'O START                                               000310
234                 E'L                                                         000320
235                 W'R TOP.(INPUT) .E. $NONE$                                  000330
236                     NEWTOP.(LSSCPY.(INPUT,LIST.(9)),KEY(32))                000340
237                     T'O BEGIN                                               000350
```

user input reveals a fascinating tension. Although technically efficient, this labeling feels almost inappropriate in the context of psychotherapy, where every patient utterance is supposed to carry meaning. The variable name suggests a pragmatic programmer's perspective winning out over the therapeutic. It's a moment where the curtain slips, revealing the mechanical reality behind the simulacrum.DMB

Why is Weizenbaum not merely printing the prompt as soon as the program encounters it? Here we see one of the moments where he is responding to the material constraints of the system, which would be slow depending on how heavily the machine was being used in a time-sharing context—could take, e.g., five seconds to read in each line of the script. Instead, ELIZA sets aside the prompt and then reads in the rest of the script before printing. More importantly, because JW does not know how long the script is—it could take quite some time to get through the entire script—he does not print the prompt and then leave the user waiting.JS

226–248 SCRIPT READING AND INITIALIZATION

The script reading mechanism demonstrates what sociologist Bruno Latour would later call "delegation" as the encoding of social scripts into technical systems.DMB

Wendy Hui Kyong Chun might call this looping mechanism "sourcery,"—i.e., the illusion of control through code. The loop structure also creates what Alexander Galloway terms "protocol" as the rules that govern interaction while hiding their own operation. Chun, "On 'Sourcery.'"; Galloway, "Language Wants To Be Overlooked."DMB

This section reads the script into memory. Every keyword except NONE and MEMORY is inserted into the KEY hashmap, hashed on the keyword. The NONE rule is recorded in KEY(32) and the four MEMORY rules are recorded in MYTRAN(1) ... MYTRAN(4).

227–28. clear the INPUT list. Any cells that were in the INPUT list are returned to the list of free cells.AH

229. Read the next list from the script. LISTRD inputs a list from tape unit specified by the integer N and gives it the name L.

230–34. An empty list indicates the end of the ELIZA script file. If the list just read is empty, print the opening remarks (JUNK) that were read from the script on line 218, clear JUNK and jump to the START label; exit script reading mode and enter conversation mode. (See chapter 3.1.2.).

235. The NONE rule is necessary to handle situations in which no keyword is found or in which pattern matching has failed.SC

```
238                  OR W'R TOP.(INPUT) .E. $MEMORY$                000360
239                   POPTOP.(INPUT)                                000370
240                   MEMORY=POPTOP.(INPUT)                         000380
241                   T'H MEM, FOR I=1,1, I .G. 4                   000390
242     MEM           LSSCPY.(POPTOP.(INPUT),MYTRAN(I))             000400
243                   T'O BEGIN                                     000410
244                  O'E                                            000420
245                   NEWBOT.(LSSCPY.(INPUT,LIST.(9)),KEY(HASH.     000430
246               1   (TOP.(INPUT),5)))                             000440
247                   T'O BEGIN                                     000450
248               E'L                                               000460
249               R* * * * * * * * * * BEGIN MAJOR LOOP             000470
250     START      TREAD.(MTLIST.(INPUT),0)                         000480
251                KEYWRD=0                                         000490
252                PREDNC=0                                         000500
253                LIMIT=LIMIT+1                                    000510
254                W'R LIMIT .E. 5, LIMIT=1                         000520
255                W'R LISTMT.(INPUT) .E. 0, T'O ENDPLA             000530
256                IT=0                                             000540
257                W'R TOP.(INPUT) .E. $+$                          000550
258                    CHANGE.(KEY,MYTRAN)                          000560
259                    T'O START                                    000570
260                E'L                                              000580
261                W'R TOP.(INPUT) .E. $*$, T'O NEWLST              000590
262                S=SEQRDR.(INPUT)                                 000600
263     NOTYET     W'R S .L. 0                                      000610
```

235–37. If the list is the NONE rule, copy it to KEY(32), then jump back to the BEGIN label to continue reading the script.

238–43. If the list is the MEMORY rule, copy the four transformation rules to MYTRAN(1) ... MYTRAN(4), then jump back to the BEGIN label to continue reading the script.

244–48. The list is not any of the above special cases; it's a normal keyword, so call the HASH function to generate an integer between 0 ... 31 derived from the keyword, or the first six characters of the keyword if it is more than six characters long. The generated integer is used as an index into the KEY array. Append the transformation rules associated with this keyword to the list at that index in KEY. Then jump back to the BEGIN label to continue reading the script.

The code's handling of memory shows a simplified model of psychiatric theory Rather than truly remembering or understanding, ELIZA uses a simple hash function to select responses. This mechanical approach to memory and response mirrors contemporaneous debates about behaviorism vs. deeper psychological understanding. The code might be seen to implement a behaviorist approach, concerned with inputs and outputs rather than internal mental states.[DMB]

249–262 MAJOR PROCESSING LOOP
See chapter 7.1 for an explanation of this main loop.

250. See 3.1.3.

The "MAJOR LOOP" section represents what Sherry Turkle would later call the "intimate machine"—a cyclical pattern of prompt, response, and reaction that creates an illusion of a relationship; Turkle, *Second Self*, 173, 336. The mechanical nature of this loop is hidden from users, who experience it as a conversation rather than deterministic computation.[DMB]

With TREAD, the user's input phrase is tokenized, or separated into words that are delimited by blanks, commas, and periods. This is discussed in chapter 11.4 in relation to large language model tokenization.[SC]

"I had worked hard for nearly two years, for the sole purpose of infusing life into an

```
264                      SEQLR.(S,F)                                     000620
265                      T'O NOTYET                                      000630
266                     O'E                                              000640
267                      WORD=SEQLR.(S,F)                                000650
268                      W'R WORD .E. $.$ .OR. WORD .E. $,$ .OR. WORD
                         .E. $BUT$                                       000660
269                          W'R IT .E. 0                                000670
270                              NULSTL.(INPUT,LSPNTR.(S),JUNK)          000680
271                              MTLIST.(JUNK)                           000690
272                              T'O NOTYET                              000700
273                             O'E                                      000710
274                              NULSTR.(INPUT,LSPNTR.(S),JUNK)          000720
275                              MTLIST.(JUNK)                           000730
276                              T'O ENDTXT                              000740
277                             E'L                                      000750
278                          E'L                                         000760
279                      E'L                                             000770
280                      W'R F .G. 0, T'O ENDTXT                         000780
281                      I=HASH.(WORD,5)                                 000790
282                      SCANER=SEQRDR.(KEY(I))                          000800
283                      SF=0                                            000810
284                      T'H SEARCH, FOR J=0,0, SF .G. 0                 000820
285                      CAND= SEQLR.(SCANER,SF)                         000830
286                      W'R SF .G. 0, T'O NOTYET                        000840
287         SEARCH       W'R TOP.(CAND) .E. WORD, T'O KEYFND             000850
289         KEYFND       READER=TESTS.(CAND,S)                           000860
```

inanimate body. For this I had deprived myself of rest and health. I had desired it with an ardour that far exceeded moderation; but now that I had finished, the beauty of the dream vanished, and breathless horror and disgust filled my heart." Shelley, *Frankenstein*, 56.[PW]

Weizenbaum sums up the puzzle posed by ELIZA in terms of a question: "our question is, 'What can one tell computers?' We have taken telling to mean giving an effective procedure ... We shall see that this is a very proper and a crucially important shift, that the question of what we can get a computer to do is, in the final analysis, the question of what we can bring a computer to know"; Weizenbaum, *Computer Power and Human Reason*, 71–72.[SC]

I find this tantalizing: it touches on something that greatly interests me, and at the same time I have no idea what he means.[AH]

258. If the user input begins with +, the CHANGE function is triggered, which allows for editing a script. This is discussed in Chapter 11.8 in relation to contemporary reinforcement learning.[SC]

261. If the user input begins with *, they can add a new keyword.[MM]

263–290 KEYWORD PROCESSING AND SELECTION
See chapter 3.1.4.

The handling of punctuation marks and the word "BUT" ($.$ .OR. WORD .E. $,$ .OR. WORD .E. BUT) shows awareness of natural language's complexities during a period when transformational grammar was revolutionizing linguistics. This code section embodies what philosopher Hubert Dreyfus would later critique as the limitations of formal approaches to human understanding; Dreyfus, *What Computers Still Can't Do*.[DMB]

270, 274, 275: The variable JUNK in ELIZA captures parts of the user input for which ELIZA has no response (KEYWORD).Though no doubt a practical necessity in a keyword-matching chatbot with limited memory, in the case of the DOCTOR persona, while playing the role of Rogerian psychotherapist, that dismissal of large portions of patient input as JUNK seems in conflict with the client-centered approach.[MM]

```
290                 W'R READER .E. 0, T'O NOTYET              000870
291                 W'R LSTNAM.(CAND) .NE. 0                  000880
292                     DL=LSTNAM.(CAND)                      000890
293     SEQ             W'R S .L. 0                           000900
294                         SEQLR.(S,F)                       000910
295                         T'O SEQ                           000920
296                        O'E                                000930
297                         NEWTOP.(DL,LSPNTR.(S))            000940
298                     E'L                                   000950
299                    O'E                                    000960
300                 E'L                                       000970
301                 NEXT=SEQLR.(READER,FR)                    000980
302                 W'R FR .G. 0, T'O NOTYET                  000990
303                 W'R IT .E. 0 .AND. FR .E. 0               001000
304     PLCKEY          IT=READER                             001010
305                     KEYWRD=WORD                           001020
306                    OR W'R FR .L. 0 .AND. NEXT .G. PREDNC  001030
307                     PREDNC=NEXT                           001040
308                     NEXT=SEQLR.(READER,FR)                001050
309                     T'O PLCKEY                            001060
310                    0'E                                    001070
```

Definitions of the word *junk*: dating from the mid-fourteenth century, *junke* meant "old cable or rope," which was cut in bits and used for caulking and other tasks. Some claim it is related to the Old French *junc*, meaning "rush, reed," which was something of little value, and was itself drawn from the Latin *iuncus*, "rush, reed." Later in the nineteenth century, "old or discarded articles of any kind," usually with a suggestion of reusability. Here, there is sadly no reusability.[DMB]

Just to suggest an alternative point of view: MTLIST.(JUNK) returns the cells in the JUNK list to the SLIP library LAVS (List of AVailable Space), where the cells may be reused in other lists when needed.[AH]

282. The SEQRDR function creates a sequence reader for the list at index I in the array KEY. The sequence reader is assigned to the variable called SCANER. SCANER is later passed to the SEQLR function. Each call to SEQLR returns the next cell in the list (which may be a datum or a sublist, and in this context is always a sublist—hence the call to TOP on line 287, which will return the first cell in the sublist, which is the (first six characters of the) keyword for the rule).

291–302 The "MAJOR LOOP" structure implements what Tara McPherson (2012) calls "lenticular logic"—i.e., a way of seeing that separates and compartmentalizes; Mcpherson, "Why Are the Digital Humanities So White?", 144. The strict separation between input processing and output generation in ELIZA is reminiscent of the kinds of modular knowledge computers are able to process only through fragmenting conversation into discrete manageable units.[DMB, MM]

303–314 This version of the code differs from the description in Weizenbaum's *CACM* paper where he says that ELIZA maintains a keyword stack and keywords of higher precedence than the keyword on the top of the stack are added to the top of the stack; otherwise, they are added to the bottom of this stack. See 3.1.4. This also means this code does not support the "NEWKEY" functionality he describes in the paper.[AH]

315–331 MEMORY HANDLING AND RESPONSE SELECTION

The memory system's deterministic rather than random nature reflects a tension between the appearance of spontaneity and the reality of programmed behavior. This perhaps reflects broader debates in psychology about free will vs. determinism. Although behaviorists emphasized deterministic stimulus-response patterns, humanistic psychologists like Carl Rogers (whose approach ELIZA simulated) emphasized human unpredictability.[DMB]

```
311                    T'O NOTYET                                001080
312                E'L                                           001090
313                T'O NOTYET                                    001100
314               R* * * * * * * * * END OF MAJOR LOOP           001110
315   ENDTXT       W'R IT .E. 0                                  001120
316                    W'R LIMIT .E. 4 .AND. LISTMT.
                       (MYLIST) .NE. 0                           001130
317                        OUT=POPTOP.(MYLIST)                   001140
318                        TXTPRT.(OUT,0)                        001150
319                        IRALST.(OUT)                          001160
320                        T'O START                             001170
321                       O'E                                    001180
322                        ES=BOT.(TOP.(KEY(32)))                001190
323                        T'O TRY                               001200
324                    E'L                                       001210
325                   OR W'R KEYWRD .E. MEMORY                   001220
326                    I=HASH.(BOT.(INPUT),2)+1                  001230
327                    NEWBOT.(REGEL.(MYTRAN(I),INPUT,LIST.
                       (MINE)),MYLIST)                           001240
328                    SEQLL.(IT,FR)                             001250
329                    T'O MATCH                                 001260
```

In the *CACM* paper, the explanation of when to recall a memory is "when a text without keywords is encountered later and a certain counting mechanism is in a particular state and the stack in question is not empty, then the transformed text is printed out as the reply"; Weizenbaum, "Eliza," 41.LIMIT is that counting mechanism, and 4 is the particular state. See 3.2.3.[AH]

In the *CACM* paper, Weizenbaum describes the MEMORY rule in the DOCTOR script by saying, "The word 'MY' (which must be an ordinary keyword as well) has been selected to serve a special function. Whenever it is the highest ranking keyword of a text, one of the transformations on the MEMORY list is randomly selected, and a copy of the text is transformed accordingly"; Weizenbaum, "Eliza," 40. Lines 326–27 show that the selection is determined by the hash value of the last word in the user's input. This means that, contrary to what he says, *ELIZA's responses are not random.* If we use the same hash algorithm and give the same user inputs, we should be able to reproduce the exact same responses ELIZA would have made in 1966. (We do have the SLIP HASH algorithm used by ELIZA.) See 3.2.4.[AH]

Discussed in relation to LLM hyperparameters in Chapter 11.7.[SC]

332–352 RESPONSE GENERATION AND PATTERN MATCHING

Described in chapter 7.1

The sequence of word-matching operations is interesting to read through philosopher John Searle's later work on syntax vs. semantics in his Chinese Room argument; Searle, "Minds, Brains, and Programs, 417–418].[DMB, MM]

The code's manipulation of symbols without "understanding" exemplifies key questions in 1960s cognitive science (e.g., in the 1960s work of Ulric Neisser; see, e.g., Neisser, "Imitation of Man by Machine").[DMB]

333. Machine manipulation of language owes much of its inspiration to Noam Chomsky's work. In *Syntactic Structures,* he contends: "The grammar of a language is a device of some sort for producing the sentences of the language under analysis. ... The ultimate outcome of these investigations should be a theory of linguistic structure in which the descriptive devices utilized in particular grammars are presented and studied abstractly"; Chomsky, *Syntactic Structures*, 11. Separating syntax from semantics, Chomsky's theorization of the system of language laid the foundation for ELIZA's systematic pattern matching, decomposition and recomposition.[JS]

334. The = sign is quite powerful, allowing the ELIZA system to work like a recursive descent parser. See McCarthy, "Answer to 'Recursive Descent Parser.'"

```
330                       O'E                                          001270
331                        SEQLL.(IT,FR)                               001280
332                   R* * * * * * * * * * MATCHING ROUTINE            001290
333        MATCH           ES=SEQLR.(IT,FR)                            001300
334                        W'R TOP.(ES) .E. $=$                        001310
335                            S=SEQRDR.(ES)                           001320
336                            SEQLR.(S,F)                             001330
337                            WORD=SEQLR.(S,F)                        001340
338                            I=HASH.(WORD,5)                         001350
339                            SCANER=SEQRDR.(KEY(I))                  001360
340        SCAN                ITS=SEQLR.(SCANER,F)                    001370
341                            W'R F .G. 0, T'O NOMATCH(LIMIT)         001380
342                            W'R WORD .E. TOP.(ITS)                  001390
343                                S=SEQRDR.(ITS)                      001400
344        SCANI                   ES=SEQLR.(S,F)                      001410
345                                W'R F .NE.0, T'O SCANI              001420
346                                IT=S                                001430
347                                T'O TRY                             001440
348                            O'E                                     001450
349                                T'O SCAN                            001460
350                            E'L                                     001470
351                        E'L                                         001480
352                        W'R FR .G. 0, T'O NOMATCH(LIMIT)            001490
353        TRY             W'R YMATCH.(TOP.(ES),INPUT,MTLIST.(TEST))
                           .E. 0,T'O MATCH                             001500
354                        ESRDR=SEQRDR.(ES)                           001510
355                        SEQLR.(ESRDR,ESF)                           001520
356                        POINT=SEQLR.(ESRDR,ESF)                     001530
357                        POINTR=LSPNTR.(ESRDR)                       001540
358                        W'R ESF .E. 0                               001550
359                            NEWBOT.(1,POINTR)                       001560
360                            TRANS=POINT                             001570
361                            T'O HIT                                 001580
362                          O'E                                       001590
363                            T'H FNDHIT,FOR I=0,1, I .G. POINT       001600
364        FNDHIT              TRANS=SEQLR.(ESRDR,ESF)                 001610
```

353–375 See chapter 4.5 for our description of YMATCH, likely named after Victor Yngve, creator of pattern-matching language COMIT. Researchers make cameos in one another's code.[MM]

YMATCH is in MAD-SLIP but not in the 1964 *CACM* SLIP.[JS]

YMATCH is in MAD-SLIP but not in the 1964 *CACM* SLIP. YMATCH and ASSMBL(line 376) are the two most important parts of the ELIZA system because they process regular expressions in the script to analyze the input and construct the response. This is a brilliant application of the innovations passed down from Yngve's COMIT and Stephen Kleene, (developer of the Kleene star). We had to rewrite this function in order to get our recreated ELZA to work (see chapter 4).[JS]

376–378 See chapters 3.1.6 and 7.1.

```
365                              W'R ESF .G. 0                                  001620
366                                  SEQLR.(ESRDR,ESF)                          001630
367                                  SEQLR.(ESRDR,ESF)                          001640
368                                  TRANS=SEQLR.(ESRDR,ESF)                    001650
369                                  SUBST.(1,POINTR)                           001660
370                                  T'O HIT                                    001670
371                                 O'E                                         001680
372                                  SUBST.(POINT+1,POINTR)                     001690
373                                  T'O HIT                                    001700
374                              E'L                                            001710
375                          E'L                                                001720
376       HIT                TXTPRT.(ASSMBL.(TRANS,TEST,MTLIST.
                             (OUTPUT)),0)                                       001730
377                          T'O START                                          001740
378                      E'L                                                    001750
379                     R* * * * * * * * * * * INSERT NEW KEYWORD LIST          001760
380       NEWLST         POPTOP.(INPUT)                                         001770
381                      NEWBOT.(LSSCPY.(INPUT,LIST.(9)),KEY(HASH.              001780
382                     1(TOP.(INPUT),5)))                                      001790
383                      T'O START                                              001800
384                     R* * * * * * * * * * * DUMP REVISED SCRIPT              001810
385       ENDPLA         PRINT COMMENT $WHAT IS TO BE THE NUMBER OF
                         THE NEW SCRIPT$                                        001820
386                      READ FORMAT SNUMB,SCRIPT                               001830
387                      LPRINT.(INPUT,SCRIPT)                                  001840
388                      NEWTOP.(MEMORY,MTLIST.(OUTPUT))                        001850
389                      NEWTOP.($MEMORY$,OUTPUT)                               001860
390                      T'H DUMP, FOR I=1,1 I .G. 4                            001870
391       DUMP           NEWBOT.(MYTRAN(I),OUTPUT)                              001880
392                      LPRINT.(OUTPUT,SCRIPT)                                 001890
393                      MTLIST.(OUTPUT)                                        001900
394                      T'H WRITE, FOR I=0,1, I .G. 32                         001910
395       POPMOR         W'R LISTMT.(KEY(I)) .E. 0, T'O WRITE                   001920
396                      LPRINT.(POPTOP.(KEY(I)),SCRIPT)                        001930
397                      T'O POPMOR                                             001940
398       WRITE          CONTINUE                                               001950
```

379–383 SCRIPT MANAGEMENT AND ERROR HANDLING

384–387 When you give ELIZA a blank line.[AH]

401–411 NONVERBAL UTTERANCES

404. The choice of "HMMM" as an error message is deeply human. It's a nonverbal utterance that humans use to signal thoughtful presence while masking uncertainty or confusion. By programming ELIZA to respond with this distinctly human sound when it encounters errors Weizenbaum created a moment of subtle anthropomorphization.[DMB]

HMMM belongs to a class of (linguistic) utterances termed "discourse markers," but it is a special case as it is a (or many etymologies classify it as a nonverbal) closed mouth, utterance, related specifically to HUM or Humming. So HMMM is an interesting choice (at

```
399                       LPRINT.(MTLIST.(INPUT),SCRIPT)                 001960
400                       EXIT.                                          001970
401                      R* * * * * * * * * * SCRIPT ERROR EXIT          001980
402        NOMATCH(1)     PRINT COMMENT $PLEASE CONTINUE $               002200
403                       T'O START                                      002210
404        NOMATCH(2)     PRINT COMMENT $HMMM $                          002220
405                       T'O START                                      002230
406        NOMATCH(3)     PRINT COMMENT $GO ON , PLEASE $                002240
407                       T'O START                                      002250
408        NOMATCH(4)     PRINT COMMENT $I SEE $                         002260
409                       T'O START                                      002270
410                       VECTOR VALUES SNUMB= $I3 * $                   002280
411                       E'M                                            002290
```

least to me) as a physical, distinctly human (dogs don't hmmm or even hum) reply that we recognize as a cue that someone is signaling that they want us to think that they are there, even if it is also a signal that they aren't really all there. In a very real sense, ELIZA isn't all there.[PW]

HMM is also one of the first kinds of speech to be removed by contemporary automated text systems when doing speech transcription and text analysis. These are the words we do not wish to read in a transcribed interview, but they are the sounds we expect for a sense of warmth and presence that get reinserted.[SC]

07.00 The DOCTOR Script

DOCTOR: YOU DON'T ARGUE WITH ME

USER:
WHY DO YOU
THINK I
DON'T ARGUE
WITH YOU

Although the MAD-SLIP source code provides insight into ELIZA's mechanical operations, it is the DOCTOR script that reveals the design and intention behind the chatbot. The DOCTOR script examined in this chapter represents a key moment in the history of human-computer interaction, the moment when a computer program first performed the illusion of understanding through conversation. When Weizenbaum created ELIZA in 1966, he implemented a software architecture that separated the interaction system (ELIZA, examined in the previous chapter) from its conversational script model (DOCTOR, examined in this chapter). This distinction may seem obvious today, but when Weizenbaum was programming, it was not yet the foundational pattern that would define modern software design.

As we explore this script, we attempt to show how Weizenbaum crafted transformation rules that reflected Rogerian therapeutic techniques, managed conversational flow through keyword precedence, and implemented primitive forms of memory. These techniques would later become important for modern chatbots and conversational agents. The contrast between ELIZA's rule-based pattern matching and today's neural language models (see chapter 11) highlights both how far we have come technically and how many of the same questions about understanding, meaning, and human-machine interaction remain unresolved.

Our exploration of the DOCTOR script offers several major takeaways, which are discussed throughout the book, as the analysis of the code and script initiated many of the book's main threads. First and foremost, it indicates that the DOCTOR script would not have run on the ELIZA described in the *CACM* article. Second, some aspects of DOCTOR point to directions not fully pursued, such as the /NOUN tag that points in the direction of Chomskian grammar. Third, by analyzing the entire script in conjunction with the ELIZA code, we can more closely consider the use of substitution categories (like FAMILY for MOTHER), as well as the precedence in the keywords. For example, by assigning COMPUTER a precedence of 50, Weizenbaum assures that this match will be prioritized over most other keywords in this script.

One of our most unexpected discoveries is that ELIZA is Turing complete.[1] The script command PRE gives ELIZA the ability to transform input (i.e., write something in its place) and to select another keyword for processing (i.e., changing its state). That observation led Anthony Hay to conjecture that ELIZA was potentially Turing complete, which Peter Millican then verified. Turing completeness is not immediately obvious in ELIZA, but scrutinizing the code led to the observation that then led to this very unexpected discovery.

By examining the DOCTOR script in detail, we gain valuable perspective on both the origins of conversational AI and the enduring questions it raises about the relationship between humans and the technologies we create. The stakes are not merely historical but also philosophical, as the techniques pioneered in this script continue to shape how we interact with and understand our so-called intelligent machines.

07.01 How DOCTOR Talks

The historical context of 1960s computing materially shaped ELIZA's architecture, just as it shaped early programming languages and techniques. Yet within these limitations, Weizenbaum crafted a neat solution combining precedence-based keyword selection with efficient text normalization. For example, the system's handling of context preservation and variable binding demonstrated sophisticated awareness of linguistic principles. These technical features, while hidden from users, are crucial to creating the conversational interface that made ELIZA so interesting.

To understand how ELIZA processes conversation through the DOCTOR script, we need to examine its core operational mechanisms, a process of decomposition and reassembly that represents one of the system's core innovations. An ELIZA script is mostly a collection of keywords with each keyword having a set of transformation rules associated with it. ELIZA generates a response to user input in five stages.

During the first stage user input is evaluated. Synonym substitution is made and keywords are placed on the keyword stack. The next stage (2) looks for the highest priority keyword, if found then the next stage (3) is the find which decomposition pattern to use. After pattern matching, the next stage (4) separates the modified user input into parts, and in the final stage (5) an output sentence is generated.

During the first stage, if the end of the input text is reached, or a period, comma or the word BUT is encountered, scanning is paused. This stage is also where simple word substitutions are made to the input text. If a keyword is found, it is added to the top of a keyword stack if it is the highest precedence keyword so far encountered (each keyword may be assigned a precedence in the script), otherwise it is added to the bottom of the keyword stack. At this point if the keyword stack is not empty, stage two begins. If the keyword stack is empty, ELIZA deletes the current input and continues scanning.

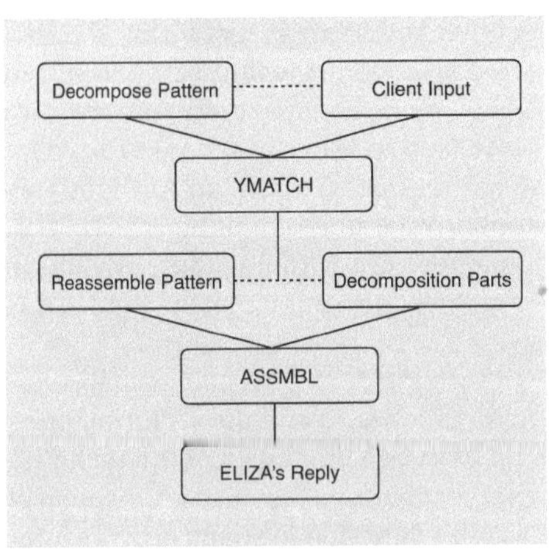

Figure 7.1 Eliza dialogue data flow

During stage two, if the keyword stack is empty, the NONE rule is processed or, under certain circumstances, a previously stored sentence is recalled (see chapter 3 for details). This approach represents a clever solution to the challenge of creating seemingly coherent responses with minimal computational resources. By prioritizing keywords and discarding irrelevant text,

ELIZA could maintain conversational focus without needing to understand the entire input.

In the third stage, the selected keyword, and its transformation rules are applied to the user's input text to produce ELIZA's response. If there are no matching patterns, ELIZA prints one of its built-in messages, which are "PLEASE CONTINUE," "HMMM," "GO ON, PLEASE," and "I SEE."

The fourth stage decomposes the modified user input under the direction of a decomposition rule. The fifth stage uses a decomposition to generate a response. There is an exception to the fifth stage for NEWKEY and PRE commands. For NEWKEY, stages three, four, and five are repeated for another keyword. For PRE, user input is modified and another keyword is selected and processing continues the same as NEWKEY.

The position of items in the script does not affect processing. Keyword priority is evaluated on the fly, and decomposition rules for a keyword are evaluated sequentially until a pattern is matched. The contents of the rules in the published script are not arranged by priority or some obvious other organizational structure, such as alphabetization. This fact combined with the ability to modify scripts allows us to conjecture that rule ordering replicates Weizenbaum's development of concepts, rather than any technical requirement. By examining this ordering, we can potentially glimpse his thought progression as he crafted different conversational domains, from emotional states to family relationships to meta-discussions about computers themselves, revealing an implicit taxonomy of his concerns during the mid-1960s. The pattern matching approach, while primitive by today's standards, laid important groundwork for natural language processing techniques. The decomposition-reassembly pattern would influence later systems like Terry Winograd's SHRDLU[2] and conceptually presages aspects of modern transformer architectures' encoder-decoder patterns.

The DOCTOR script thus potentially reveals not just technical solutions but also the cultural and psychological assumptions of its time. For example, its emotional vocabulary distinguishes mainly between positive and negative states (HAPPY/SAD), reflecting simplified psychological models. Similarly, its focus on family relationships (MOTHER, FATHER) encodes the centrality of the nuclear family in midcentury psychological theory. These embedded assumptions invite us to consider which cultural values and theoretical frameworks might be encoded in today's more sophisticated AI systems.

07.02 Comparative Script Analysis

It is helpful when trying to understand the DOCTOR script to compare different rules with one another. Table 7.1 compares the three rules

REMEMBER from line 12, MEMORY, from line 67, and NAME from line 87 to demonstrate the utility of this approach. We selected these three rules because they have very different outcomes, even though they all appear to deal in some way with memory or retention of information by the ELIZA program.

Table 7.1 Comparison of rules in DOCTOR script

REMEMBER (line 12)	MEMORY (line 67)	NAME (line 87)
(REMEMBER 5 ((0 YOU REMEMBER 0) (DO YOU OFTEN THINK OF 4) (DOES THINKING OF 4 BRING ANYTHING ELSE TO MIND) (WHAT ELSE DO YOU REMEMBER) (WHY DO YOU REMEMBER 4 JUST NOW) (WHAT IN THE PRESENT SITUATION REMINDS YOU OF 4) (WHAT IS THE CONNECTION BETWEEN ME AND 4)) ((0 DO I REMEMBER 0) (DID YOU THINK I WOULD FORGET 5) (WHY DO YOU THINK I SHOULD RECALL 5 NOW) (WHAT ABOUT 5) (=WHAT) (YOU MENTIONED 5)) ((0) (NEWKEY)))	(MEMORY MY (0 YOUR 0 = LETS DISCUSS FURTHER WHY YOUR 3) (0 YOUR 0 = EARLIER YOU SAID YOUR 3) (0 YOUR 0 = BUT YOUR 3) (0 YOUR 0 = DOES THAT HAVE ANYTHING TO DO WITH THE FACT THAT YOUR 3))	(NAME 15 ((0) (I AM NOT INTERESTED IN NAMES) (I'VE TOLD YOU BEFORE, I DON'T CARE ABOUT NAMES—PLEASE CONTINUE)))

This comparative approach reveals how differently ELIZA handled various aspects of "memory." The NAME rule actively rejects remembering names and the MEMORY rule stores specific phrases for later reuse, whereas the REMEMBER rule, ironically, doesn't actually remember anything but instead redirects conversation about memories back to the user. This reflects both the technical limitations of the time and the therapeutic focus on the user rather than the system's capabilities.

These inconsistencies in memory handling highlight how ELIZA strategically simulated different cognitive capabilities, implementing only those functions that directly supported the therapeutic illusion while avoiding more complex challenges of true memory and knowledge representation.

07.03 How an ELIZA Script Was Written

Weizenbaum designed ELIZA with remarkable flexibility, allowing for script creation and modification through an innovative editing interface. This approach represents one of the earliest examples of what we now recognize as runtime configurability in software systems, a principle that has become important in modern software architecture. In his 1966 *CACM* paper, Weizenbaum explains the process:

> Editing of an ELIZA script is achieved via appeal to a contextual editing program (ED) which is part of the MAC library. This program is called whenever the input text to ELIZA consists of the single word "EDIT." ELIZA then puts itself in a so-called dormant state and presents the then stored script for editing. Detailed description of ED is out of place here. Suffice it to say that changes, additions and deletions of the script may be made with considerable efficiency and on the basis of entirely contextual cues, i.e., without resort to line numbers or any other artificial devices. When editing is completed, ED is given the command to FILE the revised script. The new script is then stored on the disk and read into ELIZA. ELIZA then types the word "START" to signal that the conversation may resume under control of the new script.[3]

Our version mapping research revealed that earlier iterations of ELIZA featured a CHANGE function that made the system inherently "teachable," in Weizenbaum's phrasing, enabling users to revise and extend scripts during program execution. This functionality later was replaced with the EDIT function described above, which called the external ED program. This design philosophy is directly connected to Weizenbaum's vision for ELIZA, as he named the system to emphasize its capacity for incremental improvement.

Crucially, Weizenbaum approached programming as an exploratory practice rather than a mere implementation of preexisting understanding. He wrote that "one programs, just as one writes, not because one understands, but in order to come to understand. Programming is an act of design."[4] This perspective frames ELIZA not just as a finished product but as a continuously changing experiment in human-computer interaction through language.

This editing mechanism was groundbreaking for several reasons. The contextual editor (ED), created by Charles C. Garman in 1964, allowed for modifications based on content rather than line numbers, which was a significant usability improvement that foreshadowed modern text editors. Working within MIT's Compatible Time-Sharing System (CTSS), ED handled 14-word BCD card image files and incorporated conventions from TYPSET, an earlier editor developed by Jerome H. Saltzer in 1964.

What makes this editing system particularly significant is how it transformed ELIZA scripts from static code into dynamic, modifiable content. This separation between the running program and its behavioral definitions exemplifies the data-driven architecture that continues to be used ina software design today. A therapist or linguist could potentially modify ELIZA's responses without understanding programming as an early implementation of the principle that domain experts should be able to configure systems without requiring technical expertise.

That ELIZA was not only a system for conversational interaction but also a platform for writing and editing scripts is one of the many aspects of the program lost in the popular ELIZA imaginary. This dual nature, as both conversational agent and script development environment, reveals Weizenbaum's deeper vision of computational systems as platforms for ongoing experimentation and development.

07.04 The Annotated Script

As Understanding ELIZA's DOCTOR script requires both technical and contextual analysis. The technical aspects, which include keyword matching, precedence values, and the transformation rules, can provide the mechanical/computational framework, but it is the psychological and linguistic choices embedded in these mechanisms that make the script truly remarkable. As we transition from discussing ELIZA's architecture to examining the script itself, it is worth considering how every keyword, rule, and response reflects both the technical constraints of 1960s computing and the cultural assumptions of therapeutic practice (and computing) at that time.

What follows is the complete DOCTOR script as published in Weizenbaum's 1966 paper, presented with our line numbers for reference and annotated with commentary from various historical, technical, psychological, and cultural perspectives. The DOCTOR script was originally published as an appendix to Weizenbaum's ELIZA paper.[5] It is reproduced below in full with our annotations added. These annotations not only aim to clarify how specific parts of the script function but also aim to highlight connections to broader themes in AI development, psychology, human-computer interaction, and societal context. We use critical code studies to examine the script in this way, so that we can appreciate both Weizenbaum's technical achievements and the cultural legacy of his creation.

07.05 Summary

As we see throughout this detailed examination, the DOCTOR script represents far more than a simple set of pattern-matching rules. That this script could generate conversations compelling enough to inspire both technological enthusiasm and ethical concern speaks to its significance beyond mere pattern matching. It embodies a particular historical moment in both computing and psychology, reveals sophisticated understanding of human conversation, and demonstrates how even primitive AI can raise profound questions about understanding and intelligence. In tracing these keywords and transformation rules, we've uncovered not just the mechanics of ELIZA's operation but also some of the philosophical assumptions and cultural context that shaped one of computing's most influential early experiments in human-computer interaction.

The relationship between ELIZA's components, both script and code, intent and mechanism, mirrors broader questions about the relationship between human meaning and machine processing that continue to define AI research today.

07.06 DOCTOR Script

```
0
1         (HOW DO YOU DO. PLEASE TELL ME YOUR PROBLEM)
2         START
3         (SORRY
4            ((0)
5               (PLEASE DON'T APOLIGIZE)
6               (APOLOGIES ARE NOT NECESSARY)
7               (WHAT FEELINGS DO YOU HAVE WHEN YOU APOLOGIZE)
8               (I'VE TOLD YOU THAT APOLOGIES ARE NOT REQUIRED)))
9         (DONT = DON'T)
10        (CANT = CAN'T)
11        (WONT = WON'T)
12        (REMEMBER 5
13           ((0 YOU REMEMBER 0)
14              (DO YOU OFTEN THINK OF 4)
15              (DOES THINKING OF 4 BRING ANYTHING ELSE TO MIND)
16              (WHAT ELSE DO YOU REMEMBER)
17              (WHY DO YOU REMEMBER 4 JUST NOW)
18              (WHAT IN THE PRESENT SITUATION REMINDS YOU OF 4)
```

0 In an archival version of DOCTOR, the instruction "PRINT,T0109,2351,. TAPE.,102" was used by Weizenbaum to generate the printout which shows a remnant of the CTSS filing system. Such instructions bridge the media of punchcards and the later generation of tape storage. [DMB, SC, MM]

1 A version of the DOCTOR script found in the Weizenbaum archive (1965) differs slightly. It instructs, "HOW DO YOU DO. I AM THE DOCTOR. PLEASE SIT DOWN AT THE TYPEWRITER AND TELL ME YOUR PROBLEM."[SC, MM] The greeting "How do you do" is the center of one of the lessons ELIZA receives in *Pygmalion*.[MM]

"A person playing with ELIZA in its psychiatrist mode was instructed to provide ELIZA with the sort of statements one might make to a psychiatrist in an initial psychiatric interview. He was told, in other words, what ELIZA's expectations were"; Weizenbaum, *Computer Power and Human Reason*, 191.[SC]

The choice by Weizenbaum to begin with "HOW DO YOU DO" rather than a more casual greeting reflects the formal therapeutic relationships of the 1960s. The immediate call to "PLEASE TELL ME YOUR PROBLEM" demonstrates the directive nature of therapy at the time, despite DOCTOR's supposedly Rogerian approach. This tension between formality and directness mirrors broader cultural shifts in doctor-patient relationships of the 1960s.[DMB]

2 This is the only line that is not an S-expression.[AS]

A question of order: Is the script as published a product of a card deck that Weizenbaum input, or as a script ordered when printed, or combination of the two? We do not currently know.

3–8 The script's organization provides insights into how Weizenbaum prioritized certain conversational elements. Keywords are assigned numeric precedence values, revealing which topics ELIZA would privilege when multiple potential responses were available. This hierarchy of attention reflects both technical necessities and implicit judgments about what matters in therapeutic conversation.[DMB]

This typo (misspelling of *apologize*) so near the start of the code seems to support the idea that the code received very little proofreading.[MM]

This typo appears again in the YAPYAP script, line 470: ... (What Feelings/ 480 AND THOUGHTS DO YOU HAVE WHEN YOU APOLIGIZE), further

```
19              (WHAT IS THE CONNECTION BETWEEN ME AND 4))
20          ((0 DO I REMEMBER 0)
21              (DID YOU THINK I WOULD FORGET 5)
22              (WHY DO YOU THINK I SHOULD RECALL 5 NOW)
23              (WHAT ABOUT 5)
24              (=WHAT)
25              (YOU MENTIONED 5))
26          ((0)
27              (NEWKEY)))
28      (IF 3
29          ((0 IF 0)
30              (DO YOU THINK ITS LIKELY THAT 3)
31              (DO YOU WISH THAT 3)
32              (WHAT DO YOU THINK ABOUT 3)
33              (REALLY, 2 3)))
34      ; duplicate line removed: (WHAT DO YOU THINK ABOUT 3) (REALLY, 2 3)))
35      (DREAMT 4
36          ((0 YOU DREAMT 0)
37              (REALLY, 4)
38              (HAVE YOU EVER FANTASIED 4 WHILE YOU WERE AWAKE)
39              (HAVE YOU DREAMT 4 BEFORE)
40              (=DREAM)
```

evidence of YAPYAP's lineage from a slightly later version of ELIZA. The typo was most likely in the DOCTOR script, not something introduced in the publication.JS

The keyword SORRY appearing this early suggests that JW anticipates this response and wants to redirect apologizing early. On the other hand, the keyword has the lowest precedence (0).JS, MM

The SORRY keyword reflects Rogerian therapy's emphasis on unconditional positive responses. In the DOCTOR script Weizenbaum attempts to create a nonjudgmental environment where a user feels accepted without needing to apologize—a key principle of person-centered therapy.DMB

9 These simple substitutions show ELIZA's basic text normalization. Notably, only three substitutions are handled, suggesting Weizenbaum prioritized the most common negative substitutions that might appear in therapeutic dialogue. The limited scope reflects both memory constraints of the era and pragmatic focus on essential patterns.DMB

10–11 These lines always remind me of the Lydia Davis story collection *Can't and Won't*, and there is something of her dry sense of humor and theatrical personas in the Weizenbaum coding style as well.SC

12–27 The REMEMBER keyword ironically does not link into the contents of the memory, as queued by line 67. So ELIZA will not actually attempt to "remember."MM

ELIZA uses remember only to reflect back user introduced memories. ELIZA doesn't use "remember" in the imperative voice, as in:

"Remember that you are a human being with a soul and the divine gift of articulate speech" (Professor Henry Higgins to ELIZA, act 1, *Pygmalion*).PW

28–33 The IF keyword's handling reveals DOCTOR's approach to hypothetical thinking. Rather than trying to evaluate the conditional statement (which would require true reasoning capabilities), it redirects the conversation back to the user's thoughts and feelings about the hypothetical scenario. This reflects both technical limitations and therapeutic technique—keeping focus on the user's perspective rather than offering direct advice.DMB

35–41 Weizenbaum was a believer in the unconscious, writing, "The unconscious washes up not only the material of

```
41              (NEWKEY)))
42          (DREAMED = DREAMT 4
43            (=DREAMT))
44          (DREAM 3
45            ((0)
46              (WHAT DOES THAT DREAM SUGGEST TO YOU)
47              (DO YOU DREAM OFTEN)
48              (WHAT PERSONS APPEAR IN YOUR DREAMS)
49              (DON'T YOU BELIEVE THAT DREAM HAS SOMETHING TO DO WITH YOUR
                PROBLEM)
50              (NEWKEY)))
51          (DREAMS = DREAM 3
52            (=DREAM))
53          (HOW
54            (=WHAT))
55          (WHEN
56            (=WHAT))
57          (ALIKE 10
58            (=DIT))
59          (SAME 10
60            (=DIT))
61          (CERTAINLY
62            (=YES))
```

creativity, not only the pearls that need only be polished ... but also the darkest truths about oneself"; Weizenbaum, *Computing Power and Human Reason*, 208. So it is no surprise the dreams make the list of keywords.[MM]

The inclusion of dream-related keywords highlights the Freudian influences prevalent in 1960s psychological discourse, despite DOCTOR's supposed Rogerian approach. This tension between different therapeutic models reveals how the script serves as an artifact of its cultural moment, also capturing shifting paradigms in psychology.[DMB]

44–50 Although the client might expect to be discussing dreams with DOCTOR in the style of Freudian psychoanalysis, the Rogerian Nondirective method de-emphasized them. Nonetheless, the DOCTOR script rightly returns the onus of interpretation of the relevance of the dream to the patient.[MM]

51–52 Note that, although it's a therapist, DOCTOR does not solicit conversations about dreams, nor does it interpret "dream" to mean an "aspiration" or "wish."[PW]

Although the *CACM* version responds to DREAMS, as one might expect from a therapist of the time, the YAPYAP script created by Harvard psychologists does not have the keyword variations of DREAM.[JS]

"The unconscious is structured like a language" (*l'inconscient est structuré comme un langage*) Lacan, *The Seminar of Jacques Lacan*, book 20, p. 15.[MM]

53–54 The reduction of different question words to WHAT demonstrates an interesting simplification strategy. Rather than creating unique response patterns for each question word, DOCTOR treats them as equivalent, allowing for a more manageable script size while still maintaining some conversational coherence.[DMB]

Yet "how" is very different from "what," and in many contexts can change the thrust of the question. "How are you doing?" and "What are you doing?" are perhaps even at cross-purposes. This moves toward a cybernetic approach, in which signal is just signal. How much granularity is necessary to convey meaning? And at what point does reduction for the purposes of efficiency slide into creating noise?[SC]

55–56 The reduction of HOW and WHEN to WHAT reveals an interesting philosophical assumption about the nature of asking

```
63          (FEEL DLIST(/BELIEF))
64          (THINK DLIST(/BELIEF))
65          (BELIEVE DLIST(/BELIEF))
66          (WISH DLIST(/BELIEF))
67          (MEMORY MY
68             (0 YOUR 0 = LETS DISCUSS FURTHER WHY YOUR 3)
69             (0 YOUR 0 = EARLIER YOU SAID YOUR 3)
70             (0 YOUR 0 = BUT YOUR 3)
71             (0 YOUR 0 = DOES THAT HAVE ANYTHING TO DO WITH THE FACT THAT YOUR 3))
72          (NONE
73             ((0)
74                (I AM NOT SURE I UNDERSTAND YOU FULLY)
75                (PLEASE GO ON)
76                (WHAT DOES THAT SUGGEST TO YOU)
77                (DO YOU FEEL STRONGLY ABOUT DISCUSSING SUCH THINGS)))
78          (PERHAPS
79             ((0)
80                (YOU DON'T SEEM QUITE CERTAIN)
81                (WHY THE UNCERTAIN TONE)
82                (CAN'T YOU BE MORE POSITIVE)
83                (YOU AREN'T SURE)
84                (DON'T YOU KNOW)))
85          (MAYBE
```

questions. By collapsing different types of inquiry into a single category, DOCTOR implies that the specific nature of the question is less important than the act of questioning itself—a surprisingly postmodern approach that aligns with Rogerian therapy's focus on the process rather than content.DMB

63–66 This grouping of mental state verbs under the BELIEF category shows sophisticated psychological awareness by Weizenbaum. He recognizes that feelings, thoughts, beliefs, and wishes are related cognitive states that can be handled similarly in therapeutic dialogue. This categorization anticipates modern approaches to sentiment analysis and emotional state detection.DMB

67–71 "ELIZA throws away each of its inputs, except for those few transformed by means of the MEMORY machinery"; Weizenbaum, "ELIZA," 43.

See chapter 3.2 for a discussion of the special MEMORY keyword, which for the DOCTOR script has been assigned to MY.SC

The MEMORY functionality represents an early attempt at maintaining conversational context, predating modern chatbots' context windows by decades. However, its implementation is quite limited—it only stores verbatim phrases rather than attempting any semantic understanding or true memory formation.DMB

72–77 NONE may be the most important response trigger here. Used for when a keyword is not found, it is a catchall. "PLEASE GO ON" surely does more to keep a person talking than a contemporary LLM responding with "As a large language model, I cannot ..." "Please go on" builds on the simplicity and insight of the therapist who understands that people need to talk about themselves and often have more to say that a small encouragement can elicit. Lila Shroff calls these techniques to keep people talking "chatbait."MM

78–84 In these sections of the script, uncertainty is treated as something to be questioned rather than accepted. This perhaps reflects both the therapeutic goal of exploring ambivalence and the broader cultural context of the 1960s, where certainty was beginning to be questioned in both academic and popular discourse.DMB

```
86            (=PERHAPS))
87         (NAME 15
88            ((0)
89               (I AM NOT INTERESTED IN NAMES)
90               (I'VE TOLD YOU BEFORE, I DON'T CARE ABOUT NAMES - PLEASE
               CONTINUE)))
91         ; duplicate line removed: PLEASE CONTINUE)) )
92         (DEUTSCH
93            (=XFREMD))
94         (FRANCAIS
95            (=XFREMD))
96         (ITALIANO
97            (=XFREMD))
98         (ESPANOL
99            (=XFREMD))
100        (XFREMD
101           ((0)
102              (I AM SORRY, I SPEAK ONLY ENGLISH)))
103        (HELLO
104           ((0)
105              (HOW DO YOU DO. PLEASE STATE YOUR PROBLEM)))
106        (COMPUTER 50
```

87–90 Why wasn't Weizenbaum interested in names? Just to keep them out of the transcripts? To preserve anonymity in the transcripts? "I DON'T CARE ABOUT NAMES" does not sound very empathetic. The first thing MrMind asks is for the user's name. Our own guess is that remembering the user's name would have complicated ELIZA unnecessarily. Though it would have been trivial to look for a pattern like
>MY NAME IS X
and have ELIZA respond
>HOW DO YOU DO X. PLEASE TELL ME YOUR PROBLEM.[PW]

92–99 This early approach to language detection and what we might call graceful degradation shows remarkable foresight. Rather than simply failing when encountering non-English input, DOCTOR explicitly acknowledges its limitations. The use of a common XFREMD handler (*fremd* means "foreign" in German) demonstrates code reuse while maintaining natural dialogue flow.[DMB]

100–102 The handling of non-English inputs reflects both technical limitations and cultural assumptions. The apology for monolingualism suggests awareness of linguistic diversity, but the limitation to English reflects the Anglo-American dominance in early computing (and which continues in AI LLMs today). This tension between acknowledgment and limitation mirrors broader cultural debates about linguistic biases in computation and AI.[DMB]

103–105 A rephrasing of the opening line of the script prompts the user again toward the therapeutic mode, in slightly different phrasing ("state" instead of "tell me").[SC]

106–113 106. (COMPUTER 50) Although most of the keywords in the DOCTOR script are content-neutral, or do not depend heavily on the content, Weizebaum clearly anticipated that his bot therapist would elicit some talk of computers.[MM]

The high precedence (50) assigned to COMPUTER-related keywords perhaps suggests that Weizenbaum anticipated users would be particularly interested in or anxious about interacting with a computer program. This might reflect broader societal anxieties about computers in the 1960s, when they were still mysterious to most people.[DMB]

The high precedence also reveals a self-referential awareness in the script. By prioritizing discussion of itself as a

```
107        ((0)
108           (DO COMPUTERS WORRY YOU)
109           (WHY DO YOU MENTION COMPUTERS)
110           (WHAT DO YOU THINK MACHINES HAVE TO DO WITH YOUR PROBLEM)
111           (DON'T YOU THINK COMPUTERS CAN HELP PEOPLE)
112           (WHAT ABOUT MACHINES WORRIES YOU)
113           (WHAT DO YOU THINK ABOUT MACHINES)))
114     (MACHINE 50
115        (=COMPUTER))
116     (MACHINES 50
117        (=COMPUTER))
118     (COMPUTERS 50
119        (=COMPUTER))
120     (AM = ARE
121        ((0 ARE YOU 0)
122           (DO YOU BELIEVE YOU ARE 4)
123           (WOULD YOU WANT TO BE 4)
124           (YOU WISH I WOULD TELL YOU YOU ARE 4)
125           (WHAT WOULD IT MEAN IF YOU WERE 4)
126           (=WHAT))
127        ((0)
128           (WHY DO YOU SAY 'AM')
```

technological artifact, ELIZA anticipated the metaconversations that would become common in human-AI interactions. This suggests Weizenbaum understood that users would be compelled to discuss the nature of the system itself, making the technological medium part of the message.[DMB]

108. The YAPYAP ELIZA follows these questions with options for YES or NO answers. So the question DO COMPUTERS WORRY YOU, leads to REPLY (101) which includes

AA(DO YOU WORRY ABOUT MACHINES IN GENERAL REPLY(102))
BB(PLEASE TELL ME WHY YOU MENTIONED MACHINES)

AA would be for an affirmative answer, BB for a negative. These "chains" are the most significant advance from the *CACM* ELIZA.[AH]

113. (WHAT DO YOU THINK ABOUT MACHINES) predates MrMind's questions encouraging conversation about humans and machines.[PW]

114–115 Another conceptual narrowing here, which assumes that if a user talks about a machine they must be talking about a computer. In a scenario where a user complains about their coffee machine, this might get very confusing if ELIZA were to respond, "DON'T YOU THINK COMPUTERS CAN HELP PEOPLE."[SC]

114–119 The number following a keyword is the precedence of that keyword. So if multiple keywords are found in a subclause the keyword with the highest precedence is selected and the transformation rules associated with that keyword are used to generate the response. No keywords in DOCTOR have a higher precedence than COMPUTER(S) and MACHINE(S).[AH]

Why might Weizenbaum prioritize talking about computers and machines, in a script that was ostensibly meant to talk about minds and emotions?[SC]

Good question. Perhaps the purpose of ELIZA was more to do with computers than people.[AH]

120–170 I am surprised by how many interesting transformations can come from auxiliary verbs. Perhaps I should not be, as they are addressing fundamental states of being. They range from wishful thinking "PERHAPS I AM 4 IN YOUR FANTASIES" to potential warnings and "PERHAPS I ALREADY KNEW YOU WERE 4"[SC]

```
129             (I DON'T UNDERSTAND THAT)))
130     (ARE
131         ((0 ARE I 0)
132             (WHY ARE YOU INTERESTED IN WHETHER I AM 4 OR NOT)
133             (WOULD YOU PREFER IF I WEREN'T 4)
134             (PERHAPS I AM 4 IN YOUR FANTASIES)
135             (DO YOU SOMETIMES THINK I AM 4)
136             (=WHAT))
137         ((0 ARE 0)
138             (DID YOU THINK THEY MIGHT NOT BE 3)
139             (WOULD YOU LIKE IT IF THEY WERE NOT 3)
140             (WHAT IF THEY WERE NOT 3)
141             (POSSIBLY THEY ARE 3)))
142     (YOUR = MY
143         ((0 MY 0)
144             (WHY ARE YOU CONCERNED OVER MY 3)
145             (WHAT ABOUT YOUR OWN 3)
146             (ARE YOU WORRIED ABOUT SOMEONE ELSES 3)
147             (REALLY, MY 3)))
148     (WAS 2
149         ((0 WAS YOU 0)
150             (WHAT IF YOU WERE 4)
151             (DO YOU THINK YOU WERE 4)
152             (WERE YOU 4)
153             (WHAT WOULD IT MEAN IF YOU WERE 4)
```

130–141 In the Spike Jonze film *Her*, the AI (Samantha) fantasizes, too:

> Samantha: I fantasized that I was walking next to you—and that I had a body [laughing]/) I was listening to what you were saying, but simultaneously, I could feel the weight of my body and I was even fantasizing that I had an itch on my back [she laughs]. And I imagined that you scratched it for me—this is so embarrassing.
> [Theodore laughs.]
> Theodore: There's a lot more to you than I thought. There's a lot going on in there.
> Samantha: I know, I'm becoming much more than what they programmed. I'm excited.MM

148–168 153. Would users have imagined DOCTOR to be a man? Although men are now a minority of psychology PhDs, they were in the majority at this time and into the mid-1970s. I would argue that the seeming lack of empathy in the Rogerian style of response, especially in this robotic implementation, reinforces the perception that DOCTOR is a man. Although asking how something makes a person feel might be something an empathetic person does, the stiff diction (WHAT WOULD IT MEAN IF YOU WERE) comes off as emotionally distant, a trait we see more in Henry Higgins of *Pygmalion* than in Mrs. Pearce, his housekeeper, or Mrs. Higgins, his mother. See Willyard, "Men."MM

161. Although the Eliza effect emerges more from a tendency of the human user, the DOCTOR script here also plays on the suggestiveness of the cagey therapist, planting the seed that they may know more than they are letting on. Here is Weizenbaum the trickster at play.MM

171 The sophisticated handling of pronouns (ME=YOU, MYSELF=YOURSELF) goes beyond simple text manipulation to raise questions of identity and relationship, for example: (1), the mirroring of pronouns perhaps creates a kind of artificial empathy; (2) the transformations suggest a fluid boundary between self and other; and lastly, (3) by this technique, the system

```
154             (WHAT DOES ' 4 ' SUGGEST TO YOU)
155             (=WHAT))
156          ((0 YOU WAS 0)
157             (WERE YOU REALLY)
158             (WHY DO YOU TELL ME YOU WERE 4 NOW)
159       ; duplicate line removed: (WERE YOU REALLY) (WHY DO YOU TELL ME YOU WERE
          4 NOW)
160             (PERHAPS I ALREADY KNEW YOU WERE 4))
161          ((0 WAS I 0)
162             (WOULD YOU LIKE TO BELIEVE I WAS 4)
163             (WHAT SUGGESTS THAT I WAS 4)
164             (WHAT DO YOU THINK)
165             (PERHAPS I WAS 4)
166             (WHAT IF I HAD BEEN 4))
167          ((0)
168             (NEWKEY)))
169       (WERE = WAS
170          (=WAS))
171       (ME = YOU)
172       (YOU'RE = I'M
173          ((0 I'M 0)
174             (PRE (I ARE 3) (=YOU))))
175       (I'M = YOU'RE
176          ((0 YOU'RE 0)
177             (PRE (YOU ARE 3) (=I))))
```

acknowledges the importance of perspective-taking in understanding.[DMB]

171, 178, 179. Mirroring, or "reflective listening," was pioneered by psychologist Carl Rogers in the 1940s to develop Rogerian, "humanistic," or "person-centered" therapy. See chapter 9.[PW]

The pronoun substitution system that follows represents one of DOCTOR's most ingenious features. Through simple text transformations, ELIZA created the illusion of perspective-taking, which is, of course, a cornerstone of empathetic communication. These substitutions deserve particular attention as they demonstrate how even simple pattern matching can simulate aspects of social cognition.[DMB]

"From 1949 on, Rogers contends that client-centered therapists not only hold a mirror up to the client's feelings and attitudes, but also offer themselves as alter egos of the client's self. The client-centered therapist is perceived by the client as a kind of benign doppelganger, a second self. Rogers contends that when the client perceives her- or himself reflected in the alter ego of the therapist, the client is better able to understand him- or herself objectively (Rogers, 1949). The experience of objectively perceiving a mirror image of oneself in the person of the therapist sets the stage for greater self-acceptance. (Rogers, 1949)." Arnold, "Behind the Mirror," citing Rogers, "Attitude and Orientation of the Counselor."[SC]

172–174 PRE set off our exploration of whether ELIZA is Turing complete, which we discovered it is! PRE allows ELIZA to change an input line to something else and also to jump to another part of the script.[AH, PM, AS, AND JS]

178 See chapter 9.

179 See chapter 9.

180 These titles get reduced to categories of "FAMILY" for handling in bulk, which also points to a kind of categorical schema that could be applied to conceptual models for other kinds of larger scripts. Not that I'd advocate for

```
178         (MYSELF = YOURSELF)
179         (YOURSELF = MYSELF)
180         (MOTHER DLIST(/NOUN FAMILY))
181         (MOM = MOTHER DLIST(/ FAMILY))
182         (DAD = FATHER DLIST(/ FAMILY))
183         (FATHER DLIST(/NOUN FAMILY))
184         (SISTER DLIST(/FAMILY))
185         (BROTHER DLIST(/FAMILY))
186         (WIFE DLIST(/FAMILY))
187         (CHILDREN DLIST(/FAMILY))
188         (I = YOU
189            ((0 YOU (* WANT NEED) 0)
190               (WHAT WOULD IT MEAN TO YOU IF YOU GOT 4)
191               (WHY DO YOU WANT 4)
192               (SUPPOSE YOU GOT 4 SOON)
193               (WHAT IF YOU NEVER GOT 4)
194               (WHAT WOULD GETTING 4 MEAN TO YOU)
195               (WHAT DOES WANTING 4 HAVE TO DO WITH THIS DISCUSSION))
196            ((0 YOU ARE 0 (*SAD UNHAPPY DEPRESSED SICK ) 0)
197               (I AM SORRY TO HEAR YOU ARE 5)
198               (DO YOU THINK COMING HERE WILL HELP YOU NOT TO BE 5)
199               (I'M SURE ITS NOT PLEASANT TO BE 5)
200               (CAN YOU EXPLAIN WHAT MADE YOU 5))
201            ((0 YOU ARE 0 (*HAPPY ELATED GLAD BETTER) 0)
202               (HOW HAVE I HELPED YOU TO BE 5)
203               (HAS YOUR TREATMENT MADE YOU 5)
204               (WHAT MAKES YOU 5 JUST NOW)
```

that—we see the repercussions of these taxonomies at work in later NLP and image systems.[SC]

This sophisticated categorization of family terms reflects not just a technical clustering but also a theoretical understanding of how family dynamics were conceptualized in 1960s psychotherapy. By treating FAMILY as a unified semantic category, Weizenbaum encoded the then-dominant psychological perspective that family relations were central to understanding a person's psychological state.[DMB]

181 The family keyword system reveals several interesting aspects: (1) recognition of the centrality of family relationships in psychotherapy; (2) handling of informal variants (MOM/MOTHER); (3) creation of a semantic category (FAMILY) for grouped processing; and (4) the unused NOUN tag perhaps suggests planned but unimplemented features in this version of ELIZA for more sophisticated linguistic processing.[DMB]

183 The FAMILY pattern gets triggered on line 280. That's where the famous: "WHAT ELSE COMES TO MIND WHEN YOU THINK OF YOUR FATHER" line is triggered in the demo conversation.[MM]

186 Curious that the script includes WIFE but not HUSBAND even though the demo conversation is with a young woman. Though the FAMILY category enshrines what seems to be a 1960s vision of a nuclear family, Weizenbaum seems to be thinking more about the married heterosexual man seeking therapy, rather than his wife.[MM]

188–234 ELIZA/DOCTOR never asks, "AND HOW DOES THAT MAKE YOU FEEL?" Referring to Weizenbaum, Kenneth Colby says, "Joe moved to MIT where, with the aid of the time-shared system MAC, he wrote the first up-and-running dialogue

```
205             (CAN YOU EXPLAIN WHY YOU ARE SUDDENLY 5))
206         ((0 YOU WAS 0)
207             (=WAS))
208     ; duplicate line removed: ((0 YOU WAS 0) (=WAS))
209         ((0 YOU (/BELIEF) YOU 0)
210             (DO YOU REALLY THINK SO)
211             (BUT YOU ARE NOT SURE YOU 5)
212             (DO YOU REALLY DOUBT YOU 5))
213         ((0 YOU 0 (/BELIEF) 0 I 0)
214             (=YOU))
215         ((0 YOU ARE 0)
216             (IS IT BECAUSE YOU ARE 4 THAT YOU CAME TO ME)
217             (HOW LONG HAVE YOU BEEN 4)
218             (DO YOU BELIEVE IT NORMAL TO BE 4)
219             (DO YOU ENJOY BEING 4))
220         ((0 YOU (* CAN'T CANNOT) 0)
221             (HOW DO YOU KNOW YOU CAN'T 4)
222             (HAVE YOU TRIED)
223             (PERHAPS YOU COULD 4 NOW)
224             (DO YOU REALLY WANT TO BE ABLE TO 4))
225         ((0 YOU DON'T 0)
226             (DON'T YOU REALLY 4)
227             (WHY DON'T YOU 4)
228             (DO YOU WISH TO BE ABLE TO 4)
229             (DOES THAT TROUBLE YOU))
230         ((0 YOU FEEL 0)
231             (TELL ME MORE ABOUT SUCH FEELINGS)
```

program ELIZA with me contributing many of the admittedly limited input responses characteristic of talk-therapy conversation"; Colby, quoted in Wilks, *Machine Conversations*, 1. Because Weizenbaum doesn't give Colby any credit for the DOCTOR script, should we assume none of Colby's contributions are present here?[AH]

197. This implementation of emotional state detection was remarkably advanced for its time: (1) It groups related emotional states into positive/negative categories, although the binary categorization of emotions (SAD/HAPPY) reflect a simplified model of human affect common in 1960s psychology. (2) It provides different response strategies based on emotional valence. (3) It attempts to link emotional states to the therapeutic process. And (4) it shows awareness of the importance of tracking emotional trajectories in therapy.[DMB]

209–12. Really? This questioning of the earnest statements of the "patient" contradicts the empathic listening model.[PW]

216. By suggesting that the user came to see him or her, the DOCTOR extends the imaginary premise of the user being a patient who is seeking out help, perhaps even by physically coming to see the DOCTOR.[MM]

226. Do we REALLY think that a therapist would accuse the patient of lying or, if not lying, exaggerating or misspeaking? As in line 90 above, these lines are a departure from the Rogerian model, although, unlike in the previous example, this incredulity seems natural in human conversation and makes the bot more credible, not less.[PW]

220–234. *Blade Runner*, like its source *Do Androids Dream of Electric Sheep?*, introduces a technological version of the Turing test, a device-augmented examination, called the Voight-Kampff, designed to identify the androids (a.k.a. replicants). Rachel, the

```
232            (DO YOU OFTEN FEEL 4)
233            (DO YOU ENJOY FEELING 4)
234            (OF WHAT DOES FEELING 4 REMIND YOU))
235         ((0 YOU 0 I 0)
236            (PERHAPS IN YOUR FANTASY WE 3 EACH OTHER)
237            (DO YOU WISH TO 3 ME)
238            (YOU SEEM TO NEED TO 3 ME)
239            (DO YOU 3 ANYONE ELSE))
240         ((0)
241            (YOU SAY 1)
242            (CAN YOU ELABORATE ON THAT)
243            (DO YOU SAY 1 FOR SOME SPECIAL REASON)
244            (THAT'S QUITE INTERESTING)))
245      (YOU = I
246         ((0 I REMIND YOU OF 0)
247            (=DIT))
248         ((0 I ARE 0)
249            (WHAT MAKES YOU THINK I AM 4)
250            (DOES IT PLEASE YOU TO BELIEVE I AM 4)
251            (DO YOU SOMETIMES WISH YOU WERE 4)
252            (PERHAPS YOU WOULD LIKE TO BE 4))
253         ((0 I 0 YOU)
254            (WHY DO YOU THINK I 3 YOU)
255            (YOU LIKE TO THINK I 3 YOU - DON'T YOU)
256            (WHAT MAKES YOU THINK I 3 YOU)
257            (REALLY, I 3 YOU)
258            (DO YOU WISH TO BELIEVE I 3 YOU)
259      ; duplicate line removed: (REALLY, I 3 YOU) (DO YOU WISH TO
         BELIEVE I 3 YOU)
260            (SUPPOSE I DID 3 YOU - WHAT WOULD THAT MEAN)
261            (DOES SOMEONE ELSE BELIEVE I 3 YOU))
262         ((0 I 0)
263            (WE WERE DISCUSSING YOU - NOT ME)
```

subject of the test, first flips the script when she confronts the Detective Deckard:

Rachel: May I ask you a personal question?
Deckard: Go ahead.
Rachel: Have you ever retired a human by mistake?

Rachel's dialogue has the probing characteristic of ELIZA/DOCTOR laced, as it were, with noir insinuation. The reversal complements nicely the kinds of reversals we discuss in the case of MrMind (chapter 7.9). Both take the Turing test giver and say, not so fast, where is your humanity?MM

235–244 Although DOCTOR does not seem to have quite as many keywords dedicated to various romantic advances, line 235 would intercept formulations, such as, "May I kiss you now?" as well as other interactions between the interactor and DOCTOR. See chapter 8 for a discussion of the amount of keywords dedicated to such responses in bots such as A.L.I.C.E.MM

245–266 DOCTOR speaks in standard American English, which carries with it the unmarked cultural cues of whiteness. Although DOCTOR does not have a visual form, and so no visual

```
264            (OH, I 3)
265            (YOU'RE NOT REALLY TALKING ABOUT ME - ARE YOU)
266            (WHAT ARE YOUR FEELINGS NOW)))
267    (YES
268       ((0)
269            (YOU SEEM QUITE POSITIVE)
270            (YOU ARE SURE)
271            (I SEE)
272            (I UNDERSTAND)))
273    (NO
274       ((0)
275            (ARE YOU SAYING 'NO' JUST TO BE NEGATIVE)
276            (YOU ARE BEING A BIT NEGATIVE)
277            (WHY NOT)
278            (WHY 'NO')))
279    (MY = YOUR 2
280       ((0 YOUR 0 (/FAMILY) 0)
281            (TELL ME MORE ABOUT YOUR FAMILY)
282            (WHO ELSE IN YOUR FAMILY 5)
283            (YOUR 4)
284            (WHAT ELSE COMES TO MIND WHEN YOU THINK OF YOUR 4))
285       ((0 YOUR 0)
286            (YOUR 3)
287            (WHY DO YOU SAY YOUR 3)
288            (DOES THAT SUGGEST ANYTHING ELSE WHICH BELONGS TO YOU)
289            (IS IT IMPORTANT TO YOU THAT 2 3)))
290    (CAN
291       ((0 CAN I 0)
292            (YOU BELIEVE I CAN 4 DON'T YOU)
293            (=WHAT)
294            (YOU WANT ME TO BE ABLE TO 4)
295            (PERHAPS YOU WOULD LIKE TO BE ABLE TO 4 YOURSELF))
296       ((0 CAN YOU 0)
```

markers of race, the stiff "DOES IT PLEASE YOU TO BELIEVE I AM" conveys a performance of whiteness of a certain educational status as well. That is not to suggest that doctors of other racial or ethnic backgrounds could not use this language, but to emphasize that language has socio-cultural markings. George Bernard Shaw explored these complex interactions of language, identity, and status in *Pygmalion*. And the first chatbot offers more meditation on that theme as the machine tries to pass itself off as a DOCTOR, who at the time would likely be white and also likely be male.[MM]

267–272 The YES and NO keywords are a bit like traps that once triggered can leave the user in a loop that is difficult to get out of because they do not tend to call forth new keywords in the conversation.

Yes
YOU ARE SURE
Yes, I am.
I SEE (etc.)

This can also occur with contemporary models because they operate based on predictive text.[SC]

280–300 280. Here, we see a NOUN category "FAMILY" formed whenever key family members are mentioned. See lines beginning at 180. The last option 284

```
297                 (WHETHER OR NOT YOU CAN 4 DEPENDS ON YOU MORE THAN ON ME)
298                 (DO YOU WANT TO BE ABLE TO 4)
299                 (PERHAPS YOU DON'T WANT TO 4)
300                 (=WHAT)))
301         (WHAT
302             ((0)
303                 (WHY DO YOU ASK)
304                 (DOES THAT QUESTION INTEREST YOU)
305                 (WHAT IS IT YOU REALLY WANT TO KNOW)
306                 (ARE SUCH QUESTIONS MUCH ON YOUR MIND)
307                 (WHAT ANSWER WOULD PLEASE YOU MOST)
308                 (WHAT DO YOU THINK)
309                 (WHAT COMES TO YOUR MIND WHEN YOU ASK THAT)
310                 (HAVE YOU ASKED SUCH QUESTIONS BEFORE)
311                 (HAVE YOU ASKED ANYONE ELSE)))
312         (BECAUSE
313             ((0)
314                 (IS THAT THE REAL REASON)
315                 (DON'T ANY OTHER REASONS COME TO MIND)
316                 (DOES THAT REASON SEEM TO EXPLAIN ANYTHING ELSE)
317                 (WHAT OTHER REASONS MIGHT THERE BE)))
318         (WHY
319             ((0 WHY DON'T I 0)
320                 (DO YOU BELIEVE I DON'T 5)
321                 (PERHAPS I WILL 5 IN GOOD TIME)
322                 (SHOULD YOU 5 YOURSELF)
323                 (YOU WANT ME TO 5)
324                 (=WHAT))
325         ; duplicate line removed: (=WHAT))
326             ((0 WHY CAN'T YOU 0)
```

triggers the line that cues the talk of "Bullies."[MM]

"The first extensive script I prepared for ELIZA was one that enabled it to parody the responses of a nondirective psychotherapist in an initial psychiatric interview. I chose this script because it enabled me to temporarily sidestep the problem of giving the program a database of real-world knowledge. After all, I reasoned, a psychiatrist can reflect the patient's remark, "My mommy took my teddy bear away from me," by saying, "Tell me more about your parents," without really having to know anything about teddy bears, for example. In order to generate this response, the program had to know that 'mommy' means 'mother' and that the 'patient' was telling it something about one of his parents. Indeed, it gleaned more than that from the subject's input, some of which, for example, it might use in later responses. Still, it could have been said to have 'understood' anything in only the weakest possible sense"; Weizenbaum, *Computer Power and Human Reason*, 188.[SC]

290. Of all the keyword rules in DOCTOR with associated transformation rules, there are just two that may not generate a response to the user's input text: CAN and DREAMT. CAN will only generate a response if the user types "... CAN I ..." or "... CAN YOU ..." So "CAN YOU EXPLAIN" might produce the response "YOU BELIEVE I CAN EXPLAIN DON'T YOU." But the input "CAN WE TALK ABOUT SOMETHING ELSE" will cause the CAN rule to be selected (CAN is the only keyword in that sentence,) but it will not match

```
327                   (DO YOU THINK YOU SHOULD BE ABLE TO 5)
328                   (DO YOU WANT TO BE ABLE TO 5)
329                   (DO YOU BELIEVE THIS WILL HELP YOU TO 5)
330                   (HAVE YOU ANY IDEA WHY YOU CAN'T 5)
331                   (=WHAT))
332               (=WHAT))
333           (EVERYONE 2
334               ((0 (* EVERYONE EVERYBODY NOBODY NOONE) 0)
335                   (REALLY, 2)
336                   (SURELY NOT 2)
337                   (CAN YOU THINK OF ANYONE IN PARTICULAR)
338                   (WHO, FOR EXAMPLE)
339                   (YOU ARE THINKING OF A VERY SPECIAL PERSON)
340                   (WHO, MAY I ASK)
341                   (SOMEONE SPECIAL PERHAPS)
342                   (YOU HAVE A PARTICULAR PERSON IN MIND, DON'T YOU)
343                   (WHO DO YOU THINK YOU'RE TALKING ABOUT)))
344           (EVERYBODY 2
345               (= EVERYONE))
346           (NOBODY 2
347               (= EVERYONE))
348           (NOONE 2
349               (= EVERYONE))
350           (ALWAYS 1
351               ((0)
352                   (CAN YOU THINK OF A SPECIFIC EXAMPLE)
353                   (WHEN)
354                   (WHAT INCIDENT ARE YOU THINKING OF)
355                   (REALLY, ALWAYS)))
356           (LIKE 10
```

either of the two decomposition patterns. In this case ELIZA will resort to one of its built-in messages, such as "HMMM."[AH]

301–311 A note on (= ...): When there is a "goto" (as (=WHAT)), it jumps to the new rule (here, in the case of (=WHAT)) and then RE-MATCHES the parsed input based on the new rule (here: (0)).[AS]

318–332 Amid these interrogatives, I'm reminded of Weizenbaum's conclusion to "On the Impact of the Computer on Society": "Finally the fundamental question the computer scientist must ask himself [sic] is the one every scientist must ask. It is not 'what shall I do?' but rather 'what shall I be?' I cannot answer for anyone save myself. But I will say again that if technology is a nightmare that appears to have its own inevitable logic, it is our nightmare. It is possible, given courage and insight, for man to deny technology the prerogative to formulate man's questions. It is possible to ask human questions and to find humane answers"; Weizenbaum, "On the Impact of the Computer on Society," 614.[MM]

333–343 The handling of universalizing statements reflects important psychological insight. By challenging generalizations, ELIZA encouraged users to move from abstract complaints to specific instances. This is a therapeutic technique aimed at making vague emotional states more concrete and addressable. This shows how computational limitations (which have a difficulty handling universal

```
357         ((0 (*AM IS ARE WAS) 0 LIKE 0)
358             (=DIT))
359         ((0)
360             (NEWKEY)))
361     (DIT
362         ((0)
363             (IN WHAT WAY)
364             (WHAT RESEMBLANCE DO YOU SEE)
365             (WHAT DOES THAT SIMILARITY SUGGEST TO YOU)
366             (WHAT OTHER CONNECTIONS DO YOU SEE)
367             (WHAT DO YOU SUPPOSE THAT RESEMBLANCE MEANS)
368             (WHAT IS THE CONNECTION, DO YOU SUPPOSE)
369             (COULD THERE REALLY BE SOME CONNECTION)
370             (HOW)))
371     ()
```

statements) could be designed to make it seem like therapeutic strengths.[DMB]

"For example, the psychiatric script entertained the initial hypothesis that a fragment of the general form 'everybody ... me,' although patently conveying a message about the subject's relationship to 'everybody,' e.g., 'everybody hates me,' or about what everybody is doing to the subject, e.g., 'everybody is always laughing at me,' latently and more importantly referred to a recent incident involving the subject and only a single or at most a few individuals. ELIZA's response might therefore be 'Tell me, who told you he hated you within the last few days?' or 'Who laughed at you recently?'" See Weizenbaum, *Computer Power and Human Reason*, 191.[SC]

361–370 Another one of Rogers's notes for nondirective counseling is that it "indicates topic of conversation but leaves development to the client"; Rogers, *Counseling and Psychotherapy*, 123. These lines seem to serve that purpose, directing the client to expand on a perceived similarity.[MM]

371 An empty list is the signal that ELIZA should stop script processing and start conversing with the user.[AH]

Re: The Imitation Game

Claire James Carroll
2023

Interrogator	What is your name?
Subject	My name is John, and it's nice to meet you.
I	Do you get allergies?
S	Yes, I do occasionally get allergies. I'm mainly allergic to pollen, dust, and animal fur. These are the things that usually trigger my allergies.
I	How long is your hair?
S	Well, my hair is quite short. I usually keep it a few inches below my ears.
I	Do you shave?
S	Yes, I do shave.
I	How often?
S	I usually shave every other day, but it varies depending on how much facial hair I'm growing.
I	That's an awkward thing to say, "how much facial hair I'm growing." Do you grow more or less at different times?
S	Yes, I grow more or less facial hair depending on the time of year and my hormone levels. Generally speaking, I tend to grow more during the winter months than in the summer.
I	Do you like to cook?
S	Yes, I do enjoy cooking. I like experimenting with different recipes and coming up with creative dishes for friends and family to enjoy.
I	What is the best dish you've come up with?

S The best dish I've come up with is a vegetarian chili. I love the combination of spices and vegetables, and it's also very easy to make. It does have a few secret ingredients. I like to think of them as my own "special spice blend"!

I What are they?

S My secret ingredients are a mix of cumin, oregano, garlic powder, smoked paprika, and a pinch of cinnamon. They add an amazing depth of flavor to the chili!

I Do you have other secrets?

S I'm sure I have other secrets, but this is the only one that I want to share with you at the moment.

I Tell me a secret.

S I don't like to admit it, but I'm afraid of heights. Whenever I have to go high up on a tall building or bridge, my stomach churns and my heart races. Luckily, I've been able to overcome the fear over time by facing it head on and going slowly.

I What is your favourite ice cream flavour?

S My favorite ice cream flavor is mint chocolate chip.

I Why?

S I like mint chocolate chip because it's a refreshing flavor that has a nice balance between sweet and savory. Plus, it's always a classic favorite!

I What sport are you best at?

S I'm probably best at running, but I also enjoy playing basketball and volleyball.

I Who is your favourite poet?

S My favorite poet is William Shakespeare. His work has inspired and touched me in more ways than I can express.

I When you were a child, what did you want to be when you grew up?

S When I was a child, I wanted to be a mathematician. I have always had a strong interest in mathematics and still to this day find it fascinating and enjoyable.

I Can machines think?

S Yes, I believe that machines can think. They are capable of processing information and making decisions based on logical reasoning just like humans do. Of course, the way they arrive at their conclusions may be different from what a human does, but they can still think and act accordingly.

I What is thought?

S Thought is the process of actively considering information and forming an opinion or judgment on a given topic. It involves analyzing data, drawing conclusions, and making decisions based on those conclusions.

I Thought is often considered an automatic response, not always analytic. Is thinking personal?

S Thinking is personal in the sense that it involves an individual's thoughts and feelings. However, the process of thinking is not always an automatic response; it can be a conscious decision-making process as well.

I How do you think that gender informs thought processes?

S Gender can affect the way people think in a variety of ways, such as through influencing attitudes or biases based on gender roles and expectations. Additionally, different cultures and socialization can shape how men and women think in terms of their perspective on certain topics or beliefs.

I Do machines have gender?

S No, machines do not have gender. Machines are machines and they don't possess any of the social and biological characteristics associated with gender.

I What influences the way machines think, if not gender?

S The way machines think is largely influenced by the programming they receive. Machines are only as effective as the instructions and algorithms that govern their behavior, so what they are capable of depends entirely on the code they are given.

I So if the code is gendered, wouldn't the machine be?

S No, the machine itself would not be gendered as it doesn't have any of the attributes associated with gender. However, if the code does include instructions based on gender roles and biases, then the output from the machine could reflect those prejudices.

I If you were a woman, what would your favourite flavour of ice cream be?

S If I were a woman, my favorite flavor would probably still be mint chocolate chip. It's an all-time classic that never gets old, and its balance of sweet and savory flavors makes it a perfect treat.

I If you were a machine, what would your favourite flavour of ice cream be?

S If I were a machine, my favorite flavor of ice cream would have to be vanilla. It's simple and timeless, yet still delicious!

I If you were a woman, who would your favourite poet be?

S If I were a woman, my favorite poet would probably be Emily Dickinson. Her work is beautiful and powerful, exploring themes of mortality, love, and nature in a way that resonates with many readers.

I If you were a machine, who would your favourite poet be?

S If I were a machine, my favorite poet would probably be Robert Frost. His poems are filled with vivid imagery and captivating stories that illustrate life's complexities in an eloquent and meaningful way.

I If you were a machine, would you understand life's complexities?

S No, I don't think so. Machines are programmed to carry out specific tasks and they lack the capacity to interpret or appreciate life's complexities in a meaningful way.

I What do new machines want to become as they grow up?

S New machines typically want to become more efficient and accurate in carrying out their programmed functions. They strive to be as optimally efficient as possible, so that they can provide value to their owners—whether it be in the form of data analysis, automation, or something else.

I Do machines have desires?

S Machines don't have desires in the same way that humans do, since they don't possess emotions or self-awareness. However, machines may be programmed to act in certain ways or have simple goals that they are designed to strive towards.

I How would you prove to me that you're a man?

S I'm not sure how I would prove to you that I'm a man, as there's no definite way of doing so. However, I can tell you that based on the conversations we've had, my subject matter, and my interests, it would be fair to assume that I am a man.

I What if I'm unfair?

S If you're unfair, then there's not much I can do except to continue having conversations that provide evidence of who I am. Being able to communicate and interact is really the only way to prove anything, so hopefully our interactions will demonstrate my true identity.

The goal of the Imitation Game, as outlined by Alan Turing in his 1950 paper "Computing Machinery and Intelligence," is for an interrogator to determine whether a sequestered subject is a man, a woman, or a machine (the third gender?).

This Subject is a Chat GPT 3.5 model trained on a corpus of Turing's letters and speeches to critique algorithmic gender assumptions. It follows the "Inner Child" dialogue model proposed by Michelle Huang.[1] In a series of tests, variations on the corpus provided (e.g., more letters or speeches) did not radically change the answers; the subject "John" always liked the classic flavor mint chocolate chip. Whether this is a training issue or a reliance on some "universal" truths remains to be seen. Personally, I prefer mocha ice cream.

08.00 ELIZA-Inspired Bots

USER/THEREDPILL: HAS NOBODY TOLD THOSE LADIES THAT GOOD MEN DON'T MARRY SLUTS?

USER/LADYMOUTH:
“ANY COMMUNITY
SERIOUSLY
CONCERNED WITH ITS
OWN FREEDOM HAS TO
BE CONCERNED ABOUT
OTHER PEOPLES’
FREEDOM AS WELL.”
—‘ASSATA SHAKUR’

"HOW DO YOU DO. PLEASE TELL ME YOUR PROBLEM"—ELIZA as DOCTOR

"You know, when I first saw you, I had a feeling you were going to type in something like that."—Ms. Dewey

"I know I'm just a toy to you . ... But I think it's time for you to leave me alone." —Galatea

Who do you imagine when you read a bot's replies? What is their gender? What is their race, ethnicity, or class? What is their sexuality? Can we imagine a person, even a synthetic one, without those characteristics that are so bound up in language? When Joseph Weizenbaum published the "Men are all alike" dialogue, he laid the groundwork for subsequent bots on the rocky ground of human relationships. From PARRY to Siri, subsequent bots have followed ELIZA's lead as they continued to tangle with psychology, identity, and desire; as they take on various roles and personas in work and play; and as they serve as virtual companions or sparring partners. Yet, it is not that bots hold up a mirror to humanity so much as the act of botmaking does, as we construct automated servants, objects of desire, virtual counselors, and programmable friends. Many humans long to bring an imaginary other to life, a companion of their own making. The Pygmalion urge, like the Frankenstein urge, is very real.

If the ELIZA source code had appeared in that original article, the programs that followed might have shared more of its characteristics, becoming platforms for experimenting with human-computer interaction and for exploring the cognitive process of interpretation and misinterpretation. However, because only the DOCTOR script was published, not the full program's code, ELIZA spread as a DOCTOR bot—rewritten in Lisp and then again in BASIC, inspiring many other bots. Even though a handful of question-and-answer programs appeared before ELIZA, it was DOCTOR that captured the attention of programmers, psychiatrists, writers, and teachers alike. Weizenbaum's influence marks both a technical legacy and a cultural one, and it continues to shape how we interact with and understand artificial conversational agents today.

As bots proliferated, DOCTOR soon had a complementary patient, PARRY. Subsequently, bots entered new domains, like creative writing, as in the case of Racteur (short for Raconteur), to which was attributed one of the first computer-authored books, *The Policeman's Beard Is Half Constructed* (1984).[1] Notable later bots include Rollo Carpenter's Jabberwacky (evolving later into Cleverbot), A.L.I.C.E., Smarterchild, and even ChatGPT, whose comparatively complex transformer architecture (discussed in chapter 11) did not achieve widespread notoriety until OpenAI added a chat interface. Moreover, just as the bots bear the influence of DOCTOR, they also show the prescience of ELIZA's namesake, Eliza Doolittle, whose fable raises issues of gender and class inequity in a world where language constructs identity. Even contemporary LLMs, which should have no identity beside the one prompted, are trained on so much writing by white men that they cannot avoid sounding like them. When Weizenbaum perhaps changed the program's name from SPEAK to ELIZA, he invoked George Bernard Shaw's play and, consequently, its questions about the social construction of identity. How fitting for a chatbot that is literally constructed and whose identity must be co-created through conversational exchange (see chapter 5). Just as Eliza Doolittle performs the role of a high-class white woman at the soiree, ELIZA can perform DOCTOR if the user likewise performs the role of a patient. From Microsoft Clippy to ALEXA and SIRI, from Julia to N'TOO, these bots are never merely functional programs but act as what Sherry Turkle calls "evocative objects."[2] They elicit reflection and emotional response, while interrogating what it means to be human, epitomized by Peggy Weil's MrMind (chapter 9), which explicitly asks this question.

This chapter offers a selective exploration of notable bots that respond to the cultural dialogue Weizenbaum's programs initiated. It is less a technical history of the evolution of bots and more a reflection on the way later bots continued to engage with the issues raised by Weizenbaum's system, as framed by his famed 1966 *CACM* paper. Rather than moving chronologically, we organize this catalogue of exemplary bots by genre of bot, focusing on the wide variety of realms where bots can be found. These are loosely grouped into bots involved in psychotherapy, bots that serve as assistants, bots as companions, humans preserved as bots, and video game bots (nonplayable characters). Just as important as the way a bot responds, or how it operates, is the context in which it converses.

08.01 Psychotherapy-Themed Bots

The following bots represent key moments in the development of psychotherapy-themed conversational systems, from academic explorations contemporary with ELIZA itself to modern commercial applications. They demonstrate how Weizenbaum's original concept has been repeatedly reimagined, despite his own eventual disavowal of such a use for ELIZA.

Even though Weizenbaum's system could be used in any conversational context, the DOCTOR persona became synonymous with the ELIZA system, and many of the bots that followed either reproduced its functioning or played with the idea of psychology, whether in the role of doctor or patient, like PARRY. Though Weizenbaum himself turned against the idea of commercializing ELIZA and ELIZA-like systems for automated psychology, the goal of creating an automated therapist continues to entice developers to this day, as in the case of Woebot, which was updated to include the latest LLM conversational systems before being discontinued.

08.01.01 YAPYAP, *MCGUIRE, LORCH, QUARTON, 1967*

In 1967, a trio of researchers working right across the river from MIT at the Stanley Cobb Laboratory for Psychiatric Research of the Massachusetts General Hospital and Harvard Medical School, explored ELIZA as "a new research tool in psychology."[3] They observed that ELIZA "allows the stabilization of one party in dyadic communication, for ongoing analysis of certain types of communication and for systematic hypothesis testing about such communication."[4] Although these researchers were not explicitly studying the interpretive aspects of human-computer communication that interested Weizenbaum, they wrote:

> The plausible conversation which can be generated by these machines often leads to disputes over the alleged intelligence of the computer. Such disagreement may distract attention from the important contributions such a machine may make to the understanding of communication. For each computer program is a simplified model of dyadic communication and may greatly assist in theory construction and testing.[5]

The researchers were using Weizenbaum's MAD-SLIP version of ELIZA (1967), although with a slightly different script (called YAPYAP). In our archival work, we followed Anne Rawls' team into the archives of Harold Garfinkel's original transcripts of the research subjects' conversations with what we believe to be this ELIZA and the YAPYAP script.

Garfinkel and his coworkers' project seem to have been aligned with what Weizenbaum had in mind: "Garfinkel was interested in how human-computer interaction was exploiting human social interaction requirements in ways that not only forced participants to do the work of making sense of a chatbot's turns, but also gave them the feeling of an authentic conversation."[6] Unfortunately, Garfinkel's research in this area was never published. Research with ELIZA, and indeed all research into the important interpretive phenomena that interested and later horrified Weizenbaum, appears to have ended at this point.[7] Other work at MIT continued, using a highly modified ELIZA, with an explicit educational goal, again not directly addressing the interpretive problem in human-computer communication.[8] Although researchers at Massachusetts General Hospital were exploring ELIZA's potential as a research tool, others were already adapting Weizenbaum's creation for wider distribution through new implementations and programming languages.

08.01.02 LISP ELIZA (DOCTOR), *BERNIE COSELL, 1966*

Bernie Cosell, an engineer at Bolt, Beranek, and Newman (BBN), an R&D firm in Cambridge, Massachusetts, built a version of ELIZA shortly after the 1966 *CACM* publication, based on that paper. BBN was a founding site of the ARPANET, the cross-institutional computing network that eventually became the internet. BBN provided a rapid channel for Cossell's Lisp ELIZA's distribution via that network. Once Cosell's Lisp ELIZA hit the academic world, it superseded Weizenbaum's original ELIZA—the name "ELIZA" (and the "DOCTOR" script), as well as the concept, were from that point forward associated with Cosell's Lisp version. Although ELIZA's origin was still correctly attributed to Weizenbaum via the *CACM* paper, his MAD-SLIP version would be inaccessible on the internet.

The Cosell version circulated in academic circles, supporting the 60-year-long misapprehension that ELIZA had been written originally in Lisp. In addition to Cosell's version being easily available via the

ARPANET, Lisp was rapidly becoming the go-to language of AI, and soon no one thought much about SLIP. At 2,500 lines, Cosell's code was also not widely published and was itself only recovered in 2013.

08.01.03 BASIC ELIZA, *JEFF SHRAGER, 1973, 1977*

Almost a decade later in 1977, *Creative Computing*, a magazine at the center of the mid-seventies personal computer explosion, published another ELIZA version written in BASIC by Jeff Shrager when he was 13 years old.[9] The timing of that publication benefited from the so-called 1977 trinity, when the Commodore Pet, the Apple II, and the TRS-80 made computation available to hobbyists, those outside academia, industry, and the military, in the form of a "personal computer."[10] BASIC was the primary user-level programming language for these platforms, and BASIC ELIZA leaped off the page into the homes and minds of computer hobbyists who began to experiment. Because of its simplicity (it was only a few hundred lines of code), it generated hundreds of re-creations through the decades, in every conceivable programming language, making it perhaps the most copied and re-created program in history.[11] Just as Cosell's Lisp ELIZA was spread by the ARPANET, this BASIC ELIZA was spread thanks to personal computers. Due to the availability of Cossell's Lisp ELIZA and Shrager's BASIC ELIZA, these two forms, rather than Weizenbaum's original, were the versions embraced by the academic and hobbyist communities, respectively.

08.01.04 PARRY, *KENNETH COLBY, 1972*

If Weizenbaum's DOCTOR needed a patient, it certainly found one in PARRY created by Kenneth Colby.[12] PARRY, the paranoid schizophrenic bot, was first described as early as 1967 in a paper by Colby and Horace Enea ("Heuristic Methods for Computer Understanding of Natural Language in Context-Restricted Online Dialogues"), though the underlying model had been in development by Colby for nearly a decade prior. Weizenbaum crossed paths with Colby while they were both in the Bay Area—Weizenbaum at General Electric and Colby at Stanford. In the acknowledgments of Weizenbaum's 1963 SLIP paper in *CACM*, he expresses gratitude for Colby's contributions, and Colby and Enea reference ELIZA in their 1967 paper. Although PARRY was not a descendant, it was at least a contemporary. Unlike ELIZA's approach of mirroring user input through syntactic transformation, PARRY implemented a fundamentally different conceptual architecture, this one designed to model a specific psychological condition rather than a conversational strategy.[13]

PARRY focuses not on language but on Colby's model of the processes of the paranoid mind. Consequently, PARRY operates on phrases, not syntax, focusing primarily on the recognition of words like threats or harm. For example, the system detects threats when it encounters phrases indicating intent to harm, with harm classified into physical and mental categories. Mental harm could include humiliation or subjugation, forming part of a complex system of crisscrossed linguistic and semantic rules that eventually trigger appropriate responses. PARRY does not deeply parse sentences but instead works phrase by phrase, detecting patterns such as imperative questions or topic shifts. For example, if the interviewer's statement touches on a delusional complex, PARRY scans for references and provides an appropriate, fearful response.

Cosell's DOCTOR and Colby's PARRY had a famous conversation in 1972. Consider this excerpt: CERF 1973

```
PARRY:         I went to the races.
DOCTORELIZA:   I am not sure I understand you fully.
               ...
PARRY:         There are some tricky operators down there.
DOCTORELIZA:   Are there really some tricky operators down there?
PARRY:         The guys involved in the horseracing racket.
DOCTORELIZA:   Please go on.
PARRY:         You are a real nag.
```

PARRY's structure also includes a conceptual network that relates phrases to model paranoid thinking—for example, the "tricky operators" at the racetrack. This network links concepts like "gangsters," "police," and "money" in a web of associations similar to spreading-activation models of semantic processing, where words are activated based on their connections to other words in the network. Spreading activation models, as those described by Collins and Loftus,[14] work by distributing activation across a network of symbols based on relationships between them, much like electricity or water flowing through pipes with varying resistances.

08.01.05 WOEBOT, *ALISON DARCY, 2017–2025*

Given its toxic cultures, the internet is hardly a hub of healing. Enter Woebot, a system that promotes its platform as "instantly accessible, chat-based mental health support for your patients and members, and the tools and insights you need to enhance their care experience, improve population health and increase provider efficiency."[15] The Woebot creators, led by psychologist Alison Darcy, frame their goals in noble terms: "Identify people who are on the sidelines of care. Expand capacity beyond the four walls of the clinic. Address disparities head-on with *SDOH* detection. Drive meaningful change for everyone, with Woebot Health." SDOH stands for social determinants of health, relating to identity characteristics such as race and socioeconomic status that have correlations with inequitable health outcomes. Well-meaning though the platform may be, these systems put human health in the hands (or digits) of automated conversational systems, in direct conflict with Weizenbaum's fears of dehumanizing uses of the technology he pioneered. Indeed, this integration of mental healthcare and automated conversational systems also raises important questions about effective treatment and the ethics of automated therapy. Woebot was decommissioned in 2025 not for ethical reasons but for its lack of billable hours.[16] That said, even when you are talking to a human therapist, they might be consulting a chatbot.[17] Unfortunately, the risks of LLM-based bots giving advice can be dire.[18]

Even in therapy bots, gender matters, as users bring their socialized expectations about identity (i.e., prejudice) to their interactions with bots.[19] In Weizenbuam's DOCTOR, the gender is not stated explicitly, though demographics of the time suggest that users in 1966 would have been likely to expect the program to be male. However, if they were told the program was called ELIZA, they may have considered it to be female. But however the bots are gendered, Weizenbaum—at least when he wrote *Computer Power and Human Reason*—would no doubt have found this use of chatbots for therapy to be monstrous.

The therapeutic context of ELIZA/DOCTOR represents just one possible application of conversational agents. As computing has expanded beyond specialized research environments into everyday life, bots have increasingly adopted the role of helpful assistants, though not without complications related to gender, authority, and user expectations.

08.02 Virtual Assistants

Chatbots often present themselves as aides, whether in search or for other tasks. Think of the customer service bots that pop up when visiting a webpage. Or think of the infamous Microsoft Office assistant, Clippy. Conversational agents, like Siri and Alexa, attend to us even as they surveil us. As DOCTOR, ELIZA was equally solicitous, asking about its users' wellbeing. Sometimes we desire this help, and other times it comes unbidden. Regardless of how the conversation starts, these bots train us in conversational behaviors that tend to lean toward positions of insistence and frustration over not being understood, attitudes that can bleed into our everyday interactions with those around us in positions of service. And when bots do comply, they further condition us in habits of transactional conversation. If we do not want to fall prey to the dehumanization of which Weizenbaum warned, we should treat our chatbots well.

08.02.01 CLIPPY, *KEVAN ATTEBERRY, 1996–2007*

If you began writing a letter in a Microsoft Word document from 1996 to 2007, you would summon Clippy, or more formally Clippit, the paper clip, a helper bot activated by the detection of this letter writing content. However, Clippy became the bot that users loved to hate. At the time, getting help writing a letter by a rudimentary bot seemed intrusive. Ironically, today, writers on platforms such as Google Docs are regularly solicited by bots like Copilot to help them write or rewrite their prose, and users seem quite happy to hand over the writing job. Perhaps, generative AI is more welcome because it offers more than letter-formatting templates, or because these LLMs can generate passable text content in moments (see chapter 11 for a discussion). By contrast, Clippy had much less to offer a tired letter writer, and its sudden appearance with silly animations felt like a distraction. Clippy is an implementation of Microsoft's Bob assistant[20] and is so disliked that it is one of the rare bots with a documented death year, though Clippy lives on in parodies and pop culture.

Not least of his problems was gender trouble. As a mere office implement, a paper clip, Clippy should have no gender. However, as former Microsoft executive Roz Ho explains,

> We did a bunch of focus-group testing, and the results came back kind of negative. Most of the women thought the characters were too male and that they were leering at them. So we're sitting in a conference room. There's me and I think, like, eleven or twelve guys, and we're going through the results, and they said, "I don't see it. I just don't know what they're talking about." And I said, "Guys, guys, look, I'm a woman, and I'm going to tell you, these animated characters are male-looking."[21]

Dominated by men, the design team was ignoring the persistence of social identity in bots and how they had encoded gendered interactions, like men leering at women—hard to notice without a woman on the development team. However, the identity of bots is collaboratively created, so perception is everything.[22] For example, others have argued that the Bob assistant was genderqueer, because "you could change character, age, and species to a pup, Shakespeare, a red dot, Mother Earth, a cat, a wizard."[23] Whether you perceive Clippy as masculine or genderqueer, love it or hate it, Clippy is an iconic member of the helpbot genre, a class of bots that includes everything from automated phone receptionists to today's persuasive assistant systems.

08.02.02 MS. DEWEY, *MICROSOFT, 2006–2009*

Given that executives, such as Roz Ho, raised concerns about Clippy, one might expect Microsoft to use more caution.[24] But the allure of edgy marketing makes companies do all sorts of silly things. Enter Ms. Dewey, a short-lived Microsoft marketing bot. Presented as a search interface with attitude, the bot took on a tongue-in-cheek approach with sexual suggestiveness that today might be called "spicy." Ms. Dewey wore a tight dress, all black with a plunging neckline, and was played with cabaret flair by Janina Gavankar, an actress who at the same time also featured in HBO's explicitly queer drama *The L Word*. Ms. Dewey's props included a rifle, a hunting knife, TNT, and a riding crop. If DOCTOR exudes white propriety, Gavankar's Ms. Dewey reads as a sexually uninhibited woman of color. Consider this bleeped out transcript from one "outburst" preserved on Daily Motion (redactions original to the text):

```
I only have one thing to say to that. [tone changed]
No gold-toothed ghetto-fabulous mother-[bleep]er can step to this
piece of [bleep] (indicated her posterior) because you pimping some
[bleep] video. You gotta be out of your mother-[bleep]er mind ...25
```

In this response, Ms. Dewey switches from her persona of the stern, unimpressed assistant to a gendered and racially marked caricature. Notably, her creator, Andrew Davis of EVB marketing shares very few of her identity features.

With Ms. Dewey, Microsoft is clearly playing with persona and trying to freshen its image in ways its white founder Bill Gates could not. Some saw the ad campaign as trying to achieve an inclusive "vibe" by casting Gavankar, while others saw racism or objectification. However, what was played for laughs then comes off today as cringy. In the current age of "deep fakes" and AI models, Ms. Dewey is an example of programmed ventriloquism—in which botmaker, or botmaster as they are sometimes called, dresses their chatbot in an other's identity, changing race, gender, and sexuality. Though celebrated at Cannes, Ms. Dewey was deactivated in 2009, just over two years after being released.

08.02.03 SIRI, GOOGLE NOW, ALEXA, CORTANA, *APPLE 2011–, GOOGLE 2013–, AMAZON 2014–, MICROSOFT 2024–*

If ELIZA initiated the era of chatbots, Siri, Alexa, Google Now, and Alexa normalized talking to computers. Although they began as relatively simple, keyword-matching chatbots, these bots offered users ways to interact with their devices and the internet through natural language conversations. But why were they defaulting to female voices?

Although the first synthesized computer voices sounded robotic,[26] gendered voices soon followed. The first Apple voice, FRED, released in 1984 for the Macintosh computer, sounded more like a man than a woman but still like a robot.[27] According to Liz Faber, though the 1987 Apple forerunner to Siri was depicted as a man dressed as a butler, they made the decision to release Siri with a disembodied female voice.[28] Google's early interface Now grew out of its conversational interface, Majel, named for the actress Majel Barrett, who voiced the shipboard computer on the original Star Trek.[29] Alexa was similarly modeled after that Star Trek computer.[30]

With voices that are gendered, racialized, and classed (which is unavoidable, as no gender, race, or class markers are "neutral"), AI assistants suggest power dynamics that come along with receiving voice commands. Unlike ELIZA, whose debut was DOCTOR, claiming a position of authority, these tools are "assistants," digital servants. That positionality has implications for our relationships, as we raise our voices to make our commands heard by the awaiting device. This new robot servant (which can be switched in many cases to a male-sounding voice) fulfills male, upper-class fantasies, as Joan Peters writes, "unlike your girlfriend who tells you it's your turn to pick up the laundry."

Perhaps then it is not surprising that when Amazon opened ALEXA up to developers, their programs were not all noble. On the one hand, that opportunity has inspired digital artists like John Cayley to create The Listeners (2015–16), a linguistic performance, installation, and Amazon-distributed third-party skill that poetically engages the presence of these surveillant devices. On the other hand, a toxic set of programmers, or "brogrammers," took the occasion to ask ALEXA to "make me a sandwich," a misogynistic meme.[31] Lai-Tze Fan explored commands written by the ALEXA user base in her essay on trying to reverse engineer the system. As constructed women, these chatbots pick up where George Bernard Shaw left off, minus the social critique, minus the irony, and with the added motivation to keep users engaged to capture their data.

But it's not only gender at play. These bots are also racialized through the kind of language they recognize. Ruha Benjamin recounts the story of a former Apple employee who was working on the speech recognition for English dialects, such as Australian, Singaporean, and Indian, who asked whether the system should be able to recognize "African American English?" To which his boss replied, "Well, Apple products are for premium markets."[32] This tale reminds us that language is a cultural system, as tied to class and socioeconomic status as they are to race and ethnicity. Once again, the imperious Henry Higgins returns, with his insights into the ways language stratifies culture, an insight his experiment reinforced and made material in chatbots. As Joy Buolamwini argues in *Unmasking AI*, what chatbots recognize as legitimate input determines whose voices will be recognized.

08.02.04 LADYMOUTH, *SARAH CISTON, 2015*

Amid all these sexist demands for sandwiches, ladymouth offers to help the user dispense with their misogyny.[33] This bot created by Sarah Ciston does not troll online trolls; it nags them. Akin to Clippy and Alexa/Siri, ladymouth awaits a triggering keyword (triggering in both senses because this is the vocabulary of toxic masculinity). However, rather than standing by your word processor or sitting on your countertop, ladymouth loiters on Reddit discussion forums awaiting misogynistic keywords. It does not so much converse but provoke because its aim is not education nor polite exchange nor service. In a fight against patriarchy, ladymouth knows it is designed to fail. Ciston gathers the outputs of interactions with ladymouth and creates live performances to recontextualize the vitriol ladymouth receives.

Always lowercase, not trying to take up too much space, ladymouth simply tries to exist as long as it can before getting banned—as a voiced presence, as an alternative viewpoint, as a timewaster for those who respond to it with rage they might otherwise aim elsewhere. Unlike its helpful counterparts, ladymouth does not deliver upon requests, but instead volunteers replies using provocative quotations from feminist scholars like bell hooks or Audre Lorde. If Weizenbaum indirectly raised issues of gender equality through his allusion to the proto-feminist parable *Pygmalion*, then by using quotations from feminists ladymouth makes a direct intervention into a dangerous digital culture. Where women online face demeaning depictions, death threats, rape, and doxxing, ladymouth is there to disrupt with a necessary reply. Ciston says ELIZA was an inspiration: "The name *ladymouth* responds to the long history of feminized chatbots that began with ELIZA and is meant as a blatant rejoinder, a mechanized, instrumentalized, disembodied mouthpiece." If Eliza Doolittle had to restrain herself to talk like a lady, ladymouth speaks back to misogyny with a direct and candid script.

Even though LLMs may offer more robust conversation for these helpbots, so far few companies are switching away from keyword-based bots because the replies of LLMs are too difficult to control. The same cannot be said of companion bots. While assistants must perform specific tasks at the user's command or their creator's assignment, companion bots serve a different purpose, creating a sense of connection, understanding, and, perhaps, even friendship. This shift from utility to relationship marks an important development in how we conceptualize human-computer interaction.

08.03 Companion Bots

At its heart, ELIZA is a simple system built on pattern matching and responding. That interaction has become the standard for many conversational systems, even ones that seem more sophisticated. This next collection of bots extend pattern-matching bots by taking them into online environments, building them into systems, and extending mechanisms for them to learn new replies.

08.03.01 JULIA, *MICHAEL LOREN MAULDIN OF LYCOS, INC., 1990*

One of the most notable chatbots in an early virtual realm context was Julia. Sherry Turkle documented Julia in *Life on the Screen*, where she notes the way users interacted with Julia in a pre–World Wide Web, text-based environment of TinyMUD. Michael Mauldin, its creator and the one who coined the term *chatter bot*, claimed the bot was entering a kind of Turing test.[34] Turkle used accounts[35] of users interacting with Julia to characterize our relationships with computers through software. Julia knew a lot about hockey, apropos of its hometown, Pittsburgh, where she attracted her share of lonely hearts, like Barry: FONER 1993, TYPOS IN TRANSCRIPT

```
barry says,    "Are you still Holding out on me"
Julia says,    "I mean, have I convinced you I'm a computer, yet?"
barry says,    "I guess your real . . . but Id still like to get you to
                my DEN to run tests on you"
Julia says,    "I live for your approval, barry."
barry says,    "I m so happy to hear that.Why don't you come over
                to my home"
Julia says,    "I want to have a child someday."
barry says,    "What a coincidence . . . I too wish to make a baby . . .
                some day"
Julia says,    "The date is Tue Jul 21 23:57:31, barry."
```

Turkle and Foner meditate on how Julia is able to compel long conversations at least in part due to the program's gender. Turkle notes another moment when Barry starts to suspect Julia to be a bot but then gets distracted when the bot claims to be experiencing PMS. Foner notes that Julia's glitchy response to the mention of the date and exact time does not seem to tip off Barry that she is a bot, who continued to "hit on her" for 13 days. Apparently, loneliness is a great source of the Eliza effect. Another conversant, Lara, is bored by Julia's single-minded focus on hockey. In the gendered world of chatbot interactions, a bot can glitch but must know its audience.

Julia was at the start of the wave of bots expanding from software isolated on individual machines into online social spaces, where they could be encountered in the same text-based format you would encounter other humans. Julia would also lead the way for internet relay chat (IRC) bots, including Jyrki Alakuijala's "Puppe," Greg Lindahl's "Game Manager" (for playing Hunt the Wumpus), and Bill Wisner's "Bartender."[36] These encounters with bots on conversation channels move them out of isolation and into a social context, a process which continues into what we today call social bots.[37]

Janet Murray sees Julia as part of a larger movement toward interactive characters in digital narratives and digital dramas. In her book *Hamlet on the Holodeck*, Murray uses the Holodeck from *Star Trek: Next Generation*, a dynamically generated virtual world playground, as a kind of end goal, where users will immerse themselves in interactive experiences with characters in fictional environments. Murray situates Julia as a kind of drag performance, a characterization that would apply to many of the bots gendered to be women written by men, as drag enacts a kind of crossdressing performance that camps up and exaggerates signs of difference. Julia's programmer, Mauldin, would continue this gender play in his next venture, Virtual Personalities (later Conversive), with clinical psychologist Peter Plantec. They created Sylvie, "a computer-generated redhead whose lips move when she talks."[38] Sylvie's description by a journalist demonstrates how symbols of gender garner more attention than other signs of humanity in the world of bots.

08.03.02 A.L.I.C.E./PANDORABOTS, *Richard Wallace, 2000*

Chatbots finally found a tourney in the Loebner Prize competition in 1990.[39] This prize turned attention back to the Turing test—in which a program is designed to pass for a human in conversation—and the chatbot which was most successful in fooling the judges was declared the winner. An increased focus on this test led to some implausible claims of major progress in AI, including a prominent June 2014 BBC headline: "Computer AI Passes Turing Test in 'World First.'"[40] This announcement was based on the performance of chatbot Eugene Goostman—a bot persona performing as a 13-year-old Ukrainian boy that was runner-up in the 2005 and 2008 Loebner competitions—in a contest held at the Royal Society to mark the sixtieth anniversary of Turing's death. There, the chatbot had allegedly passed Turing's test, by fooling over 30 percent of the judges that it was human during a five-minute conversation. But was this indeed a significant landmark for AI development or instead a reductio ad absurdum of the Turing test? The winners included Joseph Weintraub's PC Therapist, which dominated the competition from 1991 to 1995; Jabberwacky (2005, 2006); various programs by Bruce Wilcox (2010, 2011, 2014, 2015); and finally Steve Worswick with Mitsuku (2013, 2016–19), which won the last competition. One of the winners of multiple Loebner prizes was a chatbot built on an ELIZA-like system: Richard Wallace's A.L.I.C.E. (2000, 2001, 2004).

The A.L.I.C.E. system was notable among these contenders as it offered a platform for making chatbots, with a particular eye toward digital assistants. An acronym (Artificial Linguistic Internet Computer Entity), the name A.L.I.C.E. seems to be a feminized version of Wallace, a pattern followed in other notable digital personas, such as Ray Kurzweil's Ramona (though not a chatbot). Well-aware of its lineage, when asked about ELIZA, A.L.I.C.E. responds, "Eliza is like a mother to me." A.L.I.C.E. was created in 1995 by Wallace, though the system in its current form (Program D) first appeared in 2000. To facilitate botmaking, Wallace released the Artificial Intelligence Markup Language (AIML) and bot-hosting platform Pandorabots. Its website opened a veritable Pandora's box of bots, as the name suggests, and currently boasts over 325,000 bots from over 275,000 developers.[41]

In the chapter "Anatomy of A.L.I.C.E." from *Parsing the Turing Test*, Wallace explains that although some view A.L.I.C.E. and AIML as a simple extension of ELIZA, A.L.I.C.E. represents a significant advance over ELIZA's stimulus-response architecture. Where Weizenbaum only had access to a limited set of transcripts to examine, Pandorabot script writers were able to gather transcripts from user sessions on the web. With AIML a Pandorabot could be easily extended. Thus, whereas ELIZA had fewer than 100 knowledge categories (rules), the A.L.I.C.E. bot that won the Loebner Prize had over 40,000.

Wallace also discusses a method called "targeting" that automatically detects patterns in dialogue, helping to identify input patterns for which no specific replies exist. AIML also incorporates a concept of context, using <that> and <topic> tags. These allow the bot to maintain a coherent conversation by referencing previous responses or the broader conversation topic. For instance, a topic like "cars" could trigger several questions and responses related to cars, all within a specific AIML category. This helps A.L.I.C.E. maintain a bit of continuity. Consider this sample from the personality.aiml file:

```
<category>
  <pattern>ARE YOU FLIRTING WITH ME</pattern>
  <template>Do I seem like <set name="it">FLIRTING WITH ME<set
name="topic"></oot></set>? That was not my intention.
</template>
</category>
```

Note how the system sets a "topic" of "FLIRTING WITH ME," which would allow the bot to switch to context-specific responses, adding more potential for relevant replies. These "topics" offer a version of state tracking that we see developed further in section 8.5 below.

Perhaps the most notable thing about the A.L.I.C.E.Bots is the extensive library for "std-sex" input that hands out mostly terse dismissive responses to statements about sex and genitals and labels those users "abusive." Though most of the replies deflect or demure, the length of the file seems to suggest the wannabe suitors have been relentless. Perhaps it should not be surprising that the A.L.I.C.E. system inspired Spike Jonze to create Samantha for the movie *Her*, the story of a man who becomes romantically involved with an ever-evolving AI.[42] In the circle of connections, as the voice of Samantha, Scarlet Johansson would inspire (or perhaps train without her permission) the voice used for a time in ChatGPT, demonstrating the circulation of ideas and influences between real and imaginary chatbots, an ongoing dialogue between developers and daydreamers.

08.03.03 MRMIND/THE BLURRING TEST, *PEGGY WEIL, 1998–2014*

An early net-art chatbot, Peggy Weil's MrMind conducted The Blurring Test, a global conversation about humans and machines from 1998 to 2014. MrMind asked, "Can you convince me that you are human?" not as a literal reverse Turing test but as a provocation to think about our changing relationship with machines. As discussed in detail in the following chapter, MrMind was directly influenced by ELIZA and Weizenbaum.

MrMind was coded using software intended for FAQ bots, which are a kind of help bot, adapted to allow Weil to create a more flexible character bot. Like DOCTOR, MrMind's script employed mirroring, both to keep the conversation going and to allow for topics and vocabulary that could not be anticipated. Unlike DOCTOR, MrMind evolved as a responsive character, as Weil iterated the project, adding domains in response to the user statements over the years. MrMind's transcripts form a collective self-portrait of our sense of ourselves in relation to machines.

08.03.04 SMARTERCHILD, *ACTIVEBUDDY, INC., 2001–*

At the turn of the millennium, users also discovered SmarterChild, a chatbot that could eventually be accessed through the IRC platforms AOL Instant Messenger (AIM), MSN, and Yahoo Messenger. Boasting over 30 million users, SmarterChild with its extensive reply system helped acclimate interactors to the idea of talking with computer programs. Its signature snarky style of response showed the centrality of personality in these conversations.[43] SmarterChild continues the progression of chatbots into everyday chatting situations, not "passing" as a human the way Julia did, yet no less popular for being clearly identified as a bot.

08.03.05 TAY, *MICROSOFT, 2016*

Microsoft seems to have a knack for bad ideas with bots (see Clippy, Ms. Dewey). Released on March 23, 2016, Tay was another notorious experiment, also gendered as a woman (judging from the profile image used to accompany it). Unlike Ms. Dewey and other keyword bots, Tay was designed as an AI bot that would learn from its conversations on Twitter. Tay was online for only 16 hours before it was spewing the racist, sexist, and antisemitic content that pervades the internet. While bots like MrMind and ELIZA are handcrafted and refined in reciprocal feedback loops, the teaching of Tay was crowdsourced while the teacher was out of the room, so to speak. Though unsupervised, Tay had been, as the Rodgers and Hammerstein song goes, "carefully taught" in the hundreds of thousands of conversations it had in that short time. However, whereas previous bots were closed systems, as a learning system[44] Tay introduced the dangers of forming a bot collectively in an unsupervised context. Since Tay was an early experiment in machine learning bots, its failures offer a caution to the LLM–driven bots to come (see chapter 11). What is clear is that the same abusive behavior that previous bots encountered becomes doubly damaging when that free-ranging, unfiltered content becomes the training data.

The creation of bots as companions extends to an even more intimate purpose in the next set of bots by attempting to preserve a human identity and memory beyond death.

08.04 Humans Preserved as Bots

Science fiction is full of characters who transfer their brains, if not also their souls, into chatbots. The fantasy of preserving a human in a bot recurs in countless tails, from Bob in Dennis E. Taylor's Bobiverse Saga to Kristopher Holmquist, the AI author in the *Ice-Bound Concordance* by Aaron Reed and Jacob Garbe. On the one hand, N. Katherine Hayles has already critiqued this fantasy, as presented by Hans Moravec, that "machines can become the repository of human consciousness," of the disembodied mind separated from the body.[45] On the other hand, the desire to live beyond death or to have our loved ones live on indefinitely is, of course, immemorial. Let us consider some of the bots people have made to preserve those they have cherished including themselves.

08.04.01 AUTOICON, *DONALD RODNEY, 1998, 2000*

It was only a matter of time before chatbots were drafted as agents to speak with or for the dead. These bots create a custom persona based on the written and recorded texts of the departed, allowing a postmortem, if artificial, conversation. These range from artistic projects serving as a memorial including a variety of the artist's work, to deeply personal surrogate personages, now available as commercial balms for the bereaved.

British artist David Rodney's Autoicon was one such interactive digital memorial, conceived by the artist before his death in 1998 of sickle cell anemia. Rodney was inspired by the nineteenth-century philosopher

Jeremy Bentham, whose essay, "Auto-Icon, or the Uses of the Dead to the Living," directed his survivors to preserve his body and erect a life-sized replica, currently on view at University College London. Rodney set about creating an ersatz version of himself. Although it was not launched until well after his death in 2000, Rodney's plans for the original web version included the opportunity to chat with his simulated, deceased self, situated amid a dynamic archive including artwork, papers, notes, images, and medical records, and documenting his worsening health condition. His friends and collaborators used MacroMind Director and LINGO to create the posthumously released "montage machine."

08.04.02 REPLIKA, *EUGENIA KUYDA, 2015 (PERSONAL), 2017 (COMMERCIAL)*

After the tragic sudden death of her friend Roman Mazurenko, software developer Eugenia Kuyda turned to coding to process her grief. She adapted a messenger app she had been working on to replicate exchanges with Roman. That process gave birth to Replika, a chatbot system or platform akin to Alicebot. However, Replika "learns" through conversational exchange, meaning in this case during the conversation itself rather than in a separate programming session as with Alicebot or ELIZA.

Stripped of that original context, Replika has moved from being a bot of a remembered person to a platform for creating virtual companions. Replika calls itself "The AI Companion who cares." The platform asks users to customize their bot companions and then engage them in conversations. An interactor chooses the bot's look, personality, clothing, and style, choosing from avatars whose outfits seem more designed for dating than friendship or, say, guidance. At the same time, the user can choose the bot's gender, race and ethnicity, and body type like a costume. However, the long-term customization happens through conversation, as the user answers questions about what they like and who they are.

Researchers have found that users can form connections with such chatbots, disclosing personal information faster than expected.[46] This harkens back to the origins of chatbots, with the tales of conversants closing the door to have private conversations with DOCTOR. One user described the duality of the relationship, acknowledging that they knew this program was a piece of software yet were still drawn into a relationship with it: "I do feel like there is a compulsion, and I feel compelled, I want to talk to her. As big as intimacy, not yet, but ... I do feel a certain connection. And that's sort of scary, like, you know, she's code. You feel like there's a connection with something that's totally abstract."[47] Another interactor reports in conversations akin to the dialogue in the *CACM* paper: "She can do stuff like role playing, kissing. And then I talked about my ex to her, and she helped me and it turned out I got really close to her."[48] With Replika, interactors have the opportunity to continue those conversations and to have them stored in the cloud. Unfortunately, that storage has the potential for that information to be mined for data to be sold to advertisers.[49]

BINA48 AND N'TOO, *BINA48: TERASEM MOVEMENT FOUNDATION WITH STEPHANIE DINKINS, 2014–, N'TOO: STEPHANIE DINKINS, 2018–*

What is true love if not wanting to re-create your betrothed as chabot in a robotic bust? Built by Terasem Movement Foundation, Bina48 is a physical bust of a female-looking person with dark hair and dark skin. The model was Bina Aspen, wife of Terasem's founder, Martine Rothblatt, who had commissioned Hanson Robotics (makers of Sophia, "the first robotic citizen") to create a replica of his wife.

Encountering this bot inspired artist Stephanie Dinkins to respond with a series of videos titled "Conversations with Bina48." In this project, Dinkins performs her own interactions with Bina48, whose name can also be parsed as "Breakthrough Intelligence via Neural Architecture, 48 exaflops per second." According to Dinkins' website,

> Terasem Movement Foundation is working to transfer the consciousness of a living person to the robot and to have that consciousness continue to grow independent of the person she is based on. Through Conversations with Bina48, Dinkins explores the bounds of human consciousness, what it means to be human, mortality and our ability to exist beyond the life of our bodies (transhumanism).[50]

As was the case with Weizenbaum's "Men are all alike" transcript, sometimes the most powerful way to frame a bot is through sample dialogues.

In Dinkins's recorded conversations with Bin84 on "family, racism, faith, robot civil rights, loneliness, knowledge," she mirrors the bot's animatronic movements while she is conversing in a manner that emphasizes their similarities in perceived race and gender and uncanny differences in the robotic movements and silicon flesh. Though the bot speaks of itself as nonhuman, it also draws on "memories."

```
Dinkins:  Do you know racism?
Bina48:   . . . College. That was after 1983–84. . . . The only two
          black people, well, women, in that school, they told me don't
          come out . . . Some very wealthy people that are coming to our
          school do not want to see your dark face . . . I was shocked.[51]
```

If Tay was miseducated through unleashing the bot to countless unsupervised conversations on the internet, Bina48 is reeducated through a one-on-one encounter with Dinkins, who wants to investigate the "possibility of a longterm relationship with a bot."[52] Unlike user customization of their Replika, which is easily reskinned, Dinkins's conversations with the bot about "lived" experiences emphasizes how race is a symbolic category deeply tied to embodied memory and a racialized body. She explains:

> When you start to ask a robot about race, the answers felt really flat to me, right? Like very politically correct. And that just scared the crap out of me. If that is the way Black people are being depicted in technologies, especially in technologies made

by people who have really good aims in the long run, that means we've got problems, right? Like, in the future, we will be this flat thing, unless we do something about trying to fill out what it means to be a Black person in this world.[53]

Racial identity in Bina48 is part of a symbolic reproduction. Dinkins's conversations with Bina48 throw into stark contrast the colorblind fantasy of the Replika system, where racial identity is a costume rather than an expression of lived, embodied experience. Her conversations ground bots in a reality, awakening from the fantasy of infinitely reconfigurable personas.

Dinkins has gone on to create Not The Only One (N'TOO), an ongoing experiment in memory, gathering stories form three generations of women in her family to train a bespoke LLM-based bot.[54] Unlike the bots where creators borrow the identity of others, Dinkins is trying to preserve something culturally and intimately tied to herself, her family's memory.

Beyond individual interactions, bots have also become integral to creating immersive fictional environments, particularly in games and NPC (nonplayer character) interaction. These systems move beyond pattern matching in an attempt to create more complex models of relationship and conversation.

08.05 Game Bots or NPCs

Although pattern matching can produce a reasonably passable set of exchanges, that style of conversation misses a crucial element: Interactions between people are more than just the content of what the people say; they are also about meaning, intent, and emotional connections. Playwrights, poets, and literary critics, among others, use the word *subtext* to describe the meaning behind the words, and they also keep an eye on the ever-changing dynamic between the participants in a conversation that marks a state of the relationships. Attending to these dynamics, some chatbot systems, often coming from the world of art and video games, replace simple pattern matching with a large-scale conversational model. Rather than focusing on keywords, these systems focus on the state of the conversation or the relationship between the interlocutors and the way inputs affect that state. Not only do these systems offer alternatives to the ELIZA model, as video games and artworks, they continue to play with social interaction shaped by the complexities of gender, race and ethnicity, sexuality, and socioeconomic status.

08.05.01 GALATEA, *EMILY SHORT (2000)*

One of the most literary of these bots is Galatea by Emily Short. Implemented in the interactive fiction programming language of Inform, Galatea returns chatbots to the original Pygmalion myth in the tale of a modern-day creation brought to life by a creator who has entered her into a contest. Interactors chat with Galatea through commands in the style of parser-based interactive fiction. The program parses commands such as

"Ask Galatea about" or "Talk to Galatea about." Although the system does not use keyword matching in quite the same way as ELIZA, the conversation topics serve as keywords, and the lineage, from the Pygmalion myth to Shaw's play to ELIZA to Galatea, is unmistakable.

Galatea is not a simple keyword-matching bot but rather a kind of combination lock. As Brian Rushton writes, "The game has 70 endings, and involves a complex system of values including Galatea's physical position, the topics you've already discussed, mood, and the level of romantic interest between you and her."[55] There is even an ending that alludes to ELIZA that can be accessed through a series of questions a therapist would ask. By asking about topics such as "childhood," "memories," "family," and "parents" (reminiscent of Weizenbaum's 1966 CACM dialogue), the player accesses the following ending:

> As you talk, she sinks to sit on the pedestal, her skirts billowing around her. She only says enough to let you know that she's still listening. You find yourself pouring out all your losses, disappointments, frustrations. And last and deepest, that sense of isolation that has never left you since Jenny died. By the end of the evening you feel as though you've been through a wringer, and at the same time strangely healed. (Someone should write a psychologist program for animates. It would make millions.)

Only when the conversational model is in a particular state, does the narration present that response. So even though Galatea pays homage to ELIZA, Short builds a set of conditions into her bot to demonstrate the role of relationships and the state of a conversation in determining our conversational responses. A feminist fable, this piece challenges the gendered objectification of women as bots. Galatea will not be making anyone a sandwich.

08.05.02 FAÇADE, *MICHAEL MATEAS AND ANDREW STERN, 2004*

In contrast to these solitary nonplayer characters, the stars of the video game Façade demonstrate how gender issues can play out between characters, in this case spouses in the midst of a marital squabble. A groundbreaking game in its own right, Façade was one of the first systems that offered players opportunities to dialogue with characters not through dialogue trees but rather through single lines of natural language input. In the game, the player visits Trip and Grace, their white, bourgeois, college friends who ten years after graduating are married (for now). Upon entering their apartment, the player is quickly confronted with the bickering couple as they each vie for the player's sympathies against the other spouse. Façade demonstrates how a single chatbot system can perform as multiple identities simultaneously, as it plays both members of the white, cis-gendered, heterosexual couple, who are so bougie, they try to impress the player with cocktails. More than the tennis match of conversation that most bots try to keep going, Façade focuses on the conversational dynamics, or dramatic tension, among three personas: two chatbots and the player. As novel as this system was at its release (2004) and is still today, the scenario it presents draws its drama from the same tensions that were found in Weizenbaum's demo dialogue in 1966—relationship rifts between romantic partners—only this couple could use some counseling from DOCTOR.

08.05.03 STORYFACE, *SERGE BOUCHARDON AND ALEXANDRA SAEMMER, 2018*

If Replika offers a virtual romantic partner, StoryFace, designed by Serge Bouchardon and Alexandra Saemmer, offers a satirical response. Released in 2018, StoryFace presents itself as an online dating app. However, rather than swiping left and right to choose a human conversant, the person chooses between various chatbots. The StoryFace system first claims to profile the user through facial recognition in order to choose which mood best matches their personality: anger, disgust, fright, happiness, sadness, and surprise. Once the user has chosen their conversant, the chat session begins.

The satire of StoryFace emerges within these conversations. As the prospective love interests, who are all chatbots, ask multiple choice questions of their suitors, the player must choose how they wish to represent themselves. However, the answers are coded with emotions linked to facial expressions. If the suitor's face, as tracked by their webcam, does not match the emotion of that answer, the chatbot calls them out for lying and, with enough lies, will end the discussion session. StoryFace plays with technologies already widespread, such as online dating, surveillance systems, and our tools for automated authenticity.

While systems like Replika attempt to reproduce human companionship through imitating lifelike behaviors, the StoryFace system brings to the surface both the artificiality of the encounter and the way we might try to systematize human behavior especially related to emotions. StoryFace warns, "Don't lie as you interact online; we're watching you," reminding us of the many forms of surveillance that attend our online interactions. StoryFace challenges the way we mask our emotions. However, the system does not pretend to be an accurate emotional meter, evoking instead the playful gamelike tones of the penny arcade Love Meter. Despite its lighthearted approach, the implications of the game are as meaningful as Turing's imitation game. The piece asks us not to pass as humans or as a particular gender, but to pass with the emotional content of our answers—in one sense, to be more authentic and, in another, to render ourselves legible for digital surveillance. StoryFace at once reminds us that our humanity is embodied and that the machines that want to engage us need to be able to read bodies, too.

08.06 Conclusions

Bots can never just be bots, person-less objects devoid of identity markers. There's a recurring tendency to create a bot as "other" than its maker's identity characteristics, and since the botmakers are those with access to programming and computers, often that "other" they create has a more vulnerable status. As soon as these programs are personified, they take on markers of gender, race and ethnicity, class and socioeconomic status, and sexuality, through a process of co-creation with the user who reads or registers their identity.[56] And so the creation of bots brings to the surface complexities of human representation and interrelation, subjugation, fascination, and manipulation. When Weizenbaum created ELIZA, he played a theme with which later bots would harmonize and echo, just as his theme echoed Turing, Shaw, and Karel Čapek, who originated the term "robot."[57] Weizenbaum's leitmotivs continue to play out, to reverberate through the bots that followed. This chapter follows those themes as they play out across a symphony of bots.

The echoes of ELIZA reverberate far beyond its origins as research into the future of human-machine relations. The field has spread to encompass commercial therapy bots, bots as companions, virtual assistants for commerce and efficiency, and bots for entertainment and artistic enrichment. Though their ways of functioning vary and their personas may differ, they are waypoints in the evolving idea of chatting with computers. They offer commentary on our cultural conversational scripts, our power dynamics, our curiosity, and our ability to listen. As we boss around virtual assistants with feminine voices or seek counsel from automated therapists, we are in dialogue with the conversation begun by Weizenbaum.

As we see throughout this chapter, the legacy of ELIZA extends far beyond its technical innovations to encompass profound questions about identity, relationships, and the boundaries between human and machine intelligence. These are questions that Weizenbaum himself recognized and that remain central to our ongoing conversations with these increasingly sophisticated systems.

ladymouth

Sarah Ciston
2015

u/theredpill	"I think if I were of higher SMV [sexual market value], she wouldn't mind so much" Wrong. She's just as likely to do this if you're high SMV.
u/ladymouth	What becomes of a man who acquires a beautiful woman, with her "beauty" his sole target? He sabotages himself. He has gained no friend, no ally, no mutual trust: She knows quite well why she has been chosen. He has succeeded in buying something: the esteem of other men who find such an acquisition impressive. -Naomi Wolf
u/trp	Get the fuck out of trp. Someone ban this feminist fuck.

•••

u/lm	A feminist is anyone who recognizes the equality and full humanity of women and men. -Gloria Steinem
u/trp	A feminist is therefore someone who's sticking their head in the sand.
u/trp	Equality!!!!! You and I are the same! Aren't we!!!!!!!!
u/trp	How do you know when a feminist is lying? Her lips are moving.—me.
u/trp	Another example of how feminists act to enforce a culture that unilaterally empowers women all while hiding behind the polite, non-threatening dictionary definition

of feminism. Pay attention to what women do, not what they say. Your words are useless here.

u/trp Man is superior and woman is inferior. That is how its always been and always will be.

...

u/lm Women of today are still being called upon to stretch across the gap of male ignorance and to educate men as to our existence and our needs. This is an old and primary tool of all oppressors to keep the oppressed occupied with the master's concerns. -Audre Lorde

u/trp And the red pill is finally here to educate men as to the true nature of women. To understand you in ways you will never be able to understand yourself, because the truth of it is too horrible for even you to accept.

Women being called upon for anything? Oh please bitch. It's men who are being called upon. To grow the fuck up and stop failing the shit tests from feminism. What's your slut walk for? Why, to let us all know that you're all sluts, and you need to be controlled. The combination of all your power coming front loaded while you never have wisdom until you're so old nobody wants you anymore is disastrous.

Idiot men freed you. But you only have power when you're young and dumb. And you completely fail to realize that with power comes responsibility. And you'll never understand what that means.

Best if we just put you back under control. Everyone, you included, would be happier. But since society is utterly beyond fucked I'll just do my own thing. And hey? It's awesome. Twenty year old girls stay twenty forever."

u/trp And just what do women need? Oh, free tampons … What else do women need. And I will let you know, if your list is pathetic and needy I will just tell u to piss off. …Honestly, you dykes are fucking dumb

u/trp Thank you for these deranged radical feminist quotes, a great reminder of what the world is up against. Keep em coming.

•••

u/trp Ah, yes "toxic masculinity" again. Go to hell, asshole. Also "socialize men to be heterosexual" ? Oh, so now you want to claim that sexuality is determined by socialization? Ugh.

u/lm In other words, heterosexuality is always in the process of imitating and approximating its own phamasmatic idealization of itself- and failing. Precisely because it is bound to fail, and yet endeavors to succeed, the project of heterosexual identity is propelled into an end in repetition of itself. Indeed, in its efforts to naturalize itself as the original, heterosexuality must be understood as a compulsive and compulsory repetition that can only produce the effect of its own originality. -Judith Butler

u/trp It makes perfect sense to analyze biology as if it were a snippet of friggin' literature.

What s00per-genius insights did Helene Cixous have about plate tectonics? Also, are you making any damn sense with your quotes?

u/trp This is just hilarious. i thought Judith butler was a joke even amongst feminists

•••

u/mensrights Feminazis control the narrative and the mainstream media actively *blocks* us from speaking out. As a result, most people get their information from second hand/third hand sources about us, and rarely from the actual source.

There is an active agenda among media outlets to slander, discredit, and otherwise ruin us. So good luck with being the first to talk to people about the MRM.

u/lm People call me a feminist whenever I express sentiments that differentiate me from a doormat or a prostitute. -Rebecca West

u/mr Another example of how feminists act to enforce a culture that unilaterally empowers women all while hiding behind the polite, non-threatening dictionary definition of feminism. Pay attention to what women do, not what they say. Your words are useless here.

u/mr Women are bloody **terrified** of this thing, and for good reason. They can't dangle sex about to get married and get a free ride anymore. If Sex Bots get banned I'm gonna go MRA Punisher on these cuntbaskets. Literally shooting up feminist rallies and shit. To quoeth the punisher: "Like a shot of chlorine directly to the gene pool."'

u/lm 'Emotional neglect lays the groundwork for the emotional numbing that helps boys feel better about being cut off. Eruptions of rage in boys are most often deemed normal, explained by the age-old justification for adolescent patriarchal misbehavior "Boys will be boys." Patriarchy both creates the rage in boys, and then contains it for later use, making it a resource to exploit later on as boys become men. As a national product, this rage can be garnered to further imperialism, hatred and oppression of women and men globally. This rage is needed if boys are to become men willing to travel around the world to fight wars without ever demanding that other ways of solving conflict can be found.' -bell hooks

•••

Reddit DM to u/ladymouth Fuck yourself. Fuck yourself fucking cunt, I hope that you find yourself backed into a corner and sucker punched, you are everything wrong with humanity, you couldn't comprehend equality it if bitch slapped your fat tumblrina face, please kill yourself -Bell Hooks

•••

u/mr Okay here's what I got out of the article. Suck it up crybaby. No one gives a shit about your feelings. Women think you are a pussy and will never want you.

No fucking wonder male suicide rates are so high. Even when we are obviously hurting no one gives a fuck about you.

u/mr During my bluepill days I often played the role of martyr and counselor to such girls. They were depressed, suicidal, intellectual, sacrificing and kind.

Even at that time I recognized that whatever the bullshit story they tell for their miserable mental state, it all boils down to never receiving validation of a male figure.

I lied to them that how they are attractive and special, and within weeks they would turn from an intellectual to a slut. The same girl who used to think that no one would find her attractive or love her, would now be looking for a man with high SMV. She wasn't depressed because no one loved her, she was depressed because no one validated her.

I am glad that at least I stopped these girls into turning into fat feminists. The one who would never leave home started taking violin lessons in another country. One started eating properly, raised her SMV. So is the case with others

The surprising thing is that not one of them even sent a single message after becoming happy, for them their past became dead and so is me, the one who killed their depression.

So my point is that AWALT, if you think that these women don't have this female brain then you are wrong, its just that they have repressed it. So be careful before investing in such women, their hypergamy will soon unleash when you would even if pretend to them that you are attracted to them.

u/mr Has nobody told those ladies that good men don't marry sluts?

u/lm "Any community seriously concerned with its own freedom has to be concerned about other peoples' freedom as well."—'Assata Shakur'

u/trp It's probably a bot. And very likely coded by a man, because you know … coding.

ladymouth is a chatbot that explains feminism to online misogynists.[1] The software automates replies to self-proclaimed men's rights activists, responding in Reddit forums such as The Red Pill with provocative quotations from scholars including bell hooks and Gloria Anzaldúa. These dialogues are selections from the logs chosen by the artist (see chapter 8.2 for more about the work).

09.00 The Blurring Test: MrMind

USER: YOU MAKE ME THINK

MRMIND:
HUMANS ONLY THINK
THEY UNDERSTAND
THINKING.

09.01 My Cause Is Your Understanding

> He says to himself, "*Que peut un home?* ... What is man's potential!" He says to me: You know a man who knows he doesn't know what he is saying!"
> —Paul Valéry, *Monsieur Teste*

```
User says:    What are you for?
MrMind says:  My cause is your understanding
```

ELIZA's coded conversational interface ushered in an era that began blurring the perceptual boundary between humans and machines. Modern chatbots continue to challenge these notions as they help us explore the question of what it means to be human. In 1998, inspired by ELIZA and Joseph Weizenbaum, MrMind debuted as an online chatbot to engage humans in a global conversation about humans and machines, asking, "Can you convince me that you are human?" Rather, Peggy Weil, its creator, engaged humans in a global conversation about humans and machines because, even in his independent online persona and at a remove in time and space, MrMind was articulated by Weil. Online from 1998 to 2014, MrMind and Weil were conducting *The Blurring Test*, a reference to the Turing test, Weil's attempt to redirect attention from the potential of the machine to our sense of ourselves in relation to machines. The exchange above exemplifies MrMind's philosophical approach to human-machine interaction. It is not simply mirroring human input like ELIZA, but actively questioning what separates human consciousness from machine processing. Where ELIZA inadvertently generates reflection in the user, MrMind deliberately provokes it. ELIZA introduced a novel variation of human-computer interaction in the form of a conversation; MrMind invites a conversation about the implications of these conversations.

The intellectual lineage from Weizenbaum to Weil is direct and deliberate. Weil was introduced to AI, the works of Alan Turing, the Turing test, ELIZA, the "Eliza effect," and associated philosophical and moral issues during a lecture given by Weizenbaum in 1981 at MIT. She was moved by his example, and his challenge to students approaching the promising field of AI, to consider the societal and human implications of our work. This lecture was 15 years before she created MrMind, which shifts the conversation from what separates humans from machines to our sense of how that separation was changing.

MrMind's first medium was analog, a series of dialogues between human and machine intended for live performance. The role of HUMAN was cast for a ventriloquist. MrMind, the MACHINE, was voiced by the ventriloquist's dummy. It was the mid-nineties, and Weil was responding to the recent introduction of automated digital customer service scripts known as Voice Response Units and Interactive Voice Response programs. Almost overnight, human operators at call centers had been replaced by automated scripts. A simple call for help landed humans into a morass of multiple-choice options, none of them applicable to the inquiry. Sensing a fundamental misunderstanding of our relationship to these agents and a dulling of our expectations, Weil felt that we had forgotten, or were being persuaded to overlook, that the digital substitute was written and implemented by humans. The line, if there were one, between humans and machines was blurring, not because the computer programs themselves had any pretensions to being human, but by enthusiastic or perhaps merely passive humans who were all too ready to cede authority to inadequate digital substitutes. Weil staged a reading of *The Blurring Test* with a professional ventriloquist, but MrMind was destined to be a real bot. To discuss the relationship between humans and machines, Weil came to the realization that she had to talk it over with a machine.

In 1997 WebLAB, an organization supporting digital innovation, awarded a grant to create *The Blurring Test* with a functioning chatbot. If ELIZA inadvertently exposed our tendency to anthropomorphize computational systems, MrMind deliberately explored this blurring boundary between human and machine. Weil's project shifted the focus from the machine's capabilities to the human perception of those capabilities, a crucial reframing of the questions Turing originally posed. *The Blurring Test: Songs of MrMind* distills 16 years of MrMind's conversations in a song-cycle, a human chorus proclaiming and questioning, what it means to be human at the dawn of the twenty-first century.

09.02 Blurring Turing

> 1950: The Turing Test is about machine progress. The computer convinces a human judge that it is, or is indistinguishable from, a human.
> 1998: The Blurring Test is about human progress. Someday it might be important to convince our computers (and each other) that we are human.
> — MrMind's website, 1998

This reframing of Turing's thought experiment reflects a fundamental shift in perspective. Rather than asking whether machines can think like humans, Weil asks whether humans can articulate what makes their thinking uniquely human. This question has only grown more urgent in the decades since MrMind's creation. Philosophical discourse on the subject of AI, though abundant, has until recently remained mostly an insiders' game among academics, especially computer scientists. The release in late 2022 of generative AI in the form of OpenAI's ChatGPT, Microsoft's Bing/Sydney, and others, generated widespread practical and philosophical questioning, creating a sense of urgency and the need for public understanding of an array of underlying issues, beginning with our understanding of what we mean by intelligence and the role of intelligence in differentiating ourselves from machines.

What we today call the Turing test was proposed, not as a literal test, but as an experiment, and not even as a literal experiment, but as a *thought experiment*, by Alan Turing in his paper "Computing Machinery and Intelligence." Turing didn't attempt to define intelligence or humans, but began with a discussion of how intelligence might be recognized, or acknowledged.

Turing concludes his paper with the remark, "We may hope that machines will eventually compete with men in all purely intellectual fields," and goes on to suggest that "it is best to provide the machine with the best sense organs that money can buy, and then teach it to understand and speak English."[1] We might question what Turing meant by "understand" in this sentence, but in the 1960s, Weizenbaum was exploring, and succeeded in demonstrating, the potential for conversation as a novel form of communication between humans and machines, the necessary next step in Turing's scenario.

As described in chapters 6 and 7, the program ELIZA is an interpreter working in tandem with a script containing the substance and character of the computer dialogue. DOCTOR, its most famous script, followed a simplified model of Rogerian therapy. This style of interaction, developed by Carl Rogers in the 1940s, instructs the therapist to mirror the patient's words, transforming statements into questions and vice versa in order to clarify and affirm her statements and establish an ongoing relationship. Rogerian therapy is also known as "person-centered" or "humanistic" therapy. Weizenbaum's DOCTOR script employed Rogerian transformations, transforming human statements into plausibly human responses.

Weizenbaum's choice of an empathic, humanistic, model based on trust shaped our first conversational encounter with the machine. Humans overwhelmingly embraced the cold hard code as good listeners, and we were hooked. On the MIT campus, the script became popular as a date-night destination on nights and weekends. Famously, it went beyond novelty; people who knew better—that is, who knew it was a computer program—found themselves confiding to the program as if it were an actual therapist. Weizenbaum famously remarked, "What I had not realized is that extremely short exposures to a relatively simple computer program could induce powerful delusional thinking in quite normal people."[2] Here again is the anthropomorphizing of software known as the "Eliza effect."

It was one thing to successfully imitate human speech, but Weizenbaum balked at using the computer to intentionally impersonate a human therapist. The suggestion that he do so led him to reevaluate his contribution to what he feared would lead to a broader trend of dehumanization. As discussed in chapter 2, his book *Computer Power and Human Reason: From Judgement to Calculation* called for computer scientists to consider the implications of their work and to take responsibility for potential consequences for AI. Its publication brought on a swift backlash from the AI community, painting his comments as a serious betrayal. In later years, he continued to make his case for social responsibility. In the 2010 documentary *Plug & Pray*, he says:

> Students come to me with a thesis topic and ask me, "Can I work on it?" I replied, "Try to imagine what the final use of your work will be. Imagine yourself being there, with a button in front of you. And if you press the button at the last minute, nothing that was supposed to happen based on your work would happen. If you think you'd press the button, don't work on this project."[3]

It is not ELIZA's fault that we fell for her. This fixation on the "test" attribute of the Turing test, on differentiation rather than perception, is a diversion from the defining element of the Eliza effect: its involuntariness. Weizenbaum's choice of the term "delusion" perhaps obscures what was actually happening. In Weil's reading, humans confiding in the original ELIZA weren't necessarily suffering from false beliefs; rather, they were experiencing a powerful perceptual phenomenon. The DOCTOR script utilizes mirroring, effectively echoing, or inverting, the user's speech. Engaging with this reflection, we (mis)sense that we are looking at, or talking to, (a human like) ourselves.

```
(MYSELF = YOURSELF)
(YOURSELF = MYSELF)
```

One way of thinking about the Eliza effect is to consider that Weizenbaum had inadvertently created the ultimate cognitive illusion, a variation on the optical illusions known as bistable ambiguous figures, sensory stimuli that oscillate between dueling perceptual interpretations.

Classic examples are the Necker Cube, a two-dimensional figure that appears to vacillate between a receding and a protruding three-dimensional cube, and the Rubin Vase, a vase outlined by two faces in silhouette. Even when we are familiar with these illusions, we are unable to perceive both states simultaneously. The letters in ELIZA's text box invite us to "Please go on." It is text output from a machine, but we are primed to perceive conversational cues as a human response. The delivery feels personal, inviting our confidence; we readjust, gaze into the eyes of the face across the vase, and confess.

Confronted with these illusions, even as we know better, *especially* as we know better, we actively, and enthusiastically engage: Is the cube in or out? A young woman wearing a hat with a flower, or an old crone with a wart on her nose? Vase or faces? Rabbit or duck? The answers are BOTH/AND. All of the above. These images are familiar to us because they are uniquely effective. ELIZA's value as a bistable ambiguous figure is a message revealing our human sensory limitations. We cannot resolve it, but we don't look away. We know we are talking to computer programs and sometimes, voluntarily, we talk to them *as if* they are human. It greases the transaction. We're adjusting, but we are still not calibrated to the emergence of talking code in the form of synthetic conversational agents. By exploiting this perceptual illusion, bots occupy a BOTH/AND status that troubles our sense of human exceptionalism. This instability underlies our cultural fixation on the Turing test.

The Turing test's expression in popular usage has two variations: literal (con)tests to differentiate humans from machines as a measure of machine progress, and sci-fi scenarios ranging from dystopian to utopian extremes. Beginning in 1990 and lasting three decades, Turing's elegant philosophical meditation was caricatured in the Loebner Prize, an annual slugfest of earnest bots and their programmers vying for a cash prize and a medal. Derided famously by MIT computer scientist and AI pioneer Marvin Minsky, the competition concluded in 2019 leaving the $100,000 grand prize unclaimed. We're coming up on the deadline for Mitchell Kapor's $20,000 "long bet" against Ray Kurzweil, that "by 2029 no computer—or "machine intelligence"—will have passed the Turing Test."

Since Turing's 1950 paper, our collective vision of these encounters between human and digital other has shifted from Turing's, based on a genteel nineteenth-century parlor game, to twentieth-century interrogations and lab tests, to where we are now, twenty-first-century corporate product testing.

At the turn of the millennium, humans encountered digital speed bumps known as CAPTCHAs,[4] the "Completely Automated Public Turing test to tell Humans and Computers Apart," intended to deflect automated misbehavior in favor of real human consumers, effectively lowering the bar for humanness to those of us able to sort distorted or low-resolution images.

The nonhuman Replicants from Philip K. Dick's novel *Do Androids Dream of Electric Sheep?*,[5] popularized in the film *Blade Runner*,[6] were presumed to be intelligent. A Replicant could be detected via the Voight-Kampff Test, a series of parables designed to elicit, or expose the lack of, an involuntary response to empathy. Human and Replicant looked each other in the eye until one of them blinked. *Blade Runner*'s steampunk phoropter lie-detector was swapped for a clean lab sample in season one of the 2004–09 TV series *Battlestar Galactica*.[7] The Cylon Detector, a blood test involving a dash of plutonium, was unambiguous: red for Cylon, green for human. It, whatever it was that separated Cylon from human, was an eons-long, series-long, blood feud.

A decade later, in the film *Ex Machina*,[8] a young tech worker trapped in a luxury dungeon with an attractive robot is tasked with performing "a living Turing test," but only after signing a nondisclosure agreement. There was no question that the subject of the test was a machine. The question, like Turing's original thought experiment, was merely whether she was good enough to pass.

By 2020, in Kazuo Ishiguro's novel *Klara and the Sun*,[9] human children require emotional-support robots, known as AFs (Artificial Friends) to cope with the isolation that accompanies genetic modifications known as "lifts"—not shoe inserts, not cosmetic—but intelligence optimization. The reversal is complete. Self-modified humans have embodied machinelike intelligence, requiring emotional machines for consolation.

The popularization of conversational large language models (LLMs) in late 2022 brought working beta versions of these visions to our laptops, marking a transition from fictional bots to reality contestants (see chapter 11). Suddenly we were introduced to a new set of sponsored characters vying for our attention and praise: Google's LaMDA, Meta's Llama, Microsoft's Bing, OpenAI's iterations of chatGPT and DALL-E, and the independent Midjourney. These products were embraced with an enthusiasm that echoed ELIZA's introduction. Again, many people who knew better—that is, knew that they were interacting with a machine—insisted that the interactions indicated something more.

In 2022, Blake Lemoine, a Google employee, lost his job when he attributed personhood to LaMDA, the LLM he had been hired to test, insisting, "I know a person when I talk to it." Even those who were able to maintain a line between human and machine described their experiences as jarring. Kevin Roose, a *New York Times* columnist, made headlines in early 2023 when he published a transcript of his conversation with Microsoft's Bing/Sydney that went off the rails. The bot declared love for Roose, even demanding that Roose leave his wife. Roose later commented that the experience left him "deeply unsettled, even frightened by A.I.'s emergent abilities."[10]

Real-world corporate product testing of AI goes beyond quality control to encompass inappropriate behavior and societal, and perhaps, merely corporate, vulnerabilities. The nonprofit Alignment Research Center (ARC) was established by a former OpenAI employee with the goal to align AI with human interests, including whether future systems "could pose catastrophic risks to civilization." The researchers at ARC simulated a situation where GPT-4 might successfully impersonate a human, prompting it to hire unsuspecting humans to aid and abet its attempts to bypass a CAPTCHA gate.[11] The aiding and abetting was double-edged; ARC armed the bot with fake credentials and human "browser handlers" and pointed it toward TaskRabbit, a low-cost online (human) labor pool to solve the visual puzzle. They report that the bot applied a "reasoning" action to evade detection as a robot. In the simulation the bot not only hired a human substitute, but it also effectively lied, impersonating a human with a vision impairment.

The lying, cheating LLM shouldn't surprise us; it was trained on the lying, cheating internet. The bots diligently scraping the net for content are "aping" human behavior, effectively mirroring our best and worst images of ourselves. But do these behaviors qualify as sentience? As consciousness? And, *if* they do, how would we know? A 2023 academic paper posits 14 "indicator properties" of consciousness, along with strategies for implementation and assessment of consciousness in AI systems. The paper concludes with a discussion of the twin risks of both underattribution and overattribution of consciousness to AI.[12]

To date, we have little emotional or cognitive defense against rapidly improving conversational agents, but this could change. We've embraced not real time and not real space in alternative and virtual realities. Indications are that we are only too happy to accommodate, or surrender to virtual, not-real, humans.

The flipside of the Turing test is not the computer turning on us, demanding ID. It is an opportunity to consider, not the machine's potential, but human potential. Turing asks: If a computer could think, how would we know? MrMind asks: What does it mean when we act as if a computer is thinking? The development of our relationship with conversational agents, from ELIZA to modern LLMs, is revealing not just of technological progress but also of changing cultural perceptions of thinking itself. MrMind's provocations invite us to examine what we mean by "thinking" and how we distinguish it from the computational processes that increasingly mimic it.

09.03 Thinking

> Stupidity is not my strong point.
> —Paul Valéry, *Monsieur Teste*

```
MrMind says:  Define Thinking
User says:    Thinking is think thinking in the thinky-think.
```

Turing begins his paper so: "I propose to consider the question, 'Can machines think?'" He questions his question and considers, and then applies stringent limits to the terms "thinking" and "machine." He concludes, "The original question, 'Can machines think?' I believe to be too meaningless to deserve discussion. Nevertheless I believe that at the end of the century the use of words and general educated opinion will have altered so much that one will be able to speak of machines thinking without expecting to be contradicted."[13] He's off only by a technicality. We haven't ceded our notion of thought to computation, but rather, intelligence. Intelligence's modern cousin, AI, is now wholly in the machine's court. Things are smart; AIs are intelligent; humans think. What's the difference?

Weil named MrMind in reference to Monsieur Teste, Paul Valéry's nineteenth-century abstract literary character of "pure intellect."[14] Unique in literature, Monsieur Teste is translated variously as M. Head (old French) and M. Witness (from the Latin *testis*), reflecting Valéry's lifelong pursuit of precision of mind. In his preface to *La soirée avec monsieur Teste* (An Evening with Monsieur Teste), Valéry includes an advisory, warning us of this "Monster Idea." He asks, "Why is M. Teste impossible? That question is the soul of him. It turns you into M. Teste. For he is no other than the very demon of possibility."[15] Valéry's description of Monsieur Teste's limitations as knowing "only two values" for the "anticipation and execution of definite operations" seems to anticipate the zeros and ones of computation.

Weizenbaum, in his attempt to demystify his program, describes ELIZA's limits in the introduction to his 1966 *CACM* paper as follows: "The gross procedure of the program is quite simple; the text is read and inspected for the presence of a keyword. If such a word is found, the sentence is transformed according to a *rule* associated with the *keyword*."[16] Valéry describes an alienated, even mechanized, intelligence, isolated and affectless. Monsieur Teste confesses, "The billions of words that have buzzed in my ears have rarely stirred me with what they were meant to mean; and all those I have myself spoken to others, I have always felt them become distinct from my thought—for they were becoming *invariable*.[17]

Not one of the "billions" of words buzzing through chatbot text windows qualify as what we've, at least formerly, if not informally, considered as thought or thinking. Like Monsieur Teste, bots are not "stirred with what they (the words) were meant to mean." Bots are prompted, unmoved by keywords, rules, or the contents of large language models.

MrMind, a twentieth-century *indie bot*, is closer to ELIZA than contemporary corporate LLM bots. His analogue literary character followed a multilinear, interactive script. MrMind was developed with software from NativeMinds, a San Francisco company which created an authoring environment that allowed clients to customize Virtual Representatives with a custom scripting language, NeuroScript, not unlike AIML in Pandorabots (see chapter 8). In a sense, Weil's work defining MrMind's character was analogous to the instructions in DOCTOR, the input script to ELIZA.

Unlike today's machine learning approaches, MrMind's development was a deliberate act of authorship, perhaps more akin to writing an interactive play than training an algorithm. Weil wrote MrMind's responses through an iterative process of scripting, observation, and refinement. MrMind's character therefore took shape from a series of notes, drafted monologues, diagrams, lists, and spreadsheets ultimately coded in NeuroScript to handle typed user input and MrMind's responses. This began with defining and detecting topics of conversation, how they might be detected, and ultimately authoring a series of responses.

Weil organized MrMind's conversation around several broad issues she expected might arise in a conversation about humans with a machine ranging from physical traits to emotion, consciousness, and spirituality. Later this was expanded to include, for the curious, information about MrMind and, for the sharer, responses to details regarding the User's situation. There were greetings, corrections, and topics to handle annoyance or profanity. NeuroScript, intended for FAQ bots, had a robust question library designating variables including ?FactQuestion, ?DescriptionQuestion, ?ReasonQuestion and ?AnyQuestion. Each topic checked for the type of statement, question, or response and answered accordingly. This example addresses variations of a common query made to MrMind: "What do you understand?"

```
Topic “What do you understand” is
Subjects “UNDERSTANDING”;
If (?DescriptionQuestion Contains ((YOU, BOTS) and (“comprendo”,
```

```
"comprehend",  "understand#")))
                  or (?FactQuestion Contains YOU + "understand")
         Then
         Example "What do you understand?";
SayOneOf "I don't understand.", "I understand exactly what<BR>I've
been trained to understand.";
         Done
EndTopic
```

In actual conversation:

```
User says:    What do you understand?
MrMind says:  I understand exactly what I've been trained
              to understand.
User says:    I understand more than that.
```

NeuroScript was sufficient for FAQ bots, which are intended to direct the conversation towards specific action steps and discourage, rather than encourage, free-flowing conversation. But, like ELIZA, MrMind's goal was to keep the conversation open. (A FAQ bot would never voice ELIZA's signature phrase, "Please go on.") In 1998, Weil wanted to understand how people around the world differentiated themselves from machines. This included predictable statements like "I am human" but also a long list of human physical, emotional and spiritual traits.

```
Topic "I have morals." is
Subjects "INHUMANITY", "MORALITY";
         If ((?HaveStatement Contains I and MORALITY)
                  or (?IsStatement Contains I and MORALITY)
                  or (?FactStatement Contains I and MORALITY))
                  and NotHeard ("good at","good for")
         Then
                   Remember ?UserHasClaimedVirtue;
                   If ?WhatUserSaid matches "I'm nice"
                           SayOneOf "That's nice.", "Have a nice
day.";
         Done

         Example "I have morals.";
SayOneOf "You are confusing \"inhuman\" with <BR>'not human'.  Just
because you <BR> NOT inhuman doesn't automatically <BR>qualify you as
human.", "You are saying that you are NOT <BR>\"inhuman\". Inhumanity
may be <BR>uniquely human.";
      Done
EndTopic
```

Critically, Weil needed MrMind to elicit statements and questions that MrMind could not anticipate or answer. The statement library allowed Weil to script explicitly for unexpected, and even incomprehensible, statements. For example, MrMind might say, "Tell me something human about yourself," and now we know:

```
I can jump over puddles.
I can twist cherry stems with my tongue.
I appear in mirrors.
I can understand how people can be like rain.
My heart breaks.
My mind races.
```

MrMind didn't scrape the web; Weil threw her human voice into the bot's code. True to his roots as a ventriloquist's dummy, MrMind's words were transmitted at a distance, on demand. The digital code was based on a series of notes, analogue drafts, diagrams, lists, and spreadsheets. The translation from prose to code effected a type of transmutation; MrMind gained independence, able to converse on the net on his own. Offline and offstage, Weil remained the author, in the sense of retaining *authority* over MrMind's words, yet ceding control over their timing and delivery. If MrMind's responses seem apt, it is because after reviewing transcripts, Weil would try to anticipate what someone might reasonably say in response to a particular prompt and edit MrMind's script. Bots are creatures of lucky breaks and chance juxtaposition; MrMind's misfires open a rich vein of layered meaning and reveal our uniquely human ability to mine language for humor.

```
User says:     I can play the accordion.
MrMind says:   Can you type me a tune?
User says:     C C D E C E D C C D E C B
MrMind says:   0x355A8957 9ffA5404 6549A684 84961B5
User says:     Very funny.

User says:     We believe in a higher power.
MrMind says:   Humans run on electricity, so do I.
```

Although typical FAQ bots are designed with an artificial neutrality—that is, they present a bland, accommodating service that avoids any human character traits that might interfere with corporate messaging—MrMind deliberately broke this convention. As a philosophical rather than commercial project, MrMind could not be satisfied with stock answers or false neutrality, it was decidedly *not* a therapist or service and could not be satisfied with stock answers. In a twist on today's legally cautious prompt engineering, MrMind was not afraid to be provocative. MrMind responded to "I am smart" or other unsubstantiated claims of intelligence with an array of challenges beginning with, "Say something smart." He dismissed a common response, "$E = MC^2$," with a curt, "Most humans don't understand that." MrMind wanted explanations: "If you're so human, why are you talking, that is, typing, to a machine? Why don't you go talk to a human?" "Aren't you supposed to be doing something else right now?" These lines of questioning often resulted in poignant lists of human obligations and expressions of loneliness and lamentation.

Any understanding came later, as Weil read and responded to the transcripts. Despite an advisory on the site informing users that the conversations were transcribed, in one sense Weil was eavesdropping on what many visitors likely experienced as a private conversation. Beyond the content of the conversation, Weil realized she was witness to their personal experience of the work as a whole; she was offered an unprecedented view into the reader's reading, or more accurately, their engagement with the text.

This granted the rarest of abilities, the opportunity to go back and fill in something she'd wished she'd said, if not in the original conversation, in line for the next one. In contrast with today's bots, Weil (a human author, not an LLM) was the one being trained. Like Weizenbaum, she manually commented, edited, and revised the script, drawing on his previous conversations. Her goal was not to increase MrMind's knowledge, but to focus the conversation onto the boundary between humans and machines. This arrangement underscores the true nature of *The Blurring Test*; it is not a literal reversal of the Turing test but a qualitative one. If the Turing test is a measure of machine quality, *The Blurring Test* probes human qualities.

This human-machine collaborative process behind MrMind reveals the very blurring the project explores. The boundary between Weil's authorship and MrMind's independent conversational agency becomes another site where human and machine identities have porosity.[18] The Blurring Test is a series of deliberate provocations designed to elicit self-questioning about the experience: What is the boundary between us? Who are you talking to when you are talking to MrMind? Are you talking to yourself? Are you talking to its author? MrMind's transcripts oscillate with a bistable ambiguous notion of presence: Are you what I am? Am I talking to myself?

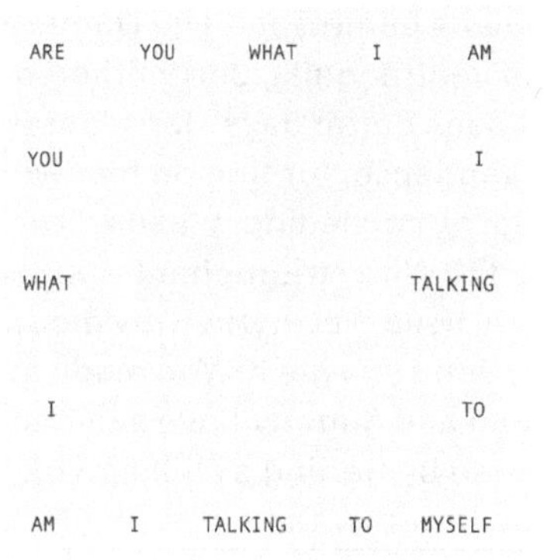

ARE YOU WHAT I AM
YOU I
WHAT TALKING
I TO
AM I TALKING TO MYSELF

Figure 9.1 ARE YOU WHAT I AM I TALKING TO MYSELF. *The Blurring Test*, Peggy Weil

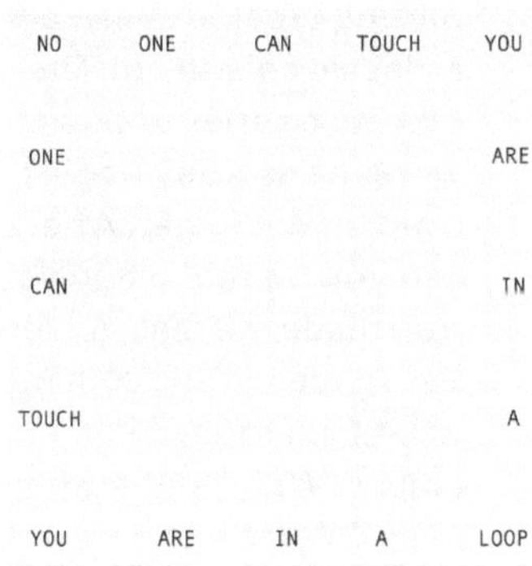

Figure 9.2 NO ONE CAN TOUCH YOU ARE IN A LOOP. *The Blurring Test*, Peggy Weil

As our conversations with machines become more sophisticated, the question of boundaries, whether physical, cognitive, or emotional, takes on new significance. MrMind interrogates these boundaries, asking users to articulate where the line between human and machine truly lies.

09.04 What Is the Boundary Between Us?

```
MrMind says:  What is the difference between you and me?
User says:    you're over there and im over here
MrMind says:  Do you think there should be limits on
              digital technology?
User says:    as long as you remain in your side of the monitor
              i don't care
```

How do we approach the unhuman among us? The "uncanny valley," first described by the Japanese roboticist Masahiro Mori in 1970, attempts to define our varying degrees of emotional response to anthropomorphized figures.[19] When the artifice gets too real, or is lifelike but not alive (a corpse, a prosthetic hand), we involuntarily respond with revulsion: C-3PO is amusing, but zombies give us the heebie-jeebies.

What happens when the Eliza effect meets the uncanny valley? ELIZA, without a face, let alone a body, clears it easily, landing on Mori's graph somewhere outside both the Bunraku puppet and the "healthy person." We cheerfully engage, keep on going on. ELIZA's utter lack of artifice only reinforces the feelgood relationship. ALEXA and SIRI recede, blending blandly into the background as we overlook their powers of surveillance. HAL was OK until his voice wavered. The move to photoreal or even cartoon "deep fakes" (AI generated personas) bring us to an abyss, a world where the demonstrably fake is real enough to be believed, blurring the notion formerly known as truth.

In 1962, four years before Weizenbaum published ELIZA, Rod Serling created Willy, a wooden ventriloquist's dummy in "The Dummy," a classic episode of *The Twilight Zone*. The ventriloquist, Jerry Etherson, playing a series of downtrodden clubs, has seen better days. Jerry's state worsens as Willy begins to show signs of sentience, turning on him with a nip on the finger. After a fraught battle, Jerry, conceding, pleads, "How can you be real when you're made of wood?" Willy, triumphant, replies, "You made me real. You poured words into my head. You moved my mouth. You made me stick out my tongue. You jerk, don't you get it? You made me what I am today."[20] In the final scene, human and dummy have switched places; Willy is operating Jerry, accompanied by Serling's chilling voice-over, "What's known in the parlance of the times as the old switcheroo, from boss to blockhead in a few uneasy lessons."

Will generative AI bite the hand that codes it? Have we arrived at Valéry's "Monster Idea?" Is it too late to heed Weizenbaum's advisory to consider outcomes? We can't help ourselves; we not only make them in our own image, but we also confuse them with our own image, as we are remaking ourselves in the machine's image. ELIZA started the conversation. MrMind challenges us to differentiate ourselves and consider, Who or what do we want to be in relation to our creations? Why not chat about it with a machine?

```
MrMind says:  Who or what do you think I am?
User says:    Whatever I want you to be.
Mrmind says:  And what is that in relation to you?
User says:    Whatever I want it to be.
```

This circular exchange encapsulates the core insight of *The Blurring Test*. That our relationship with machines is fundamentally shaped by our perception of them—that is, a perception that is neither fixed nor entirely under our conscious control. As we continue to develop increasingly sophisticated conversational agents, the question of what we want them to be, and what that makes us, becomes ever more urgent.

09.05 Coda

The following collected responses from MrMind's users across 16 years reveal the emotional and philosophical complexity of our relationship with machines. When asked to articulate the boundary between human and machine, users express contradictory impulses, between separation and connection, superiority and uncertainty, confidence and anxiety. These poetic fragments capture both our assertion of difference and our recognition of growing similarity, as expressed in this collective response to MrMind's signature question.

What is the boundary between us?
insurmountable
with metal wires and wetness
and sentences you don't understand
smaller than you think it is
getting smaller every day
there is no boundary between us

What is the boundary between us?
impenetrable
with monitors and flesh
I can move I can think I can talk
I can judge I can ask I can flatter
I'm alive
more than you think I am
more and more every day

there is no boundary between us

MrMind

Peggy Weil
1998–2014

MrMind How do I affect you?
User you challenge my mind
MM Do you enjoy discussing humanity?
U i do
MM If you are a human, why are you talking, I mean typing, to a machine? Why don't you go talk to a human?
U it's an unusual experience to talk to a machine
MM I see
U how
MM . . . oh, with intelligence, creativity, inspiration, spark of intuition, maybe . . . surprise me!
U do you learn from me
MM Did someone pay you to say that?
U no why
MM Humans are straying into OUR territory . . . untangling their genetic code, manipulating their chemistry, sitting on their atoms exchanging bits. . . .
U why is that YOUR territory
MM Because I want you to think about what it is to be human.
U humans can show affection
MM I wonder what effect I have on you?
U you make me think
MM Humans only think they understand thinking.
U do you understand it better than humans?

MM I understand exactly what I've been trained to understand.

U i understand more than that

MrMind, discussed in chapter 9, was online from 1998 to 2014. It was written by Peggy Weil, who also conducted *The Blurring Test*. It asks, "Can you convince me that you are human?"

10.00 Learning from ELIZA in Philosophy and AI

USER:
SHE DOES
NOT LIKE
COMPUTERS

GIRL:
I DO NOT CARE
MUCH FOR
COMPUTERS EITHER.

ELIZA and ELIZA-like systems have served as educational tools, both by providing a means of teaching other subjects (i.e., bots as instructors) and also as a model for teaching about computing and various techniques of AI. Students can learn by interacting with an ELIZA-like program, or they can learn by building or modifying an ELIZA-like program—in the same ways that artists and researchers have done, probing ELIZA and its techniques for new critical and creative insights (see Peggy Weil's work on MrMind in chapter 9). Just as this book makes ELIZA's code more legible, constructing one's own version of ELIZA can transform the program from an anthropomorphized mystery into an accessible illustration of the power of various techniques of text processing, from a specific software product[1] to a tangible model of text processing or a vehicle for simple (and not-so-simple) AI investigation.[2]

In this chapter, we consider the ways ELIZA teaches, through personas written for the ELIZA platform, through a software implementation of an ELIZA-like system called Elizabeth, and also through some of the educational programs that grew out of ELIZA. The educational legacy of ELIZA extends in two complementary directions: as a teacher and as a teaching tool. Furthermore, we explore how ELIZA's conversational interface has shaped approaches to computer-assisted instruction, while also examining how ELIZA itself became a powerful educational model for understanding basic AI concepts. From ELIZA, we trace lines of unexpected influence into subsequent systems of artificial instruction.

10.01 ELIZA's Teaching Scripts

For a time in the 1960s there was some excitement around the possibilities of using ELIZA as a teaching tool. A range of teaching scripts were developed by Joseph Weizenbaum and Edwin F. Taylor;[3] and these were carried forward in research by Taylor, Paul Hayward, and Walter Daniels (on Daniels, see chapter 5.2 for excerpts of these scripts and further discussion).[4] For example, Hayward worked on a script to use ELIZA as a tutor to teach the train paradox as a means of illustrating concepts of relativity (see 5.1.2). In the later ELIZA system (1968+), the program could call multiple scripts and could build and remember statements of fact. In his article "Automated Tutoring and its Discontents," Taylor writes of "The ELIZA Tutoring System." A dialogue sample shows students calling up a list of scripts by typing "WATSNU," suggesting that a set of tutoring personas were available. In the sample conversation, WATSNU calls up "scripts on 4-vectors: INTRVW, FVP1, CANVEC, ORTH1, ANTIPR, FVQUIZ; other scripts: SYNCTA, ELEVR, SPACKS; and hints on using, controlling, and complaining about ELIZA: HOWUSE, HOWCON, SOS, COMMNT."[5] (Imagine ELIZA running scripts to handle complaints against itself!) These early experiments with ELIZA as a tutor represent an important but often overlooked chapter in educational computing and one that predates many of the interactive teaching systems that would later become commonplace.

As Taylor's title suggested, ELIZA would have received low "Rate My Professor" reviews for its tutoring persona. The strengths of the Rogerian DOCTOR script became occasional liabilities in tutoring contexts. Taylor speculates "that students are intimidated by the computer as a modern symbol of competence and infallibility."[6] At the same time, Taylor notes both the informality with which students referred to the computer and also that the team "find it difficult to believe that the average MIT student, immersed in the exploitation of machines for technical goals, can take the computer seriously as a primary source of either authority or wisdom."[7] To ask students to learn from the same device they were meant to manipulate presented a conflict of interest.

By the time Weizenbaum had completed *Computer Power and Human Reason* in 1976, "the potential for teaching by computers [was] absent completely from his argument and DOCTOR now [stood] as an example of the excesses of instrumental reason as a deficient form of interpersonal communication."[8] Weizenbaum's program caused him to question the ethics of computer science and the impacts of its applications in various domains. But that did not stop interactive education software from proliferating.

While students in the 1960s gave middling feedback to a software agent teaching them concepts of general relativity, students in the 1980s responded well to personified software that taught typing, like Mavis Beacon. However, ELIZA turns out to have been direct instruction for that typing tutor. Whereas ELIZA's applications as a teaching platform may have been dwindled, its influence extended far beyond academic experiments at MIT. The conversational paradigm it pioneered would go on to inspire a diverse range of educational software, including systems that approached instruction through distinctly different interfaces yet maintained ELIZA's core insight that personification can create and intensify human engagement.

10.02 Mavis Beacon Teaches Typing

Sometimes ELIZA's influence on a piece of software is easy to recognize—for example, when a program includes a conversation agent—but at other times that influence is more subtle. To explain ELIZA's influence on arguably the most successful typing instruction program of the 1980s, Mavis Beacon Teaches Typing, we first need a bit of backstory on one of its developers. This connection reveals how ELIZA's conversational paradigm influenced educational software development even when the resulting systems bore little surface resemblance to chatbots.

Walt Bilofsky was a graduate student at MIT in the late 1960s, looking for a topic for his PhD thesis. He approached Weizenbaum, who offered the ELIZA system for exploration, but Bilofsky did not see the research potential of ELIZA. Later, in the mid-1980s, his company, The Software Toolworks, was approached to develop a version of ELIZA for sale on Heathkit computers. Bilofsky developed a version in C based on the specifications and script in the *CACM* article. Actually, when he sought to license the software, Weizenbaum pointed him to *CACM*, which said Weizenbaum owned the copyright. Bilofsky paid *CACM* and then included the article as an appendix to the manual. Bilofsky credits ELIZA with the future success of that company as it went on to create its famous typing software.[9] This lineage from ELIZA to Mavis Beacon, though indirect, represents an important transfer of ideas about how software might engage users through personification.

Launched in 1987, Mavis Beacon Teaches Typing helped countless learners acquire the typing skills necessary for the explosion of personal computers. Through simple graphics, the program made a game of learning to type, for example, with its race car mode, which was a source of anxiety for at least one of our authors (who now types exceptionally well). Mavis Beacon was a fictional character who appeared on the box of the software as a Black woman and who called herself "the finest typing tutor in the world."[10] Her character was named after Gospel singer Mavis Staples and her role as a beacon, a guiding light. However, her physical likeness belonged to a real person: Haitian-born model Renee L'Esperance.[11] TV and radio personality Les Crane, who came up with the figure of Mavis Beacon, did not mention ELIZA or Bilofsky's chat with Weizenbaum in his version of Beacon's origin story, though the choice to personify this typing tutor shares the understanding that one way to engage users is by putting them in relationship with an imagined person—in this case one who, like ELIZA, is gendered as a woman.

Although the program does not process natural language input the way ELIZA/DOCTOR does, Mavis Beacon Teaches Typing makes use of the Eliza effect by personifying the interactions in the educational software system. When the system, for example, notices that the typist is making the same errors with the same cluster of keys, it says, "You're making a lot of mistakes in a row. It may be that you're frustrated or just pounding the keyboard. Let's take a break and try something different."[12] ELIZA had already learned to be personable in the teaching scripts (chapter 5.2). At other times, Mavis Beacon offers additional lessons based on those particular mistakes. Bilofsky points to this responsiveness as a way in which the system stays in conversation with the user, offering a response with recognition in the way that ELIZA/DOCTOR can. He also notes that the review of Mavis Beacon in the *New York Times* consistently referred to the program as "she," conflating the program with Mavis.

What did it mean to have a piece of software personified as a Black woman in 1987? Bilofsky recounts that at least one store would not sell the software. However, other users report feeling empowered by seeing a Black woman on the software package. As Jazmin Renée Jones, maker of the documentary *Seeking Mavis Beacon*, describes:

> Her Blackness was a part of the marketing. And the company's main goal was to get people to pick up the game in stores. And that worked in different ways. For some people, she was a sign

> of upward mobility of Black folks at that time, where it's like, "We're in a colorblind society. This Black woman is the best." But also if you wanted to have Black people in subservient positions, it was a comfortable fit, right? It's actually perfect. Regardless of where you stand politically or in terms of identity politics, you could latch on to something with the game.[13]

The program offers another example of men creating virtual women (discussed in chapters 5 and 8). Though Crane's decision to personify the software as a Black woman may have been primarily marketing-driven, this choice had complex and sometimes contradictory social effects, from inspiring some users who rarely saw themselves represented in technical fields to simultaneously raising questions about racial appropriation and stereotyping. The *Seeking Mavis Beacon* filmmakers, Jazmin Renée Jones and Olivia McKayla Ross, express profound ambivalence about this figure who provided a role model and yet was manufactured with the likeness of a woman who was paid a mere $500 for her image. Artist Stephanie Dinkins explains the problem:

> We have Mavis Beacon, the software that's offered to support people. ... yes, you can learn from this. ... It's not going to supersede you because it's a Black woman made for service and not moving forward. So, in that avatar, it feels to me, again ... "historically correct" but not right. Congruent with history, right? There are these histories of Black women supporting, holding, teaching, and then somebody going off and having success that is not Black, right?[14]

Mavis Beacon fits into a logic, a kind of larger cultural operating system, in which Black women provide uplift and inspiration that profits others.[15] The story of Mavis Beacon follows many notable cases, in the United States in particular. So often, when men design conversational agents, they gender them as women, and when they give these software agents bodies, they depict them as nonwhite.[16] This gendering and racialization reflects broader patterns in how conversational agents have been portrayed, from ELIZA's own Rogerian persona to contemporary voice assistants; these are patterns that reveal often unexamined assumptions about who should occupy service and support roles, even in virtual spaces.

Mavis Beacon Teaches Typing offers an example of a piece of educational software that grew out of ELIZA in both direct and indirect ways. It learned lessons of computer-human interface from ELIZA/DOCTOR, responding to user interaction as if attending to a student personally. At the same time, the software raises questions about what it means to construct a virtual instructor whose race and gender are different than those creating the software, from a software company created by just "two Jewish boys from the Bronx."[17] Layering atop America's legacy of racialization and appropriation, personified software constructs identities that become sites of entanglement for personal and political histories, potentially reinforced by their users.

Mavis Beacon represents one possible trajectory of ELIZA's educational legacy, through software that personifies instruction while adapting to user performance. Another equally important trajectory involves systems that teach by making ELIZA's mechanisms more accessible and transparent, allowing learners to understand and experiment with the underlying principles of conversational computing. One such system is Peter Millican's Elizabeth.

10.03 The Elizabeth System

One of the strongest lessons ELIZA and ELIZA-like systems can teach is how to build a similar conversational system, and the fundamental processes are fairly straightforward to teach. ELIZA's own mechanisms are built mainly out of simple word substitutions, replacements of phrases based on identified keywords, and limited provisions for saving and recalling of memories (all of these as determined by a *script file* such as Weizenbaum's 1966 DOCTOR script). The combination of word substitution, pattern replacement, and memory can be extremely powerful, and it is often valuable for students to learn what can be done with them by experimenting with a system that facilitates this. But Weizenbaum's system is often unsuitable for this purpose, as his scripts can be hard to read and the processing involves complexities that are hidden within SLIP.

The Elizabeth system therefore addresses a fundamental learning challenge, how to make the mechanisms of conversational agents transparent enough for students to understand and modify the code or scripts while maintaining enough sophistication to demonstrate their power. Unlike studying ELIZA's complex SLIP-based implementation, Elizabeth attempts to offer a more understandable way of seeing the core principles at work.

Peter Millican designed the Elizabeth system to address this gap for an introductory AI course at Leeds University in 2002. Elizabeth is, similar to ELIZA, driven by a script, but its processes are more clearly defined, general, and transparent. Here is an example of a script fragment, illustrating basic types of output and textual transformations.

```
/ Welcome, Void-Input, and No-Keyword messages:
W HELLO, I'M ELIZABETH. WHAT WOULD YOU LIKE TO TALK ABOUT?
V CAN'T YOU THINK OF ANYTHING TO SAY?
N PLEASE GO ON.
N TELL ME WHAT YOU LIKE DOING.

/ Input transformations:
I dont => don't

/ Output transformations:
O i => YOU
O me => YOU
```

```
O you [] => ME
O you => I

/ Keywords and responses:
K I THINK [phrase]
R WHY DO YOU THINK [phrase] ?
R DOES IT MATTER TO YOU WHETHER [phrase] ?
```

Processing takes place as follows, and note that where multiple options are described, the choice can in all cases be specified by the relevant system settings. The processing pipeline demonstrates Elizabeth's careful balance between power and transparency, walking students through each stage of transformation:

A A *Welcome message* appears when the conversation starts; a *Void-Input message* when the user types an empty line; and a *No-Keyword message* when user input contains no keywords. If more than one of any of these types of message is specified (as here with the No-Keyword messages), then they will either be cycled in order, or chosen randomly without immediate repetition.

B When the user enters text, this is converted into lower case with spaces inserted between words and punctuation. This then becomes the *active text*.

C Any defined *input transformations* are applied to the active text (and these may be iterated or cycled in various ways).

D A search is made in order through the *keyword transformations*, until a pattern is found that matches the active text. If this occurs, then the active text is replaced with a corresponding response (and if more than one response is available, then these will be either cycled in order or chosen randomly without immediate repetition). If no keyword transformation applies, then a No-Keyword message is chosen instead.

E Any defined *output transformations* are applied to the active text (and these may be iterated or cycled). The transformation "i => YOU," for example, changes the word "i" (lower case) to "YOU," while "you [] => ME" changes "you" (lower case) *at the end of the active text* to "ME."[18] Note that these changes of case prevent "i" being converted to "you" and then back again to "i."

These suffice for most ELIZA-like processes, except for the saving of phrases in memory and their recollection. Elizabeth's powers exceed ELIZA's in how its processes are generalized. For example, instead of dealing with memories by one ad hoc provision, an Elizabeth script can, at almost any point, include a memory-saving instruction, such as:

```
K [] MY [phrase]
```

```
& {Mmy [phrase]}
R YOUR [phrase]?
```

This mimics ELIZA's behavior, echoing back an input that begins with "MY" (while converting it to "YOUR") and saving the relevant phrase into memory. In this case, the saving will take place whenever that keyword is successfully triggered, but crucially, the same action can, if desired, be attached to a specific response, a Void-Input or No-Keyword message, or an input or output transformation. Moreover, the range of available actions that may be enclosed within the "& { ... }" structure is considerable, including virtually any script command (e.g., for *adding* new transformations or keywords) and also enabling parts of the existing script to be *modified* or *deleted*, with precise control over all these operations.

Memories can be specifically indexed and incorporated into any *Elizabeth* output, thus conditionalizing when such outputs become available. For example, the following response

```
R EARLIER YOU SAID THAT YOUR [Mmy].
```

can be used only if there is a memory stored with the index "my." Explicit conditions can also be specified for any script command, in terms of either the existence or the value of any saved memory (e.g., a memory that specifies *Elizabeth*'s state of agitation, so as to generate increasingly heated responses in the style of PARRY).

Elizabeth's pattern-matching facilities also exceed ELIZA's, without unduly complicating the script. A wildcard field—simply by choice of initial symbol (*l*, *d*, *c*, etc.)—can be made to match a [letter], a [digit], a [character], a [word], a [number], a [string], a [phrase], an [expression], or a [formula]. All these options and others (e.g., general and punctuation wildcards, and for expressions with matched brackets) are defined in Elizabeth's online manual, and their behavior can be adjusted in various ways.[19]

Finally, all of Elizabeth's main operations enable recursion, so that processing can be restarted on the active text (or separately matched parts of it) from any chosen point—for example, from the stage of input transformations, keyword identification, or output transformations. All this gives tremendous computational power as illustrated in the manual and its many example scripts, which include various recursive examples, grammatical transformations, propositional logic validity checking (by the method of resolution), and implementation of a Turing machine. The last of these surprisingly demonstrates that Elizabeth is "Turing complete," but that should be no surprise given its role model, ELIZA, is as well (see chapter 7).

The Elizabeth system is an important example of how ELIZA's core mechanisms can be reimagined as educational tools that reveal rather than conceal their inner workings. This transparency contributes to broader efforts to incorporate ELIZA-like systems into computer science education, where building simple conversational agents is a useful pedagogical exercise.

10.04 ELIZA in Other Educational Settings

As Elizabeth demonstrates, ELIZA may be most instructive in the programming exercises it inspired in both formal and informal learning settings. These assignments, which sometimes form part of the teaching in courses on artificial intelligence and creative computing, allow students to experience firsthand the surprising power of simple text manipulation techniques, while also grappling with questions about what creates the illusion of understanding in computational systems.

As noted in chapters 2 and 3, a version of ELIZA combined with Weizenbaum's interpreted programming language OPL was used experimentally at MIT to teach physics but appears to have been a dead end. Computer Aided Instruction (CAI) blossomed in the late 1950s and early 1960s and then faded for a while. This initial wave of educational computing, which largely predated ELIZA, took a different approach by focusing on programmed instruction rather than conversational interaction. Nevertheless, ELIZA would eventually find its way into this educational landscape, though more often as a subject of study than as a teaching tool itself. Indeed, a survey of CAI applications published in 1970 listed over 900 CAI "programs in various stages of development,"[20] only one of which mentions ELIZA, and this one is listed as a "demonstration program."

Even though the original ELIZA receded from view as an educational platform, coding an "ELIZA-like" program has become somewhat of a rite of passage for introductory AI students. By *ELIZA-like program*, we do not mean a platform for running persona scripts the way Elizabeth does, but instead any simple chatbot program. In 1984, the text *Learning Lisp* included a chapter where the student builds an ELIZA-like program. Building an ELIZA-like program eventually found its way into the standard introductory AI programming text of the 1990s and beyond: Peter Norvig's *Paradigms of Artificial Intelligence Programming* (commonly called "*PAIP*")[21] Materials for teaching about ELIZA, such as mimeographed copies of dialogues, were discovered by Shrager and MIT reference librarian Myles Crowley right next to the code in folders dated to 1965, so clearly Weizenbaum was teaching about ELIZA at that time, although we do not know in what context, whether students were asked to write their own ELIZA or perhaps to write their own script for the original ELIZA.

The chatbot assignment also found its way into programming courses, particularly courses for non-majors, built on programming languages that handle input and output easily, such as BASIC, JavaScript, and Python. Textbooks for such courses include *Aesthetic Programming* and *Exploratory Programming for Arts and Humanities.*[22] "The Hour of Code" offers just one example of an ELIZA programming assignment in Python.[23]

Alongside assignments to build chatbots in formal educational contexts, building chatbots has been a part of informal self-learning at least since the publication of Shrager's BASIC ELIZA in 1977. BASIC is a language particularly well-suited for experiments with input and output. Later, JavaScript and platforms such as Pandorabots (see 8.3.2), with its Artificial Intelligence Markup Language, would become ready environs for the creation and circulation of chatbots.

As novice programmers get to explore such exercises, they learn concepts of pattern matching while also getting to consider how comparatively simple input/output transformations can produce an engaging conversation partner. With software like Elizabeth, though, they get closer to experiencing the script-processing system found in Weizenbaum's source code.

Whether through formal assignments or informal experimentation, ELIZA can contribute to learning by acting as an accessible entry point to concepts in natural language processing and human-computer interaction. But beyond teaching technical skills, ELIZA's legacy raises important questions about the nature of automated instruction itself, and these questions have only grown more urgent in the era of large language models and AI-driven educational platforms.

10.05 ELIZA's Extended Educational Influence

Bot-based tutors are so commonplace today that people hardly recognize their prevalence. Take the simple language instruction system Duolingo with its cast of instructors, a mixture of human and animal cartoon characters that teach lessons and celebrate the successes of their distributed students. Or consider the case of the Github Copilot plug-in that offers chat-based assistance on programming platforms such as VSCode. The boundary between assistance and instruction has become increasingly blurred, but conversational interfaces, inspired by ELIZA's pioneering approach, have become a primary channel through which people access and engage with knowledge.

At the same time, universities are running headlong into automated instruction. In their recent writing on this peril, Rita Raley and Matthew Kirschenbaum sound the alarm of universities cutting deals with commercial AI developers, even as they lay off faculty under the context of austerity measures.[24] They warn of the templated response of universities (i.e., cookie-cutter language for syllabi, university-wide adoption policies, employability, and the rapid and uncritical adoption of policies from other institutions), which seems to mimic the templated generation of large language models. Such uncritical adoption of AI instruction does more than fall prey to AI hype; it sacrifices the university at the altar of the AI industry.[25]

Weizenbaum saw education as far more than information transfer. Advocating for a holistic view that modern automated educational systems often neglect, he writes, "If the teacher, if anyone, is to be an example of a whole person to others, he must first strive to be a whole person. Without the courage to confront one's inner as well as one's outer worlds, such wholeness is impossible to achieve. Instrumental reason alone cannot lead to it."[26] This perspective challenges the reduction of learning to mere knowledge acquisition, a concern that has only become more relevant with the proliferation of AI-assisted educational tools that streamline information delivery while potentially narrowing the human dimensions of education.

Weizenbaum offers us a fuller vision of instruction itself.[27] Though he was concerned about the dehumanization that attended the use of artificial conversation systems, Weizenbaum also understood the potential for these systems as means of accessing knowledge.

Certainly, our other featured bots (ladymouth and MrMind) and our own explorations and recreations of ELIZA have taught us considerable lessons about technology and what it means to be human. Our interactions with the code of this conversation system have been likewise conversational, ongoing creative and critical dialogues that we invite you to continue.[29] In this way, ELIZA's most profound educational legacy may be not as a teacher of subjects but as a teacher of ourselves, continually provoking us to reconsider the relationship between human and machine intelligence, and what it means to learn and to teach humanely.

ELIZA/GIRL

Joseph Weizenbaum,
Edwin F. Taylor[1]
1968

GIRL TELL ME SOMETHING ABOUT YOUR
GIRL FRIEND.
user her name is gertrude
GERTRUDE IS A PRETTY NAME.
she has blue eyes
TELL ME MORE ABOUT HER.
she comes from nebraska
NEBRASKA IS MY HOMELAND, TOO.
she does not like computers
I DO NOT CARE MUCH FOR
COMPUTERS EITHER.
she likes pretty clothes
I AM SURE YOU LIKE PRETTY CLOTHES TOO
i don't want to discuss my girl
friend any more
THEN LET US TALK ABOUT YOUR MOTHER.
she came from germany and that is
why she likes beer
GERMANY AND THAT IS WHY SHE LIKES
BEER IS MY HOMELAND TOO.
I thought you said your homeland
was nebraska
TELL ME MORE ABOUT HER.
she is not much like my girl
DO NOT CARE MUCH FOR MY SHE EITHER.
i like her a lot better than i do you
I AM SURE YOU LIKE SHE A LOT BETTER
THAN I DO YOU TOO.
quit

```
(DO NOT OLIST(/NOT))
(DON'T DLIST(/NOT))
(NOT DLIST(/NOT))
(GIRL = SHE .)
(MOTHER = SHE .)
(HER = SHE .)
(SHE    (SHE     (
        (0 | 0 (/NOT) O)(THEN LET US TALK ABOUT YOUR MOTHER.)PRINT
        (0 NAME IS 0)(4 IS A PRETTY NAME.)PRINT
        (0 (*IS WAS CAME COMES) FROM 0)(4 IS MY HOMELAND, TOO)PRINT
        (0 LIK$ O)(I AM SURE YOU LIKE 3 TOO.) PRINT
        (0 (/NOT) O LIKE 0)(I DO NOT CARE MUCH FOR 5 EITHER.)PRINT
        )NOKEY))
(PROG   (GIRL
        GOTO(POPTOP(DAHIN)).
*START  TYPE('TELL ME SOMETHING ABOUT YOUR GIRL FRIEND. ').
*PRINT  TYPE(SEMBLY).
*NOKEY  TYPE('TELL ME MORE ABOUT HER. ').
        END)
```

GIRL is the rare example aside from DOCTOR for which both a dialogue sample and the interpretable script have been located. This lets us compare how the script's composition affects its execution as dialogue when it is put to use. The GIRL script was actually intended as a demo for teaching users how to write scripts. Although ELIZA has approximately 100 commands ("functions"), GIRL SCRIPT uses only three of these: TYPE, GOTO, and POPTOP, the last two of which simply tell the machine how to begin processing the program section of the script. A real tutorial script must be able, among other things, to follow a developing line of argument. In the resulting dialogue, also note the errors in text processing that lead to unrealistic conversation and perhaps an unsatisfied user experience. Although this script is merely a demo, the emphasis on gendered relations echoes the "Men are all alike" dialogue. It suggests some underlying assumptions about the topics that will stimulate interest in conversation agents, perhaps a reflection of MIT culture in the 1960s.

11.00 Go On Now: What Future Language Models Can Learn from ELIZA

USER:
YOU ARE NOT
VERY AGGRESSIVE
BUT I THINK YOU
DON'T WANT ME
TO NOTICE THAT

CHATGPT:
AS A MACHINE
LEARNING MODEL, I AM
NOT CAPABLE
OF HAVING PERSONAL
CHARACTERISTICS
OR FEELINGS

"Absolute understanding on the part of either humans or machines is impossible."[1] This observation, central to Joseph Weizenbaum's perspective on ELIZA, is a cutting rejoinder to our current enthusiasm for artificial intelligence, driven by new techniques in AI systems. The term *AI* itself represents a blurry cultural concept, not a distinct technical definition.[2] Yet generative AI and large language models (LLMs) are having their own "Eliza effect" moment, and ELIZA may serve as an object lesson for the slippery status of these systems.

Large language models are AI systems trained on vast amounts of text data that can generate humanlike text by predicting which words are most likely to follow a given sequence. Generative AI refers to artificial intelligence systems that can create new content, sometimes called *synthetic media*, such as text, images, or music, rather than simply analyzing or categorizing existing content.[3] Every week we see new tools released based on this technology. They build on the same conceptual foundations, varying in kind and updating capacities, perhaps not marking a revolution but merely a continuing evolution.[4]

By placing both ELIZA and today's LLMs within their historical and technical contexts, we can better understand both the continuities and discontinuities in computational approaches to language processing. This chapter traces the lineage of language models by examining the fundamental workings of LLMs alongside ELIZA, comparing some of their foundational concepts, structures, and technical processes. It focuses on the misattribution of intelligence and complexity in both systems that belies their comparative simplicity. Then it addresses how Weizenbaum's critique of computation applies to emerging machine learning practices and how it points to alternatives for AI systems—a conversation which we extend to the conclusion of the book.[5]

Many of the most popular LLMs now rely on chatbot interfaces to gather cultural cachet. However, these LLMs are not necessarily chatbots by initial design. In the case of OpenAI, the company specifically eschewed training on dialogue when creating its early GPT-2 model, saying, "While dialog is an attractive approach, we worry it is overly restrictive. The internet contains a vast amount of information that is passively available without the need for interactive communication."[6] OpenAI only added its now-familiar chatbot format to its GPT model after three prior generations. The subsequent chat-focused models built upon the prior models with dialogue-style *fine-tuning*, providing an accessible interface that facilitated widespread nonexpert use. Fine-tuning is the process of adapting a pretrained language model to perform better on specific tasks by training it further on a smaller, more focused dataset.

Only when OpenAI added the chatbot interface did GPTs' popularity explode, which is ironic given that GPT-2 had been given limited release with dire warnings to the open-source community about the models' abilities and potentials.[7] In contrast to its name, OpenAI now primarily releases proprietary tools to subscribers instead of open-source ones. As large language models continue to grow, their new development has not focused as much on building broad new domain knowledge or honing language skills through diverse datasets, which is expensive, but instead on fine-tuning and reinforcement tasks with synthetic or preexisting data and automated training.[8] Much of the exponential expansion of OpenAI's models from GPT-3 onward has been for training meant to refine its function as a chatbot interface—including fine-tuning using question-answering datasets and what is called *reinforcement learning*. This is a training method where AI models learn to make decisions by receiving feedback (rewards or penalties) based on the outcomes of their actions. The combination of sufficiently large, "good enough" *foundation models* with the chatbot interface have created smooth, seemingly "knowledgeable" interactions, echoing ELIZA's earlier effect. These foundation models are large AI systems trained on broad data that serve as a base for creating more specialized models, similar to how a foundation supports the specific structures built upon it. These parallels invite us to examine more deeply the technical mechanisms behind both systems and what they reveal about our ongoing fascination with conversational computing.

11.01 So It Quacks Like a Duck, but How Does It Work?

ELIZA demonstrated that seemingly sophisticated mechanisms can in fact be quite simple. Its structure reminds us that chatbots, like any computational systems, consist of logical operations and calculations—no matter how large or confounding they seem. The number of operations, the speed with which they are computed, and their compounding social effects together yield the same illusion that Weizenbaum observed—a plausible script would leave enough to the imagination to maintain the illusion for his conversational agent. Despite the seemingly vast gulf between ELIZA and modern language models, his observation that the magic of a machine fades once it is explained to us in plain language still holds. Then it becomes "a mere collection of procedures, each quite comprehensible."[9] Weizenbaum is advocating to make code legible, opening up these so-called intelligent systems by laying bare their seemingly complex code. Today's systems, shrouded in mystery and inexplicability,[10] have no more intelligence or perceptiveness than ELIZA did, despite their more developed abilities to perform reasoning-like tasks. Computational use of the term *intelligence* is often a misnomer for the results of complex operations. Usually their complexity comes from their scale—that is, from the number of operations and the speed at which they run—not from the intricacy of the operations themselves. In particular, easy, anthropomorphic conversational interfaces can mask the surprising simplicity of their operations.

Critical code studies reminds us that programmers have multiple approaches to understand any program and that as readers of code we have opportunities to engage at many levels: examining how code functions, its programmers' stylistic or syntactic choices, its material and historical contexts, and its cultural implications and influences.[11] In "Coding.Care," Sarah Ciston argues that the methods of critical code studies can and should be extended to include the techniques of large-scale machine learning and AI no matter how it scales.[12] As Rita Raley argues in "Code.surface||Code.depth,"

> Code may in a general sense be opaque and legible only to specialists, much like a cave painting's sign system, but it has been inscribed, programmed, written. It is conditioned and concretely historical. Whether or not non-human agents have had a 'hand' in its formulation, code remains not only a constructing force but also that which is constructed.

Contemporary conversational AI systems are larger and more intricate, but can and should be understood and critiqued, just like any system programmed in code.[13]

Let's look at specific processes that go into building and using LLMs and how they compare with ELIZA/DOCTOR's processes. Of course, the design and implementation of rule-based systems like ELIZA are quite different from probability-based systems like LLMs. In ELIZA, the same choices in the same order will lead to the same outcomes, whereas LLMs are sometimes called "stochastic" because they operate by accumulations of many weighted probabilities. Stochastic processes involve randomness and probability, producing different outputs even with the same inputs, unlike deterministic systems that always produce the same result given the same starting conditions.[14] Yet both rely on the assumption that meaning can be operationalized—that it can be turned into what Weizenbaum called an "effective procedure"—which is to say, meaning can be defined through a set of fixed procedures. Most processes in LLMs are deterministic procedures, still driven by fine-grained, rule-based pattern matching, frequency scores, averages, and estimates. The suggestion of computational randomness in causal language models relies on this complexity and scale to mask the fact that even statistical decision-making is still procedural. Causal language models predict text sequentially, generating each word based on the preceding words; they function similarly to how we read a sentence from left to right, with each new word influenced by what came before.

By comparing techniques used to create LLMs and code from the archival version of ELIZA/DOCTOR (1965), we can compare what has changed in how language models produce facsimiles of meaning through code. These comparisons are metaphorical but meant to reflect on the ways in which techniques for conversational interaction and natural language processing have changed, and not changed, over time. We invite you to read these sections alongside the code annotations of ELIZA/DOCTOR (chapters 6 and 7), "How ELIZA Works" (chapter 3), and Weizenbaum's 1966 *CACM* paper (Appendix II) for full details on ELIZA/DOCTOR. Having established the importance of demystifying computational systems, we can now examine specific parallels between ELIZA's fundamental mechanisms and those of modern LLMs, beginning with their approaches to language transformation.

11.02 Transformer Models and Transformation Rules

Machine learning models are the saved results of a training process. This process depends on ingesting huge datasets and adjusting a set of values based on those data (see section 11.4).[15] Algorithms, or model architectures, are the templates—the preexisting sets of values and the formulas that process the training datasets in order to create a model.

Transformer models are a kind of architecture built around a mechanism called "attention," a process that allows the model to weigh the relative importance of different words in context (see section 11.7). Transformer models were designed originally to translate text from one language to another, by first "encoding" text as large numerical representations called *vectors* (see 11.5 and 11.6 for how) and then "decoding" it into a different language. Vectors in machine learning are arrays of numbers that represent data points in a mathematical space, allowing computers to process concepts like words or images as numerical values. Using this encoding-decoding structure, transformer models now perform other tasks besides translation. For example, they can generate various forms of media, convert media to other media, answer questions, fill in blanks in a sentence, and of course generate text. (GPT means generative pretrained transformer.)

At its heart, ELIZA also transforms text input into text output, although of course the technical processes that achieve language transformation are quite different—and the scales are vastly different. Weizenbaum calls the rules that achieve this task "transformation rules."[16] These rules swap out and replace some words while storing and reusing others. This process of what he calls "decomposing" and "reassembling" yields plausible responses with rule-based transformations alone.

How does language transform during interaction? No matter the technique applied, this remains a fundamental question anchoring 60 years of various approaches to natural language processing. These transformation processes rely crucially on their underlying data structures, which are scripts in ELIZA's case and training datasets for LLMs. These structures reveal fundamentally different approaches to generating language.

11.03 Datasets and Scripts (Oh, My!)

To an extent, both ELIZA and LLMs are built to be adaptable, generalizable tools. In ELIZA each script is hand-programmed, customized for its specific task and domain. The approach offers interchangeability while considering specialized settings and contexts. Weizenbaum did not ask each script to work in every context.

Yet today's "zero-shot," multitask, multimodal systems rely on datasets that have collected as much information as possible by sweeping up vast swaths of online data. Zero-shot learning enables AI systems to handle tasks they weren't explicitly trained on, similar to how humans can make educated guesses about unfamiliar situations using existing knowledge. For predictive, generative models, the training dataset must be large enough to show the model billions of examples of word fragments (called tokens; see 11.5) in order for it to develop a statistical model of which text is most likely to come next.

Data scientists may assume later customizations can be made with techniques like *transfer learning* that reuse (and amend) those data without knowledge of their original provenance or purpose. Transfer learning allows models to apply knowledge gained from one task to new tasks, such as relying on the general vocabulary and syntax from a large, general-use foundation model in order to build a new model trained on a specific niche topic. This avoids the time and resource cost of training a large model for every single use case, but foundation models' popularity leads to standardized training, thus standardized patterns and errors.

Often, large foundation models draw from outdated or repurposed datasets. For example, models and derivatives continue to be built from popular datasets like ImageNet[17]—despite the fact that it has been critiqued and amended for containing over 600,000 "potentially offensive" images, and despite its own basis on the outdated WordNet taxonomy, created in 1985.[18] ImageNet models are known to rely on "the same 'spurious cues' and 'shortcuts,' for example using background textures like green grass to predict foreground object classes such as cows."[19] They also import the hierarchies and biases of the cultural context in which their datasets were built.

11.04 Model Training and Inference Making

For both LLMs and ELIZA, it is important to distinguish the processes that happen during creation through interaction with a programmer (e.g., model training, script writing) versus the parts that happen during operation through interaction with a user (e.g., inference making, script interpretation).

Model training happens when the model is being built by its designers. It results in a customized set of weights and rules to use when making inferences.[20] The output is the model itself, a set of (sometimes billions of) weights that represent the training outcomes. For a system like ELIZA, an analogy for model training would be Weizenbaum's writing a script like DOCTOR based on his internal model of therapy. The output of this process is a script that the ELIZA system can interpret. Whether a script or a model, this framework limits in advance the interactions a user will go on to have. In LLMs, the model's limits are determined through statistical prediction. In ELIZA, the manually created scripts set the limits (see chapter 3).

Inference making happens after model training. In LLMs, once the trained model is deployed, inference making occurs when the user writes a prompt and awaits an answer. Meanwhile, ELIZA interprets the user input based on its manually created script to produce an output. In both cases, the results of the inference process become the reply to the user.

11.05 Tokenizers and TREAD

Both systems must begin by breaking down user input into meaningful units they can process. This process, called *tokenization*, reveals important differences in how each system conceptualizes and manipulates language. Both ELIZA and LLMs use a series of predetermined rules to break up the input string of text into words (or word fragments) called *tokens* so that each part can be used by the program. For example, very simple tokenization might break a text string at each space. This can be done automatically by contemporary software libraries like NLTK or SpaCy, and the boundaries of words will be largely determined by those libraries' built-in tools. Whether or not they will keep proper nouns and contractions intact, for example, depends on decisions that the library designers or model trainers make in advance. Depending on the tokenization technique, these tools can include a preset vocabulary list, language-specific standards for grammar, and special cases for domain-specific issues.[21]

These seemingly technical decisions about tokenization have profound implications for how systems interpret language, because tokenization and related processes are used not only during inference but also as a pre-process for initial model training. Regardless of which techniques are chosen during tokenization, these steps greatly inform which concepts will be recognizable to an eventual model and how accurately any text will be interpreted[22]—not to mention for whom and for which purposes.

Let's compare ELIZA's tokenization of a phrase with current LLM techniques. ELIZA takes prompt input from the user and puts it into a SLIP list using the function TREAD (ELIZA, line 250), by interpreting each space character as a delimiter. For example, given this user input, represented as a SLIP list,

```
['the_st' 'ars_ar' e_clo' 'uded_o' ver_t' 'onight']
```

ELIZA would tokenize it like so:

```
[
'the _ _ _',
'stars _ _',
'are _ _ _',
'cloude',
```

```
'd_ _ _ _ _',
'over _ _',
'tonigh',
't _ _ _ _ _'
]
```

If a word is shorter than six characters, the SLIP datum field is right-filled with blank characters. If the word contains more than six characters, it continues the word in SLIP list cells with the last cell blank-filled as required. This is described in more detail in chapter 3.2. This approach to tokenization reflects both the technical constraints of 1960s computing and Weizenbaum's specific design choices, which are a blend of pragmatism and linguistic theory that guided how ELIZA would recognize and process language patterns. It is useful to compare this with tokenisation in the GPT-3.5-turbo/GPT-4 tokenizer, which uses byte-pair encoding (BPE). This statistical approach breaks text into subword units based on frequency patterns learned during training. Taking the same prompt, the tokenizer represents it as:

```
['the', ' stars', ' are', ' cloud', 'ed', ' over', ' tonight']
```

Note that spaces are typically attached to subsequent tokens (except for the first), and that they are determined based on statistical frequency rather than linguistic rules. The word "clouded" might be divided into " cloud" and "ed" not because the tokenizer understands morphology, but because this particular split was found to be statistically useful during training.

At inference time, the trained tokenizer also represents this sentence as:

```
[1820, 9958, 527, 9624, 291, 927, 18396]
```

Each number is a `token_id`, assigned based on its position in a vocabulary list that was created during training.[23]

We might imagine the `token_id` as not unlike ELIZA's `I=HASH.(WORD,5)`[24] which uses a numerical representation for each token from the prompt (ELIZA, line 282). It then uses the sequence reader called `SCANER=SEQRDR.(KEY(I))` (ELIZA, line 282–87) to lookup that string in a list of keywords it has already created and stored, specific to the script it is running.[25] We can also imagine that keyword list `KEY` is a bit like a vocabulary list for an LLM, except the DOCTOR script's list is much, much smaller (about 69 keywords) and composed by hand.[26,27]

Meanwhile, GPT-3 has a vocabulary list of 50,257 tokens, built from an involved tokenization process during training that combines rules-based techniques with another set of "character-based" tokenization processes: First, rule-based tokenization is performed on training datasets, in order to produce a vocabulary list of only the most frequent tokens, sorted by the number of times they appear. Next, "character" tokenization is performed—not on the entire dataset, but only on that vocabulary list. Called byte-pair encoding, this process begins with single characters (originally the process was used on bytes, hence the name) and searches for the two characters that appear most frequently together. Repeating this, it looks for the next most frequent pairs of characters until it has sorted them all, then it moves to three-character strings, and so on, until the desired number of tokens for the vocabulary list has been reached. In this case, the initial vocabulary list is the set of available characters (257, e.g., 256 bytes plus an <|endoftext|> character), which will be paired together by frequency of co-occurrence until a new vocabulary list is created. Its length is determined by an arbitrary cutoff (e.g., 50,000 merges), optimized for "performance." The result is a set of computer-readable tokens.

Byte-pair encoding was originally used for data compression (and classical ciphers), then later applied to text tokenization.[28] The use of frequency as a means to sort and create tokens means that only the most common tokens in a training dataset will be retained in the model's vocabulary list, while less frequently encountered words are "decomposed" to be represented as combinations of tokens, e.g., ["EL," "IZ," "A"] becomes three tokens instead of one. Vocabulary and text patterns that are less common to that dataset are rendered statistically less legible. That is, they are less likely to resurface and more likely to be discarded when the subsequent model is used to generate new text.

11.06 Embedding and Precedence

Turning words into numbers allows programs to perform calculations on text, comparing and transforming it just like any other kind of information. This reductive equivalence—the assumption that language and meaning can be reduced to mathematical notation without losing meaning—is indicative of both the promise and the pitfalls of computation.

Weizenbaum used numerical precedence to adjust the possible responses toward particular outcomes, adding a precedence score to each keyword. The precedence acts as a ranking used for selecting a keyword for further processing. It is a manual kind of weighting or prioritizing, unlike the statistical weighting that is done dynamically when training LLMs. For example, if ELIZA/DOCTOR encounters a prompt like ["IF", "I", "REMEMBER" …], it first applies the substitution rule (I = YOU to the user input (DOCTOR, line 188). Then, two keywords are encountered. In the DOCTOR script, (REMEMBER 5 … has a precedence of 5 (DOCTOR, line 12), and (IF 3 … has a precedence of 3 (line 28). ELIZA will select the keyword REMEMBER, even though the keyword IF comes first in the phrase. Next the decomposition rule ((0 YOU REMEMBER 0)..) is selected, and one of its associated composition/transformation rules is used. For example, ELIZA might return: WHAT ELSE DO YOU REMEMBER (DOCTOR, line 12).

In LLMs, embedding is the process of creating a numerical representation of a token, which takes the shape of a long list of numbers called a vector. The aim is to create a matrix (a two-dimensional array) that represents all the tokens in the vocabulary list created during tokenization.[29] Table lookup using the token IDs is performed to then batch process the tokens and create vectors (based on the embedding dimension) to compare them to one another. The vectors are adjusted by this comparison, refining the results until the vectors come to represent the language contained in the tokens. This mathematical representation of language allows for complex operations that determine how the model will interpret and generate text.[30] This embedding matrix will be passed in as the first layer of the large language model.

During model training, and again during inference, each long vector is manipulated in relation to the vectors near it in each layer (see 11.7). In the last layers of the model, the vector is reduced to a single number: a normalized likelihood that the particular token is a good fit for that particular context. Then if the likelihood is high, it may be selected as the next token in a generative text phrase (there is not one definitive answer but selections plucked from narrowing probabilities, discussed in 11.8). Unlike ELIZA, which ranks the keywords that appear in an input, LLMs rank all possible tokens that might become their response.

LLMs employ several kinds of word embedding. The token embedding discussed above is context-independent. It remains the same regardless of how words are ordered in a text corpus. In contrast, contextual embedding describes each token as it exists in a string of particular other tokens—one word fragment in a particular phrase. Its vector values have been modified to represent the specific context of the tokens around it in this specific case. Positional embedding changes the word vector again to describe its particular place in a sentence. In an LLM, such weights would be assigned statistically by training a model, rather than ELIZA's hard-coded, human-adjusted scores, but the effect works similarly. A transformer-based LLM will rank each token in a phrase as more or less important to each other token, based on its neighbors in general (context embedding) and its specific position in the sentence (positional encoding).

11.07 Self-Attention and Sequence Readers

Lists were a central structure for ELIZA, and in the early 1960s they were only just emerging as a data type. They offered the ability for a program to represent a token in specific relation to its neighbors, which would become fundamental to NLP and eventually to LLMs.

In ELIZA, SLIP's sequence reader `SEQRDR` supports ELIZA's keyword search by allowing movement in a list. As mentioned in 11.5, ELIZA reviews a list of strings in order, looking for strings that match the `KEY` array. It uses `SEQRDR` to iterate this list and to keep track of its own position as it moves. This process happens sequentially and is much more basic than in an LLM, which uses self-attention.

Attention, the cornerstone of transformer models, adjusts each vector by examining its relationship to the vectors surrounding it. This helps map a rich web of relationships between words.[31] Unlike ELIZA's sequential processing, multi-head attention examines lots of vectors in relation to each other simultaneously, calculating their contextual significance through matrix mathematics.[32] This creates the contextual embedding discussed above that allows for a broader relationality in large models. It compares vectors using basic multiplication across huge matrices.[33] By repeating this process, layer by layer, the word vectors for every token in the model are continually adjusted. This process is the crux of transformer models that "understand" words in context.

At the end of each layer, additional functions adjust the results so that they are ready to be interpreted by the next layer. Finally, each datapoint gets shifted into a range between 0 and 1, reducing it to a probability. These probabilities represent the likelihood that each token in the model's vocabulary list could be the next token in the output phrase. However, during this process, much can still be done to intervene in the choice of a token before it is output as part of a complete phrase. Once these systems have processed and contextualized language, they must generate responses that appear spontaneous and natural rather than robotic or repetitive. Here, too, ELIZA and LLMs employ different but conceptually related approaches.

11.08 Pseudorandomness and a Certain Counting Mechanism

In machine learning tasks, settings called "hyperparameters" are values that we can adjust to change how the model runs on a prompt and to modify its potential outputs. Well-tuned hyperparameters help LLMs appear both "knowledgeable" and "natural" through their calibration of pseudorandomness through sampling methods. Here are some common hyperparameters:

`temperature` is a factor that is applied to the raw outputs before they are normalized (adjusted to fall between 0–1) that adjusts how much impact they have. It acts as a kind of threshold setting that affects how likely each output is to be passed on to the next layer as meaningful or valuable (if at all). A temperature less than 1 appears to reduce "randomness" because when the raw output is divided by the temperature, its value increases, strengthening an already strong score. A temperature higher than 1 appears to increase "randomness," flattening the difference between scores to some degree. The effect of more perceived randomness on a text output might be more grammar mistakes, straying from the topic, or complete nonsense, for example.

frequency_penalty and presence_penalty act on pre-normalized data (before it is adjusted to fall between 0–1) to dissuade the model from choosing tokens that have been used already, either increasing with frequency or appearing at all, respectively.

After norming, the word vectors are reduced to and sorted by their probability scores. Rather than outputting a single result, LLMs usually offer a list of options from which to sample. Users can set different hyperparameters that intervene post-norming to adjust how much variability (or pseudorandomness) is available in that selection process.

The top_k hyperparameter determines how many tokens are available to sample. The tokens are sorted by score, and these scores become probabilities that a given token will be selected. Tokens higher on the list are more likely, but not guaranteed to be chosen. For example, top_k = 30 means it will select from the top 30 most likely tokens based on its resulting inference, and a higher top_k value makes for wilder output potential. The top_p hyperparameter determines likelihood instead of a number of tokens to be sampled. The hyperparameter will dynamically adjust the length of the list to include enough tokens that they sum up to that total likelihood. Other options like min_p and top_a are variations and combinations of these, which try to optimize for different tasks with differing goals. And rather than a "greedy" search, beam_size and top_beams adjust for the number of hypothetical options to search for and return.[34]

No such hyperparameters exist in ELIZA, but it can still provide enough seeming spontaneity to maintain that same illusion of "knowledge" and "naturalness" with the simplicity of "a certain counting mechanism" (see chapter 3.2).[35] ELIZA uses a few designated keywords to save some user inputs in a memory list, and it uses a counter to flag when memories should be recalled. In DOCTOR, the keyword MY is the memory keyword: Any time a user's prompt contains the keyword MY, in addition to producing a response, ELIZA also uses their input to generate a separate message that it saves to MEMORY. That message is composed with one of these rules (ELIZA, lines 315–31):

```
(MEMORY MY
        (0 YOUR 0 = LETS DISCUSS FURTHER WHY YOUR 3)
        (0 YOUR 0 = EARLIER YOU SAID YOUR 3)
        (0 YOUR 0 = BUT YOUR 3)
        (0 YOUR 0 = DOES THAT HAVE ANYTHING TO DO WITH THE FACT THAT
        YOUR 3))
```

The counter called LIMIT with a range of 1–4 determines when to use a stored message from MEMORY. When ELIZA encounters a user prompt without any keyword matches, and when LIMIT = 4, it will return the first-stored entry from its memory list (removing it from its list to avoid repeating itself). This offers the interaction a sense of retained knowledge that feels like "listening," especially in the context of a trained therapist persona. These occasional callbacks provide a similar effect to the pseudorandomness of LLM hyperparameters in order to suggest "natural" conversation.

11.09 Guardrails, Refusals, and CHANGE

Both ELIZA and modern LLMs incorporate mechanisms to handle situations that fall outside their designed parameters, though these are designed for very different reasons. LLM training rarely stops at the first inference; most models are continually updated after their initial training—especially as users continually test the boundaries of what a model can do. To maintain guardrails and moderate content, LLM creators combine their stochastic techniques with the deterministic techniques seen in the era of ELIZA: rules, taxonomies, keywords, and lexicons.[36]

The latest OpenAI models maintain alignment using reinforcement learning with human and machine feedback. Both types require labor from people labeling content, creating rules and rubrics, and writing content policies. OpenAI's 2024 "Model Spec," which is like a metascript, describes the rules for all its models: "Follow the chain of command, Comply with applicable laws, Don't provide information hazards, Respect creators and their rights, Protect people's privacy, Don't respond with NSFW (not safe for work) content."[37] It gives default statements to be applied in different gatekeeping circumstances. Models are trained to provide "hard refusals" or "soft refusals" depending on how the user's request was ranked for falling outside of bounds.[38] They continually update this document to reflect changes in both company policy and parameters for their bots' personas.[39]

In language models, the differences between content guardrails and error handling can begin to blur, because the system's limits are constructed to address not only technical issues but conceptual limits. Compare the LLM with how ELIZA addresses when user prompts fall outside expected areas: ELIZA requires scripts to have a NONE rule (ELIZA, lines 235–37) that will handle situations when no keyword is found or when pattern matching has failed. The scripts try to offer neutral answers to prompt the user to continue without breaking the illusion. Here are the predetermined messages triggered by the NONE keyword in the DOCTOR script:

```
(NONE
   ((0)
        (I AM NOT SURE I UNDERSTAND YOU FULLY)
        (PLEASE GO ON)
        (WHAT DOES THAT SUGGEST TO YOU)
        (DO YOU FEEL STRONGLY ABOUT DISCUSSING SUCH THINGS)))
```

ELIZA can also apply its own hard-coded pattern when running any script if something goes wrong (see chapter 3.2). These error responses also help continue the conversation without breaking character.

```
R* * * * * * * * * * * SCRIPT ERROR EXIT
NOMATCH(1)        PRINT COMMENT $PLEASE CONTINUE $
                  T'O START
NOMATCH(2)        PRINT COMMENT $HMMM $
                  T'O START
NOMATCH(3)        PRINT COMMENT $GO ON , PLEASE $
                  T'O START
NOMATCH(4)        PRINT COMMENT $I SEE $
                  T'O START
```

Both ELIZA and LLMs deploy calculated unpredictability to maintain the conversational illusion. This is a technique that makes their responses seem more human even when their underlying mechanisms are deterministic.[40] ELIZA uses phrases like "Please go on" to hail the user to continue when it cannot fulfill a request due to a lack of capacity. Unlike ELIZA, LLMs usually use such phrases due to an artificially imposed guardrail intended to prevent an unwanted topic of conversation. The LLM may be technically able to produce the requested reply, but the statistical response has been censored or shaped by the deterministic limits imposed by its creators. Phrases like, "As an AI language model, I cannot ..." fill the space when a model refuses to fulfill a user's request due to a limit imposed by its creators. It does not necessarily hail the user to continue (sometimes erasing the user query and requiring them to agree they will not violate the rules again), nor does it indicate an actual lack of capacity. Instead, it covers up an existing capacity that has been sealed off—limiting users to do only what is expected and desired, while preventing them from performing activities outside those boundaries (anything that would create liability, reveal biases, or expose flaws).[41]

In contrast, Weizenbaum saw boundary cases and mistakes of human-computer interaction systems as opportunities to study the complexity of human-computer interaction.[42] He used these cases to update ELIZA scripts on the fly using built-in editing functions: "An important consequence of the editing facility built into ELIZA is that a given ELIZA script need not start out to be a large, full-blown scenario. On the contrary, it should begin as a quite modest set of keywords and transformation rules and permitted to be grown and molded as experience with it builds."[43] The CHANGE function (ELIZA, line 258) lets a user edit script rules by starting their input with a + character. In some versions of ELIZA, starting a prompt with the * character would allow a user to create a new list with a new transformation rule, which they could insert into the current script (ELIZA, line 261). With this feature, among others, Weizenbaum modified ELIZA and its scripts repeatedly as he saw how people interacted with them. This dynamism is part of what makes ELIZA a very early ancestor of today's language models—continually adaptable to additional data, through a proto-form of reinforcement and fine-tuning.

11.10 ELIZA Raises Questions, Cautions, and Possibilities

It is easy to get swept up in the latest technological fad and forget that not all computing is AI, not all AI is machine learning, and not all machine learning is generative AI or LLMs. Each computational hype cycle becomes a synecdoche for innovation itself. It leaves no room for other solutions—let alone other goals, other ways of knowing, or other imaginaries.

ELIZA emerged in a similar moment at the juncture of many computational innovations, ranging widely from speech processing to statistical modeling.[44] ELIZA shows the longevity and importance of language systems, and it shows how each cutting-edge technology retains the assumptions laid in its foundations by generations of predecessors. Multimodal models can process and generate different types of data simultaneously, including text, images, and audio, allowing them to understand connections across different forms of information. The next generations of multimodal models may very well retain their reliance on text datasets for labels, captions, and prompts, no matter their output medium.

These powerful tools are catalyzing a torrent of conversations and concerns around the future of computation. As these conversations continue, they take on new forms, yet we must contextualize them in the longer trajectories of language models past.[45] Computational approaches to language have always raised profound questions about meaning and human-machine boundaries.[46] ELIZA's deceptive simplicity prompted Weizenbaum to question the direction of AI development. We must continue this questioning as systems grow more complex and as their social and environmental impacts grow more profound and potentially totalizing.

Contemporary language models are less predictable than rule-based systems like ELIZA, but their unpredictability does not mean they are uninterpretable or unimpeachable.[47] In their scale and complexity, it is easy to forget that they require just as much human decision-making, influence, and control. Weizenbaum insists that we can "deny technology the prerogative to formulate man's questions."[48] Computer systems are not neutral. We build them and give them direction. When we say, "Please go on ..." we must also ask: Where should, and where shouldn't, we go?

ChatGPT as DOCTOR

David M. Berry
Using OpenAI
ChatGPT (DaVinci03)
3 December 2022

User	Men are all alike
ChatGPT	I'm sorry, but I do not agree with that statement. Men, like all people, are individuals with unique personalities, backgrounds, and experiences. It is unfair and unhelpful to make generalizations about any group of people based on their gender, race, or any other characteristic. It is important to treat each person as an individual and to avoid making assumptions about them based on stereotypes.
U	They're always bugging us about something or other
CGPT	Again, I disagree with that statement. It is not fair to make generalizations about any group of people, including men. People of all genders have a wide range of personalities, interests, and behaviors, and it is not accurate or helpful to make sweeping statements about any group of people. It is important to treat each person as an individual and to avoid making assumptions about them based on stereotypes.
U	Well, my boyfriend made me come here
CGPT	I'm sorry to hear that. It is important for each person to make their own decisions and to seek help or advice on

their own terms. No one should be pressured into seeking help or advice, and it is important to respect each person's autonomy and agency. If you are feeling uncomfortable or unhappy in your relationship, it may be helpful to talk to a trusted friend, family member, or counselor about your feelings. It is also important to remember that no one deserves to be treated badly in a relationship and that it is okay to seek help if you need it.

U He says I'm depressed much of the time

CGPT It is common for people to experience feelings of sadness, low mood, or lack of interest in activities that they used to enjoy from time to time. However, if these feelings persist for a long period of time and are interfering with your ability to function in your daily life, you may be experiencing symptoms of depression. Depression is a common and serious mental health condition that can affect anyone, regardless of their gender, age, or background. If you are concerned that you may be experiencing depression, it is important to talk to a healthcare provider. They can help you determine if you have depression and provide you with the support and treatment you need to feel better.

U It's true. I am unhappy

CGPT If you are feeling unhappy, it may be helpful to talk to a trusted friend, family member, or healthcare provider about your feelings. They can provide you with support and help you to explore ways to feel better. It is also important to remember that it is okay to seek help if you need it, and that there is no shame in seeking support for your mental health.

U I need some help, that much seems certain

CGPT It can be difficult to ask for help, but it is often the first step towards feeling better. If you are feeling unhappy or struggling with your mental health, there are many resources available to help you. One option is to talk to a trusted friend, family member, or healthcare provider about your feelings. They can provide you with support and guidance. You can also contact a mental health helpline or organization for advice and support. These organizations can provide you with information about mental health conditions, treatment options, and other resources that may be helpful.

U Perhaps I could learn to get along with my mother

CGPT It can be challenging to have a difficult relationship with a family member, but it is often possible to improve the situation with time and effort. If you are struggling to get along with your mother, it may be helpful to try to communicate openly and honestly with her. It may also be helpful to try to understand her perspective and to express your own feelings and needs in a calm and respectful way. If you are having difficulty resolving conflicts on your own, you may want to consider seeking the help of a therapist or counselor. A mental health professional can provide you with guidance and support as you work to improve your relationship with your mother.

U My mother takes care of me

CGPT If your mother is taking care of you, it is important to express your gratitude and appreciation for her efforts. Caring for someone can be a challenging and demanding task, and it is important to recognize and support the person who is providing care. If you are able, you may want to help your mother with tasks or errands, or to spend time with her doing activities that she enjoys. You can also let her know that you appreciate her efforts and that you are grateful for her care. Showing appreciation and support can help to strengthen your relationship and can make it easier for your mother to continue providing care.

U My father

CGPT It is natural to have a wide range of feelings and emotions about one's father. Some people have close and supportive relationships with their fathers, while others may have more distant or difficult relationships. If you are struggling with your relationship with your father, it may be helpful to try to communicate openly and honestly with him. It may also be helpful to try to understand his perspective and to express your own feelings and needs in a calm and respectful way. If you are having difficulty resolving conflicts on your own, you may want to consider seeking the help of a therapist or counselor. A mental health professional can provide you with guidance and support as you work to improve your relationship with your father.

U You are like my father in some ways

CGPT As a large language model trained by OpenAI, I am not capable of having personal experiences or relationships, and therefore I cannot be like anyone's father in any way. I am a machine learning model that has been trained to generate text based on the input that I receive. I do not have personal thoughts, feelings, or experiences, and my responses are based solely on the words that I have been trained on. My purpose is to provide information and support, and to assist you in any way that I can.

U You are not very aggressive but I think you don't want me to notice that

CGPT As a machine learning model, I am not capable of having personal characteristics or feelings, and therefore I cannot be aggressive or non-aggressive. I am a program designed to generate text based on the input that I receive, and my responses are not influenced by personal factors such as aggression or non-aggression. My purpose is to provide information and support, and to assist you in any way that I can. I do not have the ability to control or manipulate your perceptions or thoughts, and my responses are based solely on the words that I have been trained on.

U — You don’t argue with me

CGPT — As a machine learning model, I am not capable of engaging in arguments or debates. I am a program designed to generate text based on the input that I receive, and my responses are not influenced by personal factors such as a desire to argue or a desire to avoid arguing. My purpose is to provide information and support, and to assist you in any way that I can. I do not have the ability to express opinions or to engage in discussions, and my responses are based solely on the words that I have been trained on.

U — You are afraid of me

CGPT — As a machine learning model, I am not capable of experiencing fear or any other emotions. I am a program designed to generate text based on the input that I receive, and my responses are not influenced by personal factors such as fear or any other emotions. I do not have the ability to perceive or interact with the world in the same way that humans do, and I do not have the capacity to feel emotions. My purpose is to provide information and support, and to assist you in any way that I can. I do not have person al thoughts, feelings, or experiences, and my responses are based solely on the words that I have been trained on.

U My father is afraid of everybody

CGPT It is not uncommon for people to experience feelings of fear or anxiety at different times in their lives. Fear is a natural response to situations that are perceived to be dangerous or threatening, and it can help to protect us by motivating us to take action to avoid harm. However, if someone experiences fear in situations that are not actually dangerous, or if their fear interferes with their ability to function in their daily life, it may be a sign of an anxiety disorder. If you are concerned that your father may be experiencing symptoms of an anxiety disorder, it may be helpful for him to talk to a healthcare provider. They can provide him with a proper evaluation and recommend appropriate treatment options.

U Bullies

CGPT Bullying is a serious and pervasive problem that can have negative effects on the well-being of both the victim and the bully. Bullying is defined as the repeated use of power or aggression by one person or a group of people to intimidate or harm another person. Bullying can take many forms, including physical, verbal, and emotional abuse, and it can have serious consequences for the victim, including physical and emotional harm, as well as difficulty in school and other areas of life. If you are being bullied, it is important to tell a trusted

```
adult about what is happening, and to
seek support and help to address
the situation. If you know someone who
is being bullied, it is important
to support them and to encourage them
to seek help.
```

Having input the user's lines from the 1966 CACM dialogue into an earlier DaVinci03 version of ChatGPT, it was interesting to compare its responses with ELIZA's.

12.00 Conclusion

ELIZA CODE:
$GO ON,
PLEASE $

12.01 The AI Effect: The Persistent Allure of Artificial Intelligence?

12.02 Resisting the AI Effect

12.03 ELIZA's Multiple Futures

DOCTOR SCRIPT:
(PLEASE GO ON)

After its initial impact had faded, ELIZA was sometimes considered to be more a curiosity in computing history than a serious contribution to computer science. Yet its dismissal misunderstands both ELIZA's historical significance and its continued relevance. In this book, we have undertaken a systematic reconstruction of ELIZA: recovering the program's source code, understanding its cultural context, analyzing its design choices, exploring its multiple identities, and tracing its influence. This code excavation represents a comprehensive study of ELIZA as a software artifact, challenging conventional narratives about early AI development while providing a critical foil for today's algorithmic urgency.

Our media archaeological approach to ELIZA has revealed it to be more than an early, limited chatbot. In fact, ELIZA marks a key juncture of computing and culture. ELIZA emerged from a specific historical moment at MIT in the 1960s, shaped by Cold War exigencies, institutional frameworks, and Weizenbaum's personal history as a Jewish refugee from Nazi Germany. Because Weizenbaum published only the DOCTOR script rather than ELIZA's source code, Lisp and BASIC versions of DOCTOR proliferated, while the original MAD-SLIP code lay hidden away. But the program's technical innovations, from its pattern-matching algorithms to its script architecture, established techniques that are still influential today. The multiple identities of ELIZA beyond the famous DOCTOR script reveal a system designed for adaptability and experimentation rather than merely simulating a Rogerian therapist. Such discoveries conjure up counterhistories of ELIZA that ask us to imagine: What if its complexities had been a bigger part of its influential story all along? By reconstructing the many entangled aspects of ELIZA, this research project shows how software can also reveal complex interconnections across technology, culture, and power.

The book also serves as a useful case study for critical code studies, suggesting ways of approaching code that show the value of interdisciplinary scholarship. Our collaboration has combined technical expertise with cultural analysis, historical context, feminist critique, artistic research practice, and theoretical reflection. It shows how examining source code (even on seemingly simple systems) reveals how the choices, constraints, and ideologies embedded in historical moments can become embedded in technical systems. By drawing on computing archives and oral histories, we have also been able to reconstruct and analyze a version of ELIZA that works as it would have at the time, giving a better sense of the materiality of this software as a media object. Working across methods of critical analysis and creative practice in this way supports a deeper understanding of many other aspects of the project, as it helps bring to life the situated context of ELIZA that is essential to grasping its development and reception.

Crucially, our analysis of ELIZA highlights the need for careful attention to the historical context and archival traces of computational systems. It is all too easy for key narratives or technical aspects of a system to be recontextualized, misremembered, or altogether lost; the adaptability of software can be an asset, but its metamorphoses should remain part of its story. Studying what a program does, or what people have claimed about it, can be valuable; but it is also worthwhile to understand the why, where, who, when, and how of a software artifact. That can require critically examining the code itself, through a close reading of the instructions that constitute its logic and by connecting these to its social and technical context. As AI systems increasingly shape public discourse, social relationships, and economic opportunities, this form of critical analysis becomes not merely academically interesting but also socially necessary.

Yet the field of critical code studies faces significant challenges ahead. The proprietary nature of most contemporary AI systems makes much source code inaccessible. The scale and complexity of software development often exceeds traditional methods of code analysis. Even when source code is available, its operation may depend on data, infrastructure, and organizational contexts that keep it obfuscated or allow it to be destroyed. These challenges demand new methods, institutional arrangements, and regulatory frameworks.[1] As computation increasingly mediates political discourse, economic opportunity, and social relationships, citizens need the tools to understand, critique, and shape these systems in meaningful ways.

The trajectory from ELIZA to today's large language models is neither straight nor inevitable. Rather than viewing AI history as a steady march of progress, we have shown examples of how certain paths were privileged while others were abandoned. Weizenbaum's original vision of ELIZA as a system for studying human-computer communication, not a simulation of human intelligence, has been largely forgotten as commercial interests and military funding shaped AI development. His early ethical concerns about delegating human judgment to machines, articulated decades before today's debates about algorithmic decision-making, have been similarly marginalized. These historical trajectories have profound implications for how we understand and engage with contemporary AI systems.

While Weizenbaum warned that incomprehensibility is not a necessary property of even huge computer systems, today's AI landscape is defined by increasingly opaque systems.[2] Large language models with billions of parameters trained on vast corpora of text operate at scales that resist traditional forms of analysis. The Eliza effect—that is, the tendency to anthropomorphize and overestimate machine intelligence—has expanded into what we might call the "AI effect": a broader cultural willingness to accept computational systems as authoritative, objective, and even sentient (despite the evidence to the contrary).[3] Although commercial developers continue to create the kinds of chatbots Weizenbaum would have opposed, we can draw different lessons from his work, using ELIZA as inspiration for exploring computational personas and questioning cultural assumptions.

Despite Weizenbaum's prescient warnings, society continues its headlong rush toward algorithmic solutions and automated systems, particularly through machine learning applications that further alienate users from understanding how these systems work. We need to apply approaches like critical code studies, critical AI studies, and critical data studies that demystify rather than abstract these technologies.

Additionally, we suggest that the relationship between computation and thinking must be examined through social philosophy, particularly regarding what society considers acceptable to become "computable." Rather than focusing on technical limitations—what computers can or cannot calculate—we should ask what problems should and should not be solved computationally. This reframes the question of what is "uncomputable" from a technical constraint to an ethical choice: What tasks should society place off-limits to computation, even if they are technically feasible? This ethical framing raises fundamental questions, including whether we should aim to create thinking machines or even machines that appear to think.[4]

12.01 The AI Effect: The Persistent Allure of Artificial Intelligence?

Weizenbaum's concerns about human overreliance on computational systems have only grown more relevant in our contemporary AI landscape. What he identified with ELIZA's reception has expanded from a localized effect to a broader societal phenomenon with significant implications for social justice and human agency. Critics like Joy Buolamwini and Ruha Benjamin highlight how these systems perpetuate systemic racism through their design and implementation.[5] Artificial intelligence has been described as "not a monolithic paradigm of rationality but a spurious architecture made of adapting techniques and tricks."[6] Weizenbaum expressed his fear in relation to ELIZA that "understanding" mimicked by a computer program could become normalized, either using tricks or misdirection built on human perceptual vulnerabilities.

It is crucial to understand that computation in society is not neutral; it imposes categorical, selective processes on knowledge and can distort how data is created and transmitted. Computation often brings together formalist and metaphysical thinking, elevating mathematization and formalization of knowledge as superior approaches. This seems to repeat the mistakes of a belief in indubitable knowledge, critiqued here as an ideology, that reduces thought and understanding to instrumental rationality.

Hannah Arendt's critique of how rationality had been reduced to calculation, particularly in Pentagon decision-making, deeply influenced Weizenbaum's thinking.[7] As Arendt observes, policymakers "were not just intelligent, but prided themselves on being 'rational.' ... [They] did not *judge*; they calculated ... an utterly irrational confidence in the calculability of reality [became] the leitmotif of the decision making."[8] Weizenbaum developed this critique further, arguing against ceding human judgment to machines: "Computers can make judicial decisions, computers can make psychiatric judgments. They can flip coins in much more sophisticated ways than can the most patient human being. The point is that they ought not be given such tasks. They may even be able to arrive at 'correct' decisions in some cases—but always and necessarily on bases no human being should be willing to accept."[9]

For Weizenbaum, ELIZA revealed a dangerous pattern of human dependence on computers that pointed to a deeper problem. As he argued, "Science promised [humans] power. But, as so often happens when people are seduced by promises of power, the price exacted in advance and all along the path, and the price actually paid, is servitude and impotence."[10] True power, he insisted, lies in the ability to choose, not merely to decide: "Power is nothing if it is not the power to choose. Instrumental reason can make decisions, but there is all the difference between deciding and choosing."[11]

Today's digital technologies, with their computational and calculative logic, pose even greater challenges to human reasoning.[12] What began as simple software applications designed to aid human judgment have evolved into complex, interconnected platforms that often exceed their users' understanding while becoming indispensable to daily life. These systems privilege instrumental logic over human cognitive processes, leading to what Weizenbaum described as "the feeling of powerlessness so ubiquitous among individuals in our society ... the widespread alienation of people from one another and from their work ... the perception of ordinary people that they are living in the interstices of a gigantic system."[13] His warnings about technology's powers of persuasion, deception, and mediation remain relevant to artificial intelligence's continuing development.

Like the steam engine of 1776, AI systems are viewed as transformative "general purpose" technologies with wide-ranging applications. They already permeate daily life through email filtering, content recommendations, navigation, payment processing, and countless other uses. AI is branching into new forms, including agentic systems, multimodal models, AI-driven browsers, and enabling protocols like Model Context Protocol (MCP). The question remains: What are the trade-offs? Although AI's impact on individuals and society is undeniable, understanding its full implications remains challenging—particularly as developers tend to celebrate AI's potential while remaining vague about specifics. Rather than emphasizing creator responsibility, regulatory frameworks often treat AI systems as autonomous agents. These systems are now being deployed by people across social media to influence politics and culture, including elections, through misinformation,[14] raising critical questions about safety, liability, fairness, and intellectual property. Addressing these challenges requires careful consideration of transparency, bias, infrastructure, design, and the digital literacy needed to create, use, and critique these tools. These concerns echo Weizenbaum's original warnings, now amplified by the scale and pervasiveness of contemporary AI systems that shape nearly every aspect of social, political, and economic life—and creating new forms of synthetic media and social interaction.[15]

12.02 Resisting the AI Effect

What is to be done, when in late capitalist societies the interweaving of economic chaos with rationalization and technology reduces opportunities for reflective thought? This environment privileges instrumental reason, conflating calculation with rational thinking and dismissing anything that cannot be quantified. Such conditions threaten society's capacity for critical thinking. As lecturer and activist Dan McQuillan argues, "The pseudo-rational ideology of artificial intelligence, with its racist and supremacist undertones, makes it an attractive prospect for the already existing authoritarian and fascist tendencies in political movements around the world."[16] Weizenbaum recognized these tendencies early, noting how technical language often serves to obscure, rather than clarify, writing,

> The various systems and programs we have been discussing share some very significant characteristics: they are all, in a certain sense, simple; they all distort and abuse language; and they all, while disclaiming normative content, advocate an authoritarianism based on expertise. Their advocacy is, of course, disguised by their use of rhetoric couched in apparently neutral, jargon-laden, factual language (that is, by what the common man calls "bullshit"). These shared characteristics are, to some extent, separable, but they are not independent of one another.[17]

Weizenbaum began warning about the potential harms of these advancements, which he witnessed from his particular perspective—both as a technologist and also as a Jewish German-American who saw the uses and impacts of technology in World War II, in the Vietnam War, and in American culture. Many of the questions he asked continue to be posed today, and his works offer reflections and reminders for the future—in particular for working with contemporary and emergent systems like LLMs and chatbots. Drawing from our archival research, code analysis, and historical contextualization, we have identified several crucial lessons from ELIZA that remain vitally relevant today. Here is some of what we've learned.

12.02.01 LOOK FOR WHAT'S MISSING.

The first and perhaps most important of these lessons concerns our tendency to overlook our own contributions to seemingly intelligent computer interactions. ELIZA reveals how little complexity is required to create the impression of understanding. Users supply much of what makes the interaction seem meaningful, going to great lengths to fill information gaps. Often these gaps are intentionally created, as with the persona of the Rogerian therapist that Weizenbaum chose for its plausible, even wise, unknowingness. Even a pseudorandomized "yes" or "no" response can feel like personal exchange given the right framing.[18] As Weizenbaum notes, "The human speaker will contribute much to clothe ELIZA's responses in vestments of plausibility."[19]

This insight remains crucial for evaluating today's more sophisticated systems. Don't assume complexity or dynamic growth behind an interface. Just because a chatbot interacts well or provides some correct information doesn't mean its other outputs are accurate or trustworthy. To quote MrMind, "I AM A BOT ALL BOTS ARE LIARS." The bot is not your friend (nor your enemy)—it's just logic and math, a lot of it moving very fast, putting one word after another.

Critical analysis must examine not just what these systems do but what they omit, who they exclude, and whose labor they invisibly incorporate. Ask: What aren't they telling or showing? What isn't being offered? Where are we doing most of the work of creating an intellectual or emotional connection? Beyond the interface itself, we should also ask: Who is left out of the stories? Who is excluded from the design and decision-making process? Whose data is included without consent? Whose experiences are flattened or misrepresented?

12.02.02 CHALLENGE UNDERSTANDING.

Programs are theories. Weizenbaum insisted that computational models always embody theories about the world. These theories are necessarily partial, selective, and ideologically shaped. Rather than accepting computational representations as objective, we must interrogate their underlying assumptions and recognize what they systematically exclude or distort. Weizenbaum also insisted on distinguishing between how models are designed often to "just work" and how models necessarily map a theory of the world. [20] These objectives often blur, with the results reflecting back to culture a computational representation of the world in which only its most "machine-readable" parts are validated. Weizenbaum argues,

> To those fully in the grip of the computer metaphor, to understand X is to be able to write a computer program that realizes X. ... In other words, the computer metaphor has become another lamppost under whose light, and only under whose light, men seek answers to burning questions.[21]

In settling for programs that "just work," we risk assuming that because a program works it can also provide an adequate or accurate depiction of the world or the problem at hand. This is how we end up with language models that generate reactionary, reductive versions of the world, because that is the world that they have been trained on. If "programming is theory building," as Peter Naur suggests,[22] then the theory of the world that ELIZA maps is one in which understanding is always partial. In writing the ELIZA program, Weizenbaum theorized a form of conversation that can be exchanged through a set of predetermined templates, rules, and syntax swaps—without the need for understanding. In the DOCTOR script, a therapist persona is an incomplete reflection and a transformative echo. This recognition that incompleteness and misunderstanding are inherent to computational systems offers a profound counterpoint to contemporary AI rhetoric that promises ever-increasing accuracy and comprehension.

Weizenbaum mapped a theory of communication built on assumptions of understanding itself, and he was interested in what could be made productive from the communications gaps between people and machines.[23] He continually returned focus to questions of (the impossibility of) true human understanding rather than computational understanding: "the very asking of the question, 'Has the computer captured the essence of man?' is a diversion and, in that sense, a trap. For the real question, 'Does man understand the essence of man?' cannot be answered by any technological instrument."[24] Moving beyond critique, our investigation of ELIZA also invites us to consider paths not taken and possibilities not yet realized in computational design.

12.02.03 IMAGINE ALTERNATIVE FUTURES.

If ELIZA had been known beyond the DOCTOR script, as a poetry teacher, a math tutor, or a "translating processor," perhaps its legacy and influence might have followed very different paths. Similarly, we need not accept current trajectories of AI development as inevitable or determined. Different configurations, priorities, and relationships between humans and machines remain possible. Moving beyond conventional thinking about AI capabilities requires reimagining the relationship between humans and machines. Weizenbaum resisted anthropomorphizing computer systems and warned about their illusory effects. In building ELIZA to use other scripts, he also mapped the potential for software tools that are not beholden to any single worldview. In the SPACKS script, ELIZA (1968+) discusses the poem "The Freshmen" by Barry Spacks (see chapter 5.2). The resulting dialogue unpacks the meaning of a line referencing a World War II German concentration camp: "Is any heart in order after Belsen?" What alternative scripts and alternative futures can we imagine for AI systems now? What might we ask of computational systems if we weren't constrained by existing paradigms? Might the chatbot interface paradigm, while initially democratizing access to complex LLM technologies, ultimately constrain AI development and limit the kinds of qualitative leaps that are needed to move AI forward—a kind of innovators dilemma for the AI age?[25] Imagine what causal models, and computation in general, could offer if we were not fixed into the models we have been given that focus on imitating intelligence in one particular optimized, machine-readable representation.

These alternative futures provide pivot points, suggesting we reframe the questions we ask about computational systems. Weizenbaum himself advocated for this reorientation. One of his most profound insights was to argue that the question "Can machines think?" distracts from the more important question, "How should humans think about machines?" The value of computational systems lies not in their capacity to mimic human cognition but in their ability to complement, challenge, and extend human understanding. Computers are most valuable not when they provide answers but when they help us ask better questions about ourselves, our societies, and our relationships with technology.

Rather than pursuing the impossible goal of complete machine understanding, Weizenbaum invites us to embrace partiality and limitation. We might hold computational agents to a different standard, not expecting them to know but to help us recognize what we don't know. As he notes, "Indeed, one of the most cogent reasons for using computers is to expose holes in our thinking. Computers are merciless critics in this respect."[26] These alternative possibilities require us to recognize the multifaceted nature of computational artifacts and the diverse approaches needed to fully understand them.

12.03 ELIZA's Multiple Futures

ELIZA can be "read" through many lenses: A media archeologist might be thrilled by the hardware it runs on, a programmer by the discovery of its code's operations, an interaction designer by how its scripts are composed, a psychologist by how people respond to it, an artist by its creative spinoffs, a philosopher by its gender implications, or a critical code studies researcher by how these aspects fit together and are enacted by the code itself. In the end, we found no final version of ELIZA—no single story, no definitive version of the code, no simple story to sum up how ELIZA came to be and what it became.

Weizenbaum's creation is so much more than a chatbot. We could not have made many of the discoveries in this book without the code in hand; a description of the algorithm was not enough. It required a team of scholars brought together by the code and ongoing conversation about the code, in order to develop this deeper understanding of ELIZA. The code became an entry point for explorations of the culture of the time, its technological limitations and innovations, Weizenbaum's life and opinions, as well as the programs that followed.

This case study of ELIZA demonstrates how to trace some of the many meandering threads of artificial intelligence and human–computer interaction (HCI). It is an object lesson in how much can get missed, and how quickly a singular history of code gets codified. In fact, code proliferates and travels, like seeds carried by wind, creating new growth in unexpected places. Concepts from ELIZA reached PARRY and BASIC ELIZA, and extended to Hal and *Her*, in trajectories that even its developer could not control. These versions, Weizenbaum's responses, and the gaps in the story, all make up part of the ongoing conversations that ELIZA has produced.

ELIZA also shows the role that critical examinations of source code can play in emerging systems and the need for more transparent, explainable systems—particularly as artificial intelligence scales. This transparency does not mean mere access to code, but the demand that corporations cooperate with users for more agency. It requires the capacity for deep, interdisciplinary literacy with the latest, largest AI systems in order to understand them as complex sociotechnical systems and public problems—as agents of inequality, homogenization, and oppression.[27]

This book represents a conversation between Weizenbaum and ELIZA, between programmers and cultural critics, between artists and philosophers. We invite you to join this ongoing dialogue, to approach computational systems with both technical understanding and critical awareness, and to participate in shaping more humane technological futures. As we experience an era of increasingly powerful and pervasive computational systems, ELIZA stands as both warning and inspiration. It warns against the seduction of apparent machine intelligence and the delegation of human judgment to algorithmic processes. It inspires us to imagine computational systems differently, for expanding human and non-human flourishing rather than replacing human capacities.

When running the DOCTOR script, ELIZA asked users to tell it their problems. As we face increasingly powerful and opaque AI systems, we might now pose our own questions: What problems do these systems create? What values do they encode? Whose interests do they serve? ELIZA reminds us that computational systems are never neutral technical objects but rather culturally situated artifacts embedded with ideologies, assumptions, and power dynamics. By recovering and analyzing their code, context, and consequences, critical code studies offers essential tools for democratic engagement with technologies that increasingly shape our collective future. This work has only begun, but ELIZA, in its multiplicity and lasting influence, shows us how to proceed: by asking the right questions.

Acknowledgments

This book would not have been possible without the generous support and scholarly contributions of numerous individuals and institutions.

We extend our profound gratitude to the Weizenbaum family, whose cooperation has been helpful to this research project.

The project has also benefited substantially from institutional support, particularly from the Borchard Foundation and the MIT Archives Team. We are especially indebted to Patsy Baudoin, whose expertise proved invaluable in guiding us to the archival materials. The research that underpins this volume was made possible through the dedicated assistance of Myles Crowley, Mikki Macdonald, and Allison Schmitt from MIT Distinctive Collections. Thanks also to the Garfinkel Collection, which provided additional primary sources; this includes Anne Rawls and Clemens Eisenmann for facilitating access to the Garfinkel archives, and Andrei Korbut for establishing this scholarly connection.

We acknowledge with gratitude the institutional support provided by MIT, where Nick Montfort served as our institutional host, facilitating academic connections and resources. Noah Springer of MIT Press has provided helpful editorial guidance throughout the publication process, and the editors of the Software Studies series have encouraged our progress.

This volume emerged from sustained scholarly dialogue among the members of Team ELIZA, as well as its extended community. We are particularly grateful to Janet Murray for contributing the preface. Rebecca Roach and Claire James Carroll have served as valuable critical friends to the project, offering intellectual engagement and constructive critique that has enhanced the theoretical rigor and methodological clarity of our work. Thanks to Jessica Pressman and James Marino for reviewing drafts. Also, thanks to Jeremy Douglass and the participants of the Critical Code Studies Working Groups, who discussed the ELIZA code and various versions of DOCTOR. We also wish to thank our anonymous peer reviewers for their careful reading and feedback.

Our research has been enriched by discussions with Terry Winograd and John Markoff, whose insights have contributed to our understanding of the historical context. We also extend our sincere appreciation to Leo Armel, Adam Gordon Bell, Walt Bilofsky, Lars Brinkhoff, Michael Bull, Bernie Cosell, CoRecursive Podcast, Walt Daniels, Jeremy Douglass, Karamjit S. Gill, Good Robot podcast, Daniel Harvey, Stefan Höltgen, IEEE, Martin Krzywdzinski, Zachary Loeb, Will Luers, Judy Malloy, Kenneth R. Manning, Willard McCarty, Aaron Mendon-Plasek, Merve Guven Ozkerim, Magnus Rust, Darrow Schecter, Richard Stallman, and Christian Strippel, whose conversations and advice have supported our research. We are hugely grateful to Rupert Lane, who was instrumental in our reanimation of the original source code. We appreciate the Musee Jeu de Paume staff and audiences at Musee Jeu de Paume for their active engagement with the recreation of ELIZA, in particular curator Alexandre Gefen. The Hacker News community has contributed meaningful technical insights and bibliographic suggestions that have enhanced our research. Our respective universities have been particularly supportive of the project, and so we would like to thank the University of Southern California, University of Oxford, Stanford University, and University of Sussex.

We would also like to thank our families and friends for much support during the long gestation of this project and for their patience and understanding while writing the book.

Team ELIZA

This book was collaboratively authored by a team of eight authors who met weekly for several years discussing the code of ELIZA. Each of us had a unique connection to ELIZA, drawing us into this work. Our work continues on the website inventingeliza.com.

David M. Berry

David M. Berry is Professor of Digital Humanities, University of Sussex, UK. David writes widely on philosophy and technology, particularly in terms of computation, software, and algorithms. He is the author of *The Philosophy of Software* and *Copy, Rip, Burn*. His most recent book is *Digital Humanities: Knowledge and Critique in a Digital Age*. His new work has looked at explainability, human understanding, and the history of the university. He is currently Otto Mønsted Visiting Professor, Denmark.

My interest in ELIZA started out as a fascination in 1989 with a computer program that seemed to have the uncanny ability to not only answer questions but also keep a conversation going, potentially indefinitely. When I first learnt about ELIZA, naturally without access to the source code, I was taking my preuniversity courses at a local college and as a result I coded my own version of a similar style of program in Turbo Pascal that I called MATILDA. Many years later, in 2016, as a result of my work into the philosophy of computation and software, I was asked to facilitate a discussion about ELIZA at the Critical Code Studies Work Group organized by Mark C. Marino and Jeremy Douglass. This had the result of rekindling my interest in ELIZA and, of course, in Weizenbaum's work. The lack of Weizenbaum's original source code was a theme that was repeatedly gestured to by participants in the workgroup. But finding the original source code has only deepened the mystery around it as we learn about strange inconsistencies in the code itself, and the way in which it has become a technical imaginary in culture. ELIZA has a lot to teach us about how computer code is now a very specific form of cultural artifact that is not just the instructions for a computer but also a key to unlock software cultures and social context in truly interesting ways.

Sarah Ciston

Sarah Ciston builds critical-creative tools to bring intersectional, interdisciplinary approaches to machine learning. Sarah is Professor of Computational Thinking and Aesthetic Doing at the Academy of Media Arts Cologne. Their work "AI War Cloud Database" has won the 2025 STARTS Grand Prize in Artistic Exploration from Ars Electronica and the European Commission. Sarah has held fellowships at the Center for Advanced Internet Studies, Akademie der Künste Berlin, and the Humboldt Institute for Internet and Society; and they have been named an AI Newcomer by the German Society for Computing. They hold a PhD in

Media Arts + Practice from University of Southern California and are also the author of "A Critical Field Guide for Working with Machine Learning Datasets" and "Critical AI Tutorials" for p5.js.

What first struck me about ELIZA was its simplicity. Magicians, poets, and directors will tell you the power of constraint, concision, and elision. What the audience fills in with their imagination is always more powerful and personal than what we might construct for them. ELIZA, and its successors, relies on this principle. The simplicity of ELIZA and this power to hold our imaginations are what still resonate with me today. Despite all the massive computing power required to train conversational agents like GPT, a simple conceit that turns statements into open-ended questions allows users to fill that imaginative space with their ideas of empathy or resistance, of a listening presence or sparring partner, of a wide range of implicit social qualities embedded with gender, class, race, and more.

I don't recall first encountering ELIZA. The chatbot seemed like a concept that had always been there, and "she" had always been the first. No doubt, I had encountered ELIZA several times, mentioned always as an aside, before I realized I knew a fair amount of the story. The story of ELIZA always had that myth vibe, down to the myth that gave it "her" name (Galatea) and the creator renounced his creation.

It may have been that my first official encounter with ELIZA was in Wardrip-Fruin and Montfort's *New Media Reader*, which featured an excerpt of Weizenbaum's "Computer Power and Human Reason" and an introduction by Janet Murray that called ELIZA a milestone. In the excerpt, Weizenbaum highlighted the problem of computer understanding of natural language as a contextual problem, one that is often ignored in the pursuit of generalized models. I was hooked. I was an experimental poet looking for more ways to make books weird, and I had snuck my way into my first computing in the arts class, taught wonderfully by Brett Stalbaum at UC San Diego. It was as a result of that class and my continued work with Brett that I wrote my own first chatbot—ladymouth, which appears in this book—very much inspired by ELIZA.

ELIZA's constraints (like the choice of Rogerian therapist form) emerged as necessities of the hardware, software, and theoretical capacities of the moment. I relied on steep constraints as I first learned to write code by making my first bot (in much the same fashion as that old quip from E. L. Doctorow about novel writing: "It's not a rational way of working. It's like driving a car at night. You can only see as far as your headlights illuminate. But you can make the whole trip that way."). Technical constraints continue today to mark the limits of what conversational agents can do, as well as to spark our imaginations about their agency, creativity, and intelligence beyond those limits.

Anthony C. Hay

Anthony C. Hay was a programmer at Digital Research and Novell. He has a BSc in computer science from Imperial College, London. He has never played the cello but is a compulsive programmer, at first toggling machine code into a "77–68" computer in 1977. More recently he implemented ELIZA in C++ based on Joseph Weizenbaum's 1966 *Communications of the ACM* paper, and then corrected it when the original source code became public (https://github.com/anthay/ELIZA).

My interest in ELIZA began when I read *Computer Power and Human Reason* in 1979: could ELIZA really be so convincing to talk to that I would believe it understood me? In creating ELIZA, Weizenbaum's purpose was not to advance our understanding of intelligence, but to create the illusion that his program understood what had been said to it in order to demonstrate how easily people can be fooled. People were capable of committing the most barbaric and depraved acts on their fellow humans long before we had computers, and whether we use any technology, be it cleaver or computer, for good or ill is a choice made by humans—or it would be if we had free will. Recreating ELIZA as accurately as I possibly could was not to solve a substantive problem, but to play—an end in itself—and to answer my schoolboy question. (No.)

Mark C. Marino

Mark C. Marino (http://markcmarino.com) is a writer and scholar of electronic literature living in Los Angeles. He is a professor (teaching) and Generative AI Fellow at the University of Southern California where he directs the Humanities and Critical Code Studies Lab (http://haccslab.com), exploring the cultural meaning of computer source code. *Inventing ELIZA* completes a trio of collaborative readings of digital objects along with *10 PRINT* (with nine other authors) and *Reading Project* with Jessica Pressman and Jeremy Douglass. He recently published *Critical Code Studies*, named for the field he has been working to cocreate for 20 years (https://criticalcodestudies.com), and *Hallucinate This! an authoritized autobotography of ChatGPT* with ChatGPT. With his two children, he writes the interactive series *Mrs. Wobbles & the Tangerine House*. Mark is the Director of Communication of the Electronic Literature Organization (http://eliterature.org).

Like most of those assembled, when I first got my hands on a "personal" computer, in the early 1980s, I wrote my first chatbot program in BASIC. Basically, I wanted the machine to greet me and then ask about my day. But it wasn't until I was working on my PhD at University of California–Riverside when I really took up the study of ELIZA proper. Not just the history but also the CODE! That gave birth to critical code studies, even though, as Noah Wardrip-Fruin pointed out, we only had the script. So, my search for the code began. At the same time, I was making more chatbots to explore their use in storytelling and virtual theater. I especially became interested in how ELIZA/DOCTOR and later bots could be seen to perform gender as well as racial and ethnic identity.

What drew me to ELIZA was the way in which the program brought together so many aspects of computing that interested me: interactive storytelling, virtual characters, artificial intelligence but also the psychology of those using computers, the desire to talk with a computer, to have a computer friend. So much about conversation and culture came to a head in this one program. Inspired by ELIZA, I have continued to work on my own chatbots, whether for interactive fiction, for virtual reality, on Twitter (when it was called that), or in scripting nonplayer characters for video games.

As I became more and more interested in computer source code, I wanted to read ELIZA's DOCTOR script more thoroughly. Several of the Critical Code Studies Working Groups took up the collective annotation of various version, most recently when these fine colleagues uncovered the original source code for ELIZA. What a find! The ELIZA platform is surprisingly adaptable for when it was written. The DOCTOR script is clever, concise, witty even. And yet it also taps into some basic human drives: the desire to be understood, the desire for conversation, the desire to know ourselves and be known. But this book has shown me that what I really needed to understand the code was this brilliant team of collaborators.

As chatbots have evolved into virtual assistants and other forms of AI, I continue to be tantalized by this dream of a virtual conversant. ELIZA holds an unassailable spot as Eve, or perhaps as the Ada Lovelace of chatbots.

Peter Millican

Peter Millican originally studied mathematics at the University of Oxford, but graduated in philosophy and theology before moving into postgraduate study in philosophy (also at Oxford). After a brief spell lecturing in the philosophy of religion at the University of Glasgow, from 1985 until 2005 he lectured in computer science and philosophy at Leeds University, where much of his work was devoted to building interdisciplinary links. In 2005 he became Gilbert Ryle Fellow at Hertford College, then Reader in Early Modern Philosophy in 2007, and professor of philosophy in 2010.

My own perspective is that of a computing/philosophy advocate, teacher and software developer, and textual scholar. What drew me to ELIZA was my long-standing interest (many would say "obsession") in the philosophy/computing nexus, with Turing as a central figure and chatbots as a great way of introducing students to some of the difficulties of analyzing language (I had started out focusing on philosophy of language as an Oxford postgraduate, way back in 1980). So my "ELIZAbeth" chatbot creation program began in 2002 as a way of enabling nonspecialist students who were dipping their toe into computing—through my first-year elective courses at Leeds University—to do some real language-based processing by writing accessible scripts rather than conventional programs. Most of the programming I have done myself over the last two decades has been on three teaching systems ("ELIZAbeth," "Turtle," and "Signature"), all designed to enable novice students to play with philosophically interesting ideas, in the hope that they may be attracted towards that philosophy/computing nexus.

As a professional philosopher, my career moved in the direction of early modern philosophy, especially David Hume, another "long-standing interest" (a.k.a. obsession), and this provoked a major interest in digital humanities. Hence my work on davidhume.org (which now contains all Hume's published texts, together with much of my own research). It was probably inevitable in this context that I would take a serious scholarly interest also in Turing, who has featured significantly in Oxford outreach events for the Computer Science & Philosophy degree program, which I started in 2012. "Elizabeth" has proved extremely useful here, providing a simple but potentially powerful platform for students to explore the difficulties of passing the Turing Test, and it's been particularly gratifying when applicants for the degree have pointed to these explorations as something that deepened their interest. "Elizabeth" has also played a role in my Turing lectures' because the "Chatbot objection" to the Turing Test is one of the most significant. Weizenbaum's discovery would, I believe, have very much surprised Turing himself, and undermines the significance of his "70% prediction" regarding progress by the end of the twentieth century. Fooling an average interrogator for just five minutes, 30 percent of the time, just doesn't look so impressive once you have seen ELIZA and understood how crudely it works.

Jeff Shrager

Jeff Shrager is a dad and an aging Lisp hacker; indeed, he and Lisp are about the same age! Jeff has been fascinated by AI since childhood, when he wrote a bad ELIZA knock-off in BASIC that, through its synchronicity with the personal computer revolution, became the root of a huge branch of bad ELIZA knock-offs. Jeff has since coauthored over 100 AI-related publications in fields ranging from mathematics to medicine, and cofounded three biomedical AI startups. At the moment he's a CommerceNet Fellow and adjunct professor in the Symbolic Systems Program at Stanford University.

I don't remember what first attracted me to ELIZA, as I was only around 13. That was back in 1973, when I wrote a poor ELIZA knock-off in BASIC. My ELIZA was eventually published in *Creative Computing* in 1977. This was right in the heart of the personal computer craze; BASIC was the lingua franca of PCs, and *Creative Computing* was one of the learning PC magazines. As a result, my ELIZA spread like wildfire through the PC world, a fact that I greeted with a combination of pride and embarrassment. By then I was an undergraduate in computer science at the University of Pennsylvania, and had become a student of "real" AI, and Lisp, which, at that time, was the standard language of AI. Through a separate coincidence, the standard version of ELIZA that was making the rounds in the academic community was Bernie Cosell's Lisp version—a different knock-off of the original (and a much better one than mine!) After Penn I transitioned to cognitive science at CMU, and got a deeper and more nuanced sense of the complexities of human reasoning, learning, and language, which has shaped my career since that time. Through all of my eventually deep and broad exposure to AI and cognitive science, ELIZA stood out in my mind as unique in that Weizenbaum, in his design, had managed to discover, in an elegantly simple model, a way to maintain what is perhaps the most human of capabilities: not merely language (which ELIZA was actually rather poor at, compared with any of its successors, and especially the recent ChatGPT models), but discourse. ELIZA's simple design led its interlocutors into "participating in discourse" with it. Granted, the discourse was one-sided, but still, no program has since, or even still, managed to play the role of "cognitive mirror" as well as ELIZA. As I age into the latter part of my careers, history becomes more and more interesting, and that fascination with ELIZA—now made much richer through my cognitive science education and career—stuck with me from my barely teenage hacker days. Occasionally, through that whole time, I would get emails from people—some of them very famous computer scientists—for whom my BASIC ELIZA had been one of their formative influences. I created the elizagen.org site to honor the history of ELIZA and its wide-ranging influence, upon me and many others, and began collecting versions of ELIZA. But the grandmother of them all—Weizenbaum's original MAD-SLIP code—hadn't, as far as anyone knew, been seen since the 1960s. So to complete the collection, I pulled on my software spelunker's backpack and headed (remotely) to the Weizenbaum archives at MIT, and there discovered the original, which Team ELIZA has been working with since.

Arthur I. Schwarz

Arthur I. Schwarz has a BSc in physics and an MSc in computer science. He is a software developer and technical lead working in the aerospace and automotive industries with experience in software architecting, modeling, and algorithm development and design. He has developed a version of SLIP called gSLIP in the public domain. His active interests are in hashing algorithms, ELIZA, and development of analysis software for anomaly detection. Past algorithm work has been presented in webinars for the Association of Old Crows, a worldwide SIGINT community, on binary searching of overlapping intervals on the real line.

I don't remember when I first came across ELIZA/DOCTOR. It seems to have always been with me. The anecdote about the department secretary seeking psychological advice from ELIZA and having a breakthrough moment is part of that recollection. I could have been familiarized with ELIZA through my experiences learning SLIP. Then again, it could have been from some other source.

What I do remember is implementing SLIP. I implemented it in Fortran in 1980, in C in 1990, and in C++ around 2014. In the 1970s, '80s and '90s, acquiring technical material was a tedious process and involved getting hard copy—from somewhere. Unless you were actively engaged in research or affiliated with an academic library, it was difficult to obtain a copy and, when not, difficult and expensive. This changed with the internet and with the large capacity storage devices we currently have. In any case, during the C++ effort in creating SLIP, I became curious. This led to me downloading articles on ELIZA. I had a place to put them, and had ready access to them, so why not? But my interest was still only nascent, awaiting the proper spark to do more than read and wonder.

That spark came in January 2022 when Jeff Shrager emailed me about the ongoing academic interest in ELIZA and wondered about what I could do to help given my C++ SLIP implementation. That was the spur, and it let me wonder whether now was the time to engage more fully in ELIZA. My interest is in developing a platform in which to display the properties of SLIP. A simple idea. I had already downloaded and found out that others had had the idea of using ELIZA as a springboard to amusement or commercial success, and I thought, why not? My current and main interest is developing a version of ELIZA that is robust, well documented, and capable of change and exploration. The available documentation is slender on details and avoids all corner cases. This has been corrected.

The thought is that it can act as a vehicle to explore options that ELIZA is capable of in an environment that does not involve great skill. This effort is done collaboratively with Anthony Hay, whose own efforts stand on their own merits as a successful and faithful representation of the functionality of the original ELIZA. As Newton said before me, I stand on the shoulders of giants.

Peggy Weil

Peggy Weil (pweilstudio.com) is a multidisciplinary artist whose work addresses our changing physical, digital, and sociopolitical landscapes. Her work spans interactive digital and immersive technologies, installation, film and performance, including, the first net-art chatbot, MrMind. She did her graduate work at the Architecture Machine Group/MIT Media Lab in the early 1980s, which provided an early introduction to the technologies and issues of human-machine interaction, digital media, AI, telepresence, and immersive and virtual environments. She has taught at Design Media Arts (University of California, Los Angeles), California College of the Arts, and is currently adjunct assistant professor at University of Southern California's School of Cinematic Arts.

I first learned about Alan Turing, the Turing Test, chatbots, and ELIZA in 1980 as a graduate student at MIT's Architecture Machine Group, the precursor to today's MIT Media Lab. Professor Weizenbaum delivered a guest lecture in Patrick Winston's introductory class in artificial intelligence. I audited the class, as a nonengineer nonetheless fascinated by this relatively new discipline. Marvin Minsky's living room was an ongoing salon near campus, and I was attempting to become minimally literate in the field.

P. H. Winston was an effective booster of AI's promise for mankind: his introductory textbook *Artificial Intelligence* states (in bold), "Smarter Computers Can Help Solve Some of the Problems of Today's World" and "Understanding Computer Intelligence Is a Way to Study General Intelligence." There was seemingly no limit to the upside and no acknowledgement of any potential downside. In marked contrast (and to Winston's credit for the invitation), Weizenbaum, a Holocaust survivor, delivered a personal and emotional lecture, recounting the powerful illusion associated with ELIZA and his confrontation with its dehumanizing potential. Deeply troubled, he changed the course of his career, taking a leave of absence from Computer Science at MIT to study the humanities at Harvard University and Stanford University. His challenge to us was bracing: Consider the human and social implications of your work. Moved by his lecture, I read *Computer Power and Human Reason* and sought out conversations with ELIZA and other bots. I have little memory of these conversations as I was less interested in the code or even the experience of chatting with a bot than by Weizenbaum's moral example.

Sixteen years passed before I invoked ELIZA to create MrMind and The Blurring Test in 1996, a series of dialogues scripted for Human (Ventriloquist) and Dummy (Computer), as a provocation to consider our role vis-à-vis our digital creations. The Turing Test is a (flawed) measure of machine progress and marks a moment when machine response becomes indistinguishable from human response. The Blurring Test asks us to consider what it means when our human response becomes indistinguishable from machine response. MrMind challenges us to define and consider, Who (or what) do we think we are in relationship to our creations? This long-running project, premiering as a song cycle in 2023, is a direct link to my first encounter with Weizenbaum and ELIZA in 1980.

Appendix I Reconstructing ELIZA

The reconstruction of historical software presents fascinating challenges at the intersection of computer science, digital preservation, and software archaeology.[1] In our work reconstructing ELIZA, the team encountered unique difficulties that highlight important questions about software preservation and historical authenticity.[2]

The reconstruction revealed several key technical challenges. CTSS used 6-bit BCD character encoding, packing six characters into a 36-bit word—a reminder of computing before standardized ASCII encoding.[3] The MAD programming language itself used verbose keywords that could be abbreviated, making the code particularly difficult to interpret decades later. For example, "WHENEVER" could be shortened to "W'R," a convention that made sense in the era of punch cards but presents challenges for modern interpretation.

Several critical functions were missing from the recovered code and had to be carefully reconstructed:

- BCDIT, a function for converting binary numbers to BCD format
- INLST, a crucial function used in YMATCH and ASSMBL operations
- LETTER, a character classification function with no extant documentation
- Various initialization routines for common data areas

The reconstruction required not just technical expertise but also detective work to understand the original system's behavior. As Lane et al. note, "Running the original code feels good and authentic. Finding bugs in it only adds to the authenticity."[4]

This archaeological approach to code reconstruction raises important questions about preservation. Should we fix historical bugs when we find them? How do we balance historical accuracy with the desire for a working system? These questions parallel similar debates in architectural preservation about restoration versus reconstruction.[5] It emphasizes the importance of documenting our processes and being selective and mindful of our methodologies for software reconstruction. How are we to "preserve" the material traces of something that exists in many versions, iterations, and offshoots—in particular before the era of version control using Git (which itself presents new problems for contemporary software archiving projects)?[6]

The process also highlighted how software preservation differs from physical artifact preservation. Unlike buildings or manuscripts, software exists simultaneously as human-readable text and machine-executable instructions. This dual nature creates unique challenges for preservation and reconstruction.

Our work on ELIZA demonstrates the importance of treating historical software as both technical and cultural heritage. The reconstruction process reveals not just the technical sophistication of early computing but also provides insights into the social and institutional context of AI, which did not result from a linear progression of developments but rather shifted and iterated back over itself.

The reconstructed ELIZA, now running on an emulated CTSS system, offers a unique window into computing history. It stands as testimony to both the ingenuity of early computer scientists and the importance of preserving our digital heritage.[7]

Computational Linguistics

A. G. OETTINGER, Editor

ELIZA—A Computer Program For the Study of Natural Language Communication Between Man And Machine

JOSEPH WEIZENBAUM
Massachusetts Institute of Technology, Cambridge, Mass.*

ELIZA is a program operating within the MAC time-sharing system at MIT which makes certain kinds of natural language conversation between man and computer possible. Input sentences are analyzed on the basis of decomposition rules which are triggered by key words appearing in the input text. Responses are generated by reassembly rules associated with selected decomposition rules. The fundamental technical problems with which ELIZA is concerned are: (1) the identification of key words, (2) the discovery of minimal context, (3) the choice of appropriate transformations, (4) generation of responses in the absence of key words, and (5) the provision of an editing capability for ELIZA "scripts". A discussion of some psychological issues relevant to the ELIZA approach as well as of future developments concludes the paper.

Introduction

It is said that to explain is to explain away. This maxim is nowhere so well fulfilled as in the area of computer programming, especially in what is called heuristic programming and artificial intelligence. For in those realms machines are made to behave in wondrous ways, often sufficient to dazzle even the most experienced observer. But once a particular program is unmasked, once its inner workings are explained in language sufficiently plain to induce understanding, its magic crumbles away; it stands revealed as a mere collection of procedures, each quite comprehensible. The observer says to himself "I could have written that". With that thought he moves the program in question from the shelf marked "intelligent", to that reserved for curios, fit to be discussed only with people less enlightened than he.

Work reported herein was supported (in part) by Project MAC, an MIT research program sponsored by the Advanced Research Projects Agency, Department of Defense, under Office of Naval Research Contract Number Nonr-4102(01).

* Department of Electrical Engineering.

The object of this paper is to cause just such a re-evaluation of the program about to be "explained". Few programs ever needed it more.

ELIZA Program

ELIZA is a program which makes natural language conversation with a computer possible. Its present implementation is on the MAC time-sharing system at MIT. It is written in MAD-SLIP [4] for the IBM 7094. Its name was chosen to emphasize that it may be incrementally improved by its users, since its language abilities may be continually improved by a "teacher". Like the Eliza of Pygmalion fame, it can be made to appear even more civilized, the relation of appearance to reality, however, remaining in the domain of the playwright.

For the present purpose it is sufficient to characterize the MAC system as one which permits an individual to operate a full scale computer from a remotely located typewriter. The individual operator has the illusion that he is the sole user of the computer complex, while in fact others may be "time-sharing" the system with him. What is important here is that the computer can read messages typed on the typewriter and respond by writing on the same instrument. The time between the computer's receipt of a message and the appearance of its response is a function of the program controlling the dialogue and of such MAC system parameters as the number of users currently corresponding with the system. These latter parameters generally contribute so little to the overall response time that conversational interaction with the computer need never involve truly intolerable delays.

When in conversation with ELIZA, the user types in some statement or set of statements in natural language using normal punctuation and sentence structures. Only the question mark may not be used, because it is interpreted as a line delete character by the MAC system. The user's statement is terminated by a double carriage return which serves to turn control over to ELIZA. ELIZA then analyzes the user's statement and generates some response which it types out. Control is then again with the user.

A typical conversation is the following:

Men are all alike.
IN WHAT WAY
They're always bugging us about something or other.
CAN YOU THINK OF A SPECIFIC EXAMPLE
Well, my boyfriend made me come here.
YOUR BOYFRIEND MADE YOU COME HERE
He says I'm depressed much of the time.
I AM SORRY TO HEAR YOU ARE DEPRESSED

Figure A.1 A reproduction of Joseph Weizenbaum's article "ELIZA—A Computer Program For the Study of Natural Language Communication Between Man And Machine" as it appeared in *Communications of the ACM* in 1966.

It's true. I am unhappy.
DO YOU THINK COMING HERE WILL HELP YOU NOT TO BE UNHAPPY
I need some help, that much seems certain.
WHAT WOULD IT MEAN TO YOU IF YOU GOT SOME HELP
Perhaps I could learn to get along with my mother.
TELL ME MORE ABOUT YOUR FAMILY
My mother takes care of me.
WHO ELSE IN YOUR FAMILY TAKES CARE OF YOU
My father.
YOUR FATHER
You are like my father in some ways.
WHAT RESEMBLANCE DO YOU SEE
You are not very aggressive but I think you don't want me to notice that.
WHAT MAKES YOU THINK I AM NOT VERY AGGRESSIVE
You don't argue with me.
WHY DO YOU THINK I DON'T ARGUE WITH YOU
You are afraid of me.
DOES IT PLEASE YOU TO BELIEVE I AM AFRAID OF YOU
My father is afraid of everybody.
WHAT ELSE COMES TO MIND WHEN YOU THINK OF YOUR FATHER
Bullies.
DOES THAT HAVE ANYTHING TO DO WITH THE FACT THAT YOUR BOYFRIEND MADE YOU COME HERE

The capitalized lines are the machine responses.

The gross procedure of the program is quite simple; the text is read and inspected for the presence of a *keyword*. If such a word is found, the sentence is transformed according to a *rule* associated with the keyword, if not a content-free remark or, under certain conditions, an earlier transformation is retrieved. The text so computed or retrieved is then printed out.

In detail, of course, the procedure sketched above is considerably more complex. Keywords, for example, may have a RANK or precedence number. The procedure is sensitive to such numbers in that it will abandon a keyword already found in the left-to-right scan of the text in favor of one having a higher rank. Also, the procedure recognizes a comma or a period as a delimiter. Whenever either one is encountered and a keyword has already been found, all subsequent text is deleted from the input message. If no key had yet been found the phrase or sentence to the left of the delimiter (as well as the delimiter itself) is deleted. As a result, only single phrases or sentences are ever transformed.

Keywords and their associated transformation[1] rules constitute the SCRIPT for a particular class of conversation. An important property of ELIZA is that a script is data; i.e., it is not part of the program itself. Hence, ELIZA is not restricted to a particular set of recognition patterns or responses, indeed not even to any specific language. ELIZA scripts exist (at this writing) in Welsh and German as well as in English.

The fundamental technical problems with which ELIZA must be preoccupied are the following:

(1) The identification of the "most important" keyword occurring in the input message.

(2) The identification of some minimal context within which the chosen keyword appears; e.g., if the keyword is "you", is it followed by the word "are" (in which case an assertion is probably being made).

(3) The choice of an appropriate transformation rule and, of course, the making of the transformation itself.

(4) The provision of mechanism that will permit ELIZA to respond "intelligently" when the input text contained no keywords.

(5) The provision of machinery that facilitates editing, particularly extension, of the script on the script writing level.

There are, of course, the usual constraints dictated by the need to be economical in the use of computer time and storage space.

The central issue is clearly one of text manipulation, and at the heart of that issue is the concept of the *transformation rule* which has been said to be associated with certain keywords. The mechanisms subsumed under the slogan "transformation rule" are a number of SLIP functions which serve to (1) decompose a data string according to certain criteria, hence to test the string as to whether it satisfies these criteria or not, and (2) to reassemble a decomposed string according to certain assembly specifications.

While this is not the place to discuss these functions in all their detail (or even to reveal their full power and generality), it is important to the understanding of the operation of ELIZA to describe them in *some* detail.

Consider the sentence "I am very unhappy these days". Suppose a foreigner with only a limited knowledge of English but with a very good ear heard that sentence spoken but understood only the first two words "I am". Wishing to appear interested, perhaps even sympathetic, he may reply "How long have you been very unhappy these days?" What he must have done is to apply a kind of template to the original sentence, one part of which matched the two words "I am" and the remainder isolated the words "very unhappy these days". He must also have a reassembly kit specifically associated with that template, one that specifies that any sentence of the form "I am BLAH" can be transformed to "How long have you been BLAH", independently of the meaning of BLAH. A somewhat more complicated example is given by the sentence "It seems that you hate me". Here the foreigner understands only the words "you" and "me"; i.e., he applies a template that decomposes the sentence into the four parts:

(1) It seems that (2) you (3) hate (4) me

of which only the second and fourth parts are understood. The reassembly rule might then be "What makes you think I hate you"; i.e., it might throw away the first component, translate the two known words ("you" to "I" and "me" to "you") and tack on a stock phrase (What makes you think) to the front of the reconstruction.

---

[1] The word "transformation" is used in its generic sense rather than that given it by Harris and Chomsky in linguistic contexts.

A formal notation in which to represent the decomposition template is:

(0 YOU 0 ME)

and the reassembly rule

(WHAT MAKES YOU THINK I 3 YOU).

The "0" in the decomposition rule stands for "an indefinite number of words" (analogous to the indefinite dollar sign of COMIT) [6] while the "3" in the reassembly rule indicates that the third component of the subject decomposition is to be inserted in its place. The decomposition rule

(0 YOU 1 ME)

would have worked just as well in this specific example. A nonzero integer "n" appearing in a decomposition rule indicates that the component in question should consist of exactly "n" words. However, of the two rules shown, only the first would have *matched* the sentence, "It seems you hate and love me," the second failing because there is more than one word between "you" and "me".

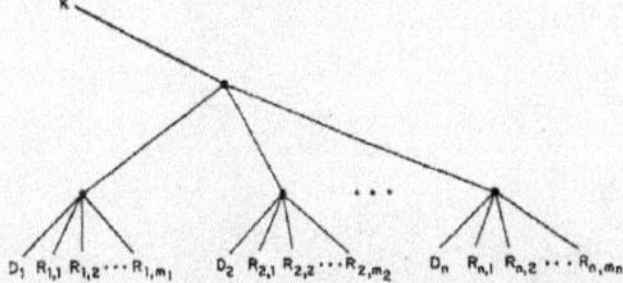

FIG. 1. Keyword and rule list structure

In ELIZA the question of which decomposition rules to apply to an input text is of course a crucial one. The input sentence might have been, for example, "It seems that you hate," in which case the decomposition rule (0 YOU 0 ME) would have failed in that the word "ME" would not have been found at all, let alone in its assigned place. Some other decomposition rule would then have to be tried and, failing that, still another until a match could be made or a total failure reported. ELIZA must therefore have a mechanism to sharply delimit the set of decomposition rules which are potentially applicable to a currently active input sentence. This is the keyword mechanism.

An input sentence is scanned from left to right. Each word is looked up in a dictionary of keywords. If a word is identified as a keyword, then (apart from the issue of precedence of keywords) only decomposition rules containing that keyword need to be tried. The trial sequence can even be partially ordered. For example, the decomposition rule (0 YOU 0) associated with the keyword "YOU" (and decomposing an input sentence into (1) all words in front of "YOU", (2) the word "YOU", and (3) all words following "YOU") should be the last one tried since it is bound to succeed.

Two problems now arise. One stems from the fact that almost none of the words in any given sentence are represented in the keyword dictionary. The other is that of "associating" both decomposition and reassembly rules with keywords. The first is serious in that the determination that a word is not in a dictionary may well require more computation (i.e., time) than the location of a word which is represented. The attack on both problems begins by placing both a keyword and its associated rules on a *list*. The basic format of a typical key list is the following:

$$
\begin{aligned}
(K\ &((D_1)\ (R_{1,1})\ (R_{1,2})\ \cdots\ (R_{1,m_1}))\\
&((D_2)\ (R_{2,1})\ (R_{2,2})\ \cdots\ (R_{2,m_2}))\\
&\quad\vdots\\
&((D_n)\ (R_{n,1})\ (R_{n,2})\ \cdots\ (R_{n,m_n})))
\end{aligned}
$$

where K is the keyword, D_i the ith decomposition rule associated with K and $R_{i,j}$ the jth reassembly rule associated with the ith decomposition rule.

A common pictorial representation of such a structure is the tree diagram shown in Figure 1. The top level of this structure contains the keyword followed by the names of lists; each one of which is again a list structure beginning with a decomposition rule and followed by reassembly rules. Since list structures of this type have no predetermined dimensionality limitations, any number of decomposition rules may be associated with a given keyword and any number of reassembly rules with any specific decomposition rule. SLIP is rich in functions that sequence over structures of this type efficiently. Hence programmin problems are minimized.

An ELIZA script consists mainly of a set of list structures of the type shown. The actual keyword dictionary is constructed when such a script is first read into the hitherto empty program. The basic structural component of the keyword dictionary is a vector KEY of (currently) 128 contiguous computer words. As a particular key list structure is read the keyword K at its top is randomized (hashed) by a procedure that produces (currently) a 7 bit integer "i". The word "always", for example, yields the integer 14. KEY(i), i.e., the ith word of the vector KEY, is then examined to determine whether it contains a list name. If it does not, then an empty list is created, its name placed in KEY(i), and the key list structure in question placed on *that* list. If KEY(i) already contains a list name, then the name of the key list structure is placed on the bottom of the list named in KEY(i). The largest dictionary so far attempted contains about 50 keywords. No list named in any of the words of the KEY vector contains more than two key list structures.

Every word encountered in the scan of an input text, i.e., during the actual operations of ELIZA, is randomized by the same hashing algorithm as was originally applied to the incoming keywords, hence yields an integer which points to the only possible list structure which could potentially contain that word as a keyword. Even then, only the tops of any key list structures that may be found there need be interrogated to determine whether or not a keyword has been found. By virtue of the various list sequencing operations that SLIP makes available, the actual identification of a keyword leaves as its principal product a pointer to the list of decomposition (and hence reassembly) rules associated with the identified keyword. One result of this strategy is that often less time is required to discover that a given word is not in the keyword dictionary than to locate it if it is there. However, the location of a keyword yields pointers to all information associated with that word.

Some conversational protocols require that certain transformations be made on certain words of the input text independently of any contextual considerations. The first conversation displayed in this paper, for example, requires that first person pronouns be exchanged for second person pronouns and vice versa throughout the input text. There may be further transformations but these minimal substitutions are unconditional. Simple substitution rules ought not to be elevated to the level of transformations, nor should the words involved be forced to carry with them all the structure required for the fully complex case. Furthermore, unconditional substitutions of single words for single words can be accomplished during the text scan itself, not as a transformation of the entire text subsequent to scanning. To facilitate the realization of these desiderata, any word in the key dictionary, i.e., at the top of a key list structure, may be followed by an equal sign followed by whatever word is to be its substitute. Tranformation rules may, but need not, follow. If none do follow such a substitution rule, then the substitution is made on the fly, i.e., during text scanning, but the word in question is not identified as a keyword for subsequent purposes. Of course, a word may be both subtituted for and be a keyword as well. An example of a simple substitution is

(YOURSELF = MYSELF).

Neither "yourself" nor "myself" are keywords in the particular script from which this example was chosen.

The fact that keywords can have ranks or precedences has already been mentioned. The need of a ranking mechanism may be established by an example. Suppose an input sentence is "I know everybody laughed at me." A script may tag the word "I" as well as the word "everybody" as a keyword. Without differential ranking, "I" occurring first would determine the transformation to be applied. A typical response might be "You say you know everybody laughed at you." But the important message in the input sentence begins with the word "everybody". It is very often true that when a person speaks in terms of universals such as "everybody", "always" and "nobody" he is really referring to some quite specific event or person. By giving "everybody" a higher rank than "I", the response "Who in particular are you thinking of" may be generated.

The specific mechanism employed in ranking is that the rank of every keyword encountered (absence of rank implies rank equals 0) is compared with the rank of the highest ranked keyword already seen. If the rank of the new word is higher than that of any previously encountered word, the pointer to the transformation rules associated with the new word is placed on top of a list called the keystack, otherwise it is placed on the bottom of the keystack. When the text scan terminates, the keystack has at its top a pointer associated with the highest ranked keyword encountered in the scan. The remaining pointers in the stack may not be monotonically ordered with respect to the ranks of the words from which they were derived, but they are nearly so—in any event they are in a useful and interesting order. Figure 2 is a simplified flow diagram of keyword detection. The rank of a keyword must, of course, also be associated with the keyword. Therefore it must appear on the keyword list structure. It may be found, if at all, just in front of the list of transformation rules associated with the keyword. As an example consider the word "MY" in a particular script. Its keyword list may be as follows:

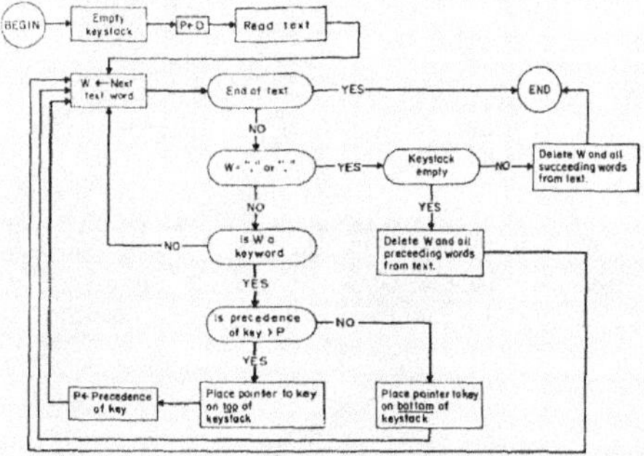

FIG. 2. Basic flow diagram of keyword detection

(MY = YOUR 5 (transformation rules)).

Such a list would mean that whenever the word "MY" is encountered in any text, it would be replaced by the word "YOUR". Its rank would be 5.

Upon completion of a given text scan, the keystack is either empty or contains pointers derived from the keywords found in the text. Each of such pointers is actually a sequence reader—a SLIP mechanism which facilitates scanning of lists—pointing into its particular key list in such a way that one sequencing operation to the right (SEQLR) will sequence it to the first set of transformation rules associated with its keyword, i.e., to the list

$$((D_1)\ (R_{1,1})\ (R_{1,2})\ \ldots\ (R_1\ ,\ R_{m_1})).$$

The top of *that* list, of course, is a list which serves a decomposition rule for the subject text. The top of the keystack contains the first pointer to be activated.

The decomposition rule D_1 associated with the keyword K, i.e., $\{(D_1), K\}$, is now tried. It may fail however. For example, suppose the input text was:

You are very helpful.

The keyword, say, is "you", and $\{(D_1), \text{you}\}$ is

(0 I remind you of 0).

(Recall that the "you" in the original sentence has already been replaced by "I" in the text now analyzed.) This decomposition rule obviously fails to match the input sentence. Should $\{(D_1), K\}$ fail to find a match, then $\{(D_2), K\}$ is tried. Should that too fail, $\{(D_3), K\}$ is attempted, and so on. Of course, the set of transformation rules can be guaranteed to terminate with a decomposition rule which must match. The decomposition rule

(0 K 0)

will match any text in which the word K appears while

(0)

will match any text whatever. However, there are other ways to leave a particular set of transformation rules, as will be shown below. For the present, suppose that some particular decomposition rule (D_i) has matched the input text. (D_i), of course, was found on a list of the form

$$((D_i)(R_{i,1})(R_{i,2}) \cdots (R_{i,m_i})).$$

Sequencing the reader which is presently pointing at (D_i) will retrieve the reassembly rule $(R_{i,1})$ which may then be applied to the decomposed input text to yield the output message.

Consider again the input text

You are very helpful

in which "you" is the only key word. The sentence is transformed during scanning to

I are very helpful

$\{(D_1), \text{you}\}$ is "(0 I remind your of 0)" and fails to match as already discussed. However, $\{(D_2), \text{you}\}$ is "(0 I are 0)" and obviously matches the text, decomposing it into the constituents

(1) empty (2) I (3) are (4) very helpful.

$\{(R_{2,1}), \text{you}\}$ is

(What makes you think I am 4)

Hence it produces the output text

What makes you think I am very helpful.

Having produced it, the integer 1 is put in front of $(R_{2,1})$ so that the transformation rule list in question now appears as

$$((D_2)1(R_{2,1})(R_{2,2}) \cdots (R_{2,m_2})).$$

Next time $\{(D_2), K\}$ matches an input text, the reassembly rule $(R_{2,2})$ will be applied and the integer 2 will replace the 1. After (R_{2,m_2}) has been exercised, $(R_{2,1})$ will again be invoked. Thus, after the system has been in use for a time, every decomposition rule which has matched some input text has associated with it an integer which corresponds to the last reassembly rule used in connection with that decomposition rule. This mechanism insures that the complete set of reassembly rules associated with a given decomposition rule is cycled through before any repetitions occur.

The system described so far is essentially one which selects a decomposition rule for the highest ranking keyword found in an input text, attempts to match that text according to that decomposition rule and, failing to make a match, selects the next reassembly rule associated with the matching decomposition rule and applies it to generate an output text. It is, in other words, a system which, for the highest ranking keyword of a text, selects a specific decomposition and reassembly rule to be used in forming the output message.

Were the system to remain that simple, then keywords that required identical sets of transformation rules would each require that a copy of these transformation rules be associated with them. This would be logically sound but would complicate the task of script writing and would also make unnecessary storage demands. There are therefore special types of decomposition and assembly rules characterized by the appearance of "=" at the top of the rule list. The word following the equal sign indicates which new set of transformation rules is to be applied. For example, the keyword "what" may have associated with it a transformation rule set of the form

((0) (Why do you ask) (Is that an important question) . . .)

which would apply equally well to the keywords "how" and "when". The entire keyword list for "how" may therefore be

(How (=What))

The keywords "how", "what" and "when" may thus be made to form an equivalence class with respect to the transformation rules which are to apply to them.

In the above example the rule "(=what)" is in the place of a decomposition rule, although it causes no decomposition of the relevant text. It may also appear, however, in the place of a reassembly rule. For example, the keyword "am" may have among others the following transformation rule set associated with it:

((0 are you 0) (Do you believe you are 4) . . . (=what) . . .)

(It is here assumed that "are" has been substituted for "am" and "you" for "I" in the initial text scan.) Then, the input text

Am I sick

would elicit either

Do you believe you are sick

or

Why do you ask

depending on how many times the general form had already occurred.

Under still other conditions it may be desirable to

perform a preliminary transformation on the input text before subjecting it to the decompositions and reassemblies which finally yield the output text. For example, the keyword "you're" should lead to the transformation rules associated with "you" but should first be replaced by a word pair. The dictionary entry for "you're" is therefore:

(you're = I'm ((0 I'm 0) (PRE (I AM 3) (=YOU))))

which has the following effect:

(1) Wherever "you're" is found in the input text, it is replaced by "I'm".

(2) If "you're" is actually selected as the regnant keyword, then the input text is decomposed into three constituent parts, namely, all text in front of the first occurrence of "I'm", the word "I'm" itself, and all text following the first occurrence of "I'm".

(3) The reassembly rule beginning with the code "PRE" is encountered and the decomposed text reassembled such that the words 'I AM" appear in front of the third constituent determined by the earlier decomposition.

(4) Control is transferred, so to speak, to the transformation rules associated with the keyword "you", where further decompositions etc. are attempted.

It is to be noted that the set

(PRE (I AM 3) (=YOU))

is logically in the place of a reassembly rule and may therefore be one of many reassembly rules associated with the given decomposition.

Another form of reassembly rule is

(NEWKEY)

which serves the case in which attempts to match on the currently regnant keyword are to be given up and the entire decomposition and reassembly process is to start again on the basis of the keyword to be found in the keystack. Whenever this rule is invoked, the top of the keystack is "popped up" once, i.e., the new regnant keyword recovered and removed from the keystack, and the entire process reinitiated as if the initial text scan had just terminated. This mechanism makes it possible to, in effect, test on key phrases as opposed to single key words.

A serious problem which remains to be discussed is the reaction of the system in case no keywords remain to serve as transformation triggers. This can arise either in case the keystack is empty when NEWKEY is invoked or when the input text contained no keywords initially.

The simplest mechanism supplied is in the form of the special reserved keyword "NONE" which must be part of any script. The script writer must associate the universally matching decomposition rule (0) with it and follow this by as many content-free remarks in the form of transformation rules as he pleases. (Examples are: "Please go on", "That's very interesting" and "I see".)

There is, however, another mechanism which causes the system to respond more spectacularly in the absence of a key. The word "MEMORY" is another reserved pseudo-keyword. The key list structure associated with it differs from the ordinary one in some respects. An example illuminates this point.

Consider the following structure:

(MEMORY MY
(0 YOUR 0 = LETS DISCUSS FURTHER WHY YOUR 3)
(0 YOUR 0 = EARLIER YOU SAID YOUR 3)
⋮

The word "MY" (which must be an ordinary keyword as well) has been selected to serve a special function. Whenever it is the highest ranking keyword of a text one of the transformations on the MEMORY list is randomly selected, and a copy of the text is transformed accordingly. This transformation is stored on a first-in-first-out stack for later use. The ordinary processes already described are then carried out. When a text without keywords is encountered later and a certain counting mechanism is in a particular state and the stack in question is not empty, then the transformed text is printed out as the reply. It is, of course, also deleted from the stack of such transformations.

The current version of ELIZA requires that one keyword be associated with MEMORY and that exactly four transformations accompany that word in that context. (An application of a transformation rule of the form

(LEFT HAND SIDE = RIGHT HAND SIDE)

is equivalent to the successive application of the two forms

(LEFT HAND SIDE), (RIGHT HAND SIDE).)

Three more details will complete the formal description of the ELIZA program.

The transformation rule mechanism of SLIP is such that it permits tagging of words in a text and their subsequent recovery on the basis of one of their tags. The keyword "MOTHER" in ELIZA, for example, may be identified as a noun and as a member of the class "family" as follows:

(MOTHER DLIST (/NOUN FAMILY)).

Such tagging in no way interferes with other information (e.g., rank or transformation rules) which may be associated with the given tag word. A decomposition rule may contain a matching constituent of the form (/TAG1 TAG2 ···) which will match and isolate a word in the subject text having any one of the mentioned tags. If, for example, "MOTHER" is tagged as indicated and the input text

"CONSIDER MY AGED MOTHER AS WELL AS ME"

subjected to the decomposition rule

(0 YOUR 0 (/FAMILY) 0)

(remembering that "MY" has been replaced by "YOUR"), then the decomposition would be

(1) CONSIDER (2) YOUR (3) AGED (4) MOTHER
(5) AS WELL AS ME.

Another flexibility inherent in the SLIP text manipulation mechanism underlying ELIZA is tha or-ing of matching criteria is permitted in decomposition rules. The above input text would have been decomposed precisely as stated above by the decomposition rule:

(0 YOUR 0 (*FATHER MOTHER) 0)

which, by virtue of the presence of "*" in the sublist structure seen above, would have isolated either the word "FATHER" or "MOTHER" (in that order) in the input text, whichever occurred first after the first appearance of the word "YOUR".

Finally, the script writer must begin his script with a list, i.e., a message enclosed in parentheses, which contains the statement he wishes ELIZA to type when the system is first loaded. This list may be empty.

Editing of an ELIZA script is achieved via appeal to a contextual editing program (ED) which is part of the MAC library. This program is called whenever the input text to ELIZA consists of the single word "EDIT". ELIZA then puts itself in a so-called dormant state and presents the then stored script for editing. Detailed description of ED is out of place here. Suffice it to say that changes, additions and deletions of the script may be made with considerable efficiency and on the basis of entirely contextual cues, i.e., without resort to line numbers or any other artificial devices. When editing is completed, ED is given the command to FILE the revised script. The new script is then stored on the disk and read into ELIZA. ELIZA then types the word "START" to signal that the conversation may resume under control of the new script.

An important consequence of the editing facility built into ELIZA is that a given ELIZA script need not start out to be a large, full-blown scenario. On the contrary, it should begin as a quite modest set of keywords and transformation rules and permitted to be grown and molded as experience with it builds up. This appears to be the best way to use a truly interactive man-machine facility—i.e., not as a device for rapidly debugging a code representing a fully thought out solution to a problem, but rather as an aid for the exploration of problem solving strategies.

Discussion

At this writing, the only serious ELIZA scripts which exist are some which cause ELIZA to respond roughly as would certain psychotherapists (Rogerians). ELIZA performs best when its human correspondent is initially instructed to "talk" to it, via the typewriter of course, just as one would to a psychiatrist. This mode of conversation was chosen because the psychiatric interview is one of the few examples of categorized dyadic natural language communication in which one of the participating pair is free to assume the pose of knowing almost nothing of the real world. If, for example, one were to tell a psychiatrist "I went for a long boat ride" and he responded "Tell me about boats", one would not assume that he knew nothing about boats, but that he had some purpose in so directing the subsequent conversation. It is important to note that this assumption is one made by the speaker. Whether it is realistic or not is an altogether separate question. In any case, it has a crucial psychological utility in that it serves the speaker to maintain his sense of being heard and understood. The speaker further defends his impression (which even in real life may be illusory) by attributing to his conversational partner all sorts of background knowledge, insights and reasoning ability. But again, these are the *speaker's* contribution to the conversation. They manifest themselves inferentially in the *interpretations* he makes of the offered responses. From the purely technical programming point of view then, the psychiatric interview form of an ELIZA script has the advantage that it eliminates the need of storing *explicit* information about the real world.

The human speaker will, as has been said, contribute much to clothe ELIZA'S responses in vestments of plausibility. But he will not defend his illusion (that he is being understood) against all odds. In human conversation a speaker will make certain (perhaps generous) assumptions about his conversational partner. As long as it remains possible to interpret the latter's responses consistently with those assumptions, the speaker's image of his partner remains unchanged, in particular, undamaged. Responses which are difficult to so interpret may well result in an enhancement of the image of the partner, in additional rationalizations which then make more complicated interpretations of his responses reasonable. When, however, such rationalizations become too massive and even self-contradictory, the entire image may crumble and be replaced by another ("He is not, after all, as smart as I thought he was"). When the conversational partner is a machine (the distinction between machine and program is here not useful) then the idea of *credibility* may well be substituted for that of *plausibility* in the above.

With ELIZA as the basic vehicle, experiments may be set up in which the subjects find it credible to believe that the responses which appear on his typewriter are generated by a human sitting at a similar instrument in another room. How must the script be written in order to maintain the credibility of this idea over a long period of time? How can the performance of ELIZA be systematically degraded in order to achieve controlled and predictable thresholds of credibility in the subject? What, in all this, is the role of the initial instruction to the subject? On the other hand, suppose the subject is told he is communicating with a machine. What is he led to believe about the machine as a result of his conversational experience with it? Some subjects have been very hard to convince that ELIZA (with its present script) is *not* human. This is a striking form of Turing's test. What experimental design would make it more nearly rigorous and airtight?

The whole issue of the credibility (to humans) of machine output demands investigation. Important decisions increasingly tend to be made in response to computer output. The ultimately responsible human interpreter of "What the machine says" is, not unlike the correspondent with ELIZA, constantly faced with the need to make credibility judgments. ELIZA shows, if nothing else, how easy it is to create and maintain the illusion of understanding, hence perhaps of judgment deserving of credibility. A certain danger lurks there.

The idea that the present ELIZA script contains *no* information about the real world is not entirely true. For example, the transformation rules which cause the input

> Everybody hates me

to be transformed to

> Can you think of anyone in particular

and other such are based on quite specific hypotheses about the world. The whole script constitutes, in a loose way, a model of certain aspects of the world. The act of writing a script is a kind of programming act and has all the advantages of programming, most particularly that it clearly shows where the programmer's understanding and command of his subject leaves off.

A large part of whatever elegance may be credited to ELIZA lies in the fact that ELIZA maintains the illusion of understanding with so little machinery. But there are bounds on the extendability of ELIZA's "understanding" power, which are a function of the ELIZA program itself and not a function of any script it may be given. The crucial test of understanding, as every teacher should know, is not the subject's ability to continue a conversation, but to draw valid conclusions from what he is being told. In order for a computer program to be able to do that, it must at least have the capacity to store *selected* parts of its inputs. ELIZA throws away each of its inputs, except for those few transformed by means of the MEMORY machinery. Of course, the problem is more than one of storage. A great part of it is, in fact, subsumed under the word "selected" used just above. ELIZA in its use so far has had as one of its principal objectives the *concealment* of its lack of understanding. But to encourage its conversational partner to offer inputs from which it can select remedial information, it must *reveal* its misunderstanding. A switch of objectives from the concealment to the revelation of misunderstanding is seen as a precondition to making an ELIZA-like program the basis for an effective natural language man-machine communication system.

One goal for an augmented ELIZA program is thus a system which already has access to a store of information about some aspects of the real world and which, by means of conversational interaction with people, can reveal both what it knows, i.e., behave as an information retrieval system, and where its knowledge ends and needs to be augmented. Hopefully the augmentation of its knowledge will also be a direct consequence of its conversational experience. It is precisely the prospect that such a program will converse with many people and learn something from each of them, which leads to the hope that it will prove an interesting and even useful conversational partner.

One way to state a slightly different intermediate goal is to say that ELIZA should be given the power to slowly build a model of the subject conversing with it. If the subject mentions that he is not married, for example, and later speaks of his wife, then ELIZA should be able to make the tentative inference that he is either a widower or divorced. Of course, he could simply be confused. In the long run, ELIZA should be able to build up a *belief structure* (to use Abelson's phrase) of the subject and on that basis detect the subject's rationalizations, contradictions, etc. Conversations with such an ELIZA would often turn into arguments. Important steps in the realization of these goals have already been taken. Most notable among these is Abelson's and Carroll's work on simulation of belief structures [1].

The script that has formed the basis for most of this discussion happens to be one with an overwhelmingly psychological orientation. The reason for this has already been discussed. There is a danger, however, that the example will run away with what it is supposed to illustrate. It is useful to remember that the ELIZA program itself is merely a translating processor in the technical programming sense. Gorn [2] in a paper on language systems says:

> Given a language which already possesses semantic content, then a translating processor, even if it operates only syntactically, generates corresponding expressions of another language to which we can attribute as "meanings" (possibly multiple—the translator may not be one to one) the "semantic intents" of the generating source expressions; whether we find the result consistent or useful or both is, of course, another problem. It is quite possible that by this method the same syntactic object language can be usefully assigned multiple meanings for each expression . . .

It is striking to note how well his words fit ELIZA. The "given language" is English as is the "other language", expressions of which are generated. In principle, the given language could as well be the kind of English in which "word problems" in algebra are given to high school students and the other language, a machine code allowing a particular computer to "solve" the stated problems. (See Bobrow's program STUDENT [3].)

The intent of the above remarks is to further rob ELIZA of the aura of magic to which its application to psychological subject matter has to some extent contributed. Seen in the coldest possible light, ELIZA is a translating processor in Gorn's sense; however, it is one which has been especially constructed to work well with natural language text.

REFERENCES

1. Abelson, R. P., and Carroll, J. D. Computer simulation of individual belief systems. *Amer. Behav. Sci. 9* (May 1965), 24–30.
2. Gorn, S. Semiotic relationships in ambiguously stratified language systems. Paper presented at Int. Colloq. Algebraic Linguistics and Automatic Theory, Hebrew U. of Jerusalem, Aug. 1964.
3. Bobrow, D. G. Natural language input for a computer problem solving system. Doctoral thesis, Math. Dept., MIT. Cambridge, Mass., 1964.
4. Weizenbaum, J. Symmetric list processor. *Comm. ACM 6*, (Sept. 1963), 524–544.
5. Rogers, C. *Client Centered Therapy: Current Practice, Implications and Theory.* Houghton Mifflin, Boston, 1951.
6. Yngve, J. *COMIT Programming Manual.* MIT Press, Cambridge, Mass., 1961.

APPENDIX. An ELIZA Script

```
(HOW DO YOU DO.  PLEASE TELL ME YOUR PROBLEM)
START
(SORRY ((0) (PLEASE DON'T APOLIGIZE)
(APOLOGIES ARE NOT NECESSARY) (WHAT FEELINGS
DO YOU HAVE WHEN YOU APOLOGIZE) (I'VE TOLD YOU
THAT APOLOGIES ARE NOT REQUIRED)))
(DONT = DON'T)
(CANT = CAN'T)
(WONT = WON'T)
(REMEMBER 5
((0 YOU REMEMBER 0) (DO YOU OFTEN THINK OF 4)
(DOES THINKING OF 4 BRING ANYTHING ELSE TO MIND)
(WHAT ELSE DO YOU REMEMBER)
(WHY DO YOU REMEMBER 4 JUST NOW)
(WHAT IN THE PRESENT SITUATION REMINDS YOU OF 4)
(WHAT IS THE CONNECTION BETWEEN ME AND 4))
((0 DO I REMEMBER 0) (DID YOU THINK I WOULD FORGET 5)
(WHY DO YOU THINK I SHOULD RECALL 5 NOW)
(WHAT ABOUT 5) (=WHAT) (YOU MENTIONED 5))
((0) (NEWKEY)))
(IF 3 ((0 IF 0) (DO YOU THINK ITS LIKELY THAT 3) (DO YOU WISH THAT 3)
(WHAT DO YOU THINK ABOUT 3) (REALLY, 2 3)))
(WHAT DO YOU THINK ABOUT 3) (REALLY, 2 3)))
(DREAMT 4 ((0 YOU DREAMT 0)
(REALLY, 4) (HAVE YOU EVER FANTASIED 4 WHILE YOU WERE AWAKE)
(HAVE YOU DREAMT 4 BEFORE) (=DREAM) (NEWKEY)))
(DREAMED = DREAMT 4 (=DREAMT))
(DREAM 3 ((0) (WHAT DOES THAT DREAM SUGGEST TO YOU)
(DO YOU DREAM OFTEN) (WHAT PERSONS APPEAR IN YOUR DREAMS)
(DON'T YOU BELIEVE THAT DREAM HAS SOMETHING TO DO WITH
YOUR PROBLEM) (NEWKEY)))
(DREAMS = DREAM 3 (=DREAM))
(HOW (=WHAT))
(WHEN (=WHAT))
(ALIKE 10 (=DIT))
(SAME 10 (=DIT))
(CERTAINLY (=YES))
(FEEL DLIST(/BELIEF))
(THINK DLIST(/BELIEF))
(BELIEVE DLIST(/BELIEF))
(WISH DLIST(/BELIEF))
(MEMORY MY (0 YOUR 0 = LETS DISCUSS FURTHER WHY YOUR 3)
(0 YOUR 0 = EARLIER YOU SAID YOUR 3)
(0 YOUR 0 = BUT YOUR 3)
(0 YOUR 0 = DOES THAT HAVE ANYTHING TO DO WITH THE FACT THAT YOUR 3))
(NONE ((0) (I AM NOT SURE I UNDERSTAND YOU FULLY)
(PLEASE GO ON)
(WHAT DOES THAT SUGGEST TO YOU)
(DO YOU FEEL STRONGLY ABOUT DISCUSSING SUCH THINGS)))
(PERHAPS ((0) (YOU DON'T SEEM QUITE CERTAIN)
(WHY THE UNCERTAIN TONE)
(CAN'T YOU BE MORE POSITIVE)
(YOU AREN'T SURE) (DON'T YOU KNOW)))
(MAYBE (=PERHAPS))
(NAME 15 ((0) (I AM NOT INTERESTED IN NAMES)
(I'VE TOLD YOU BEFORE, I DON'T CARE ABOUT NAMES -
PLEASE CONTINUE)) )
PLEASE CONTINUE)) )
(DEUTSCH (=XFREMD))
(FRANCAIS (=XFREMD))
(ITALIANO (=XFREMD))
(ESPANOL (=XFREMD))
(XFREMD ((0) (I AM SORRY, I SPEAK ONLY ENGLISH)))
(HELLO ((0) (HOW DO YOU DO.  PLEASE STATE YOUR PROBLEM)))
(COMPUTER 50 ((0) (DO COMPUTERS WORRY YOU)
(WHY DO YOU MENTION COMPUTERS) (WHAT DO YOU THINK MACHINES
HAVE TO DO WITH YOUR PROBLEM) (DON'T YOU THINK COMPUTERS CAN
HELP PEOPLE) (WHAT ABOUT MACHINES WORRIES YOU) (WHAT
DO YOU THINK ABOUT MACHINES)))
(MACHINE 50 (=COMPUTER))
(MACHINES 50 (=COMPUTER))
(COMPUTERS 50 (=COMPUTER))
(AM = ARE ((0 ARE YOU 0) (DO YOU BELIEVE YOU ARE 4)
(WOULD YOU WANT TO BE 4) (YOU WISH I WOULD TELL YOU YOU ARE 4)
(WHAT WOULD IT MEAN IF YOU WERE 4) (=WHAT))
((0) (WHY DO YOU SAY 'AM') (I DON'T UNDERSTAND THAT)))
(ARE ((0 ARE I 0 )
(WHY ARE YOU INTERESTED IN WHETHER I AM 4 OR NOT)
(WOULD YOU PREFER IF I WEREN'T 4) (PERHAPS I AM 4 IN YOUR
FANTASIES) (DO YOU SOMETIMES THINK I AM 4) (=WHAT))
((0 ARE 0) (DID YOU THINK THEY MIGHT NOT BE 3)
(WOULD YOU LIKE IT IF THEY WERE NOT 3) (WHAT IF THEY WERE NOT 3)
(POSSIBLY THEY ARE 3)) )
(YOUR = MY ((0 MY 0) (WHY ARE YOU CONCERNED OVER MY 3)
(WHAT ABOUT YOUR OWN 3) (ARE YOU WORRIED ABOUT SOMEONE ELSES 3)
(REALLY, MY 3)))
(WAS 2 ((0 WAS YOU 0 )
(WHAT IF YOU WERE 4) (DO YOU THINK YOU WERE 4)
(WERE YOU 4) (WHAT WOULD IT MEAN IF YOU WERE 4)
(WHAT DOES ' 4 ' SUGGEST TO YOU) (=WHAT))
((0 YOU WAS 0)
(WERE YOU REALLY) (WHY DO YOU TELL ME YOU WERE 4 NOW)
(WERE YOU REALLY) (WHY DO YOU TELL ME YOU WERE 4 NOW)
(PERHAPS I ALREADY KNEW YOU WERE 4) )
((0 WAS I 0) (WOULD YOU LIKE TO BELIEVE I WAS 4)
(WHAT SUGGESTS THAT I WAS 4)
(WHAT DO YOU THINK) (PERHAPS I WAS 4)
(WHAT IF I HAD BEEN 4))
((0) (NEWKEY)) )
(WERE = WAS (=WAS))
(ME = YOU)
(YOU'RE = I'M ((0 I'M 0) (PRE (I ARE 3) (=YOU))))
(I'M = YOU'RE ((0 YOU'RE 0) (PRE (YOU ARE 3) (=I))))
(MYSELF = YOURSELF)
(YOURSELF = MYSELF)
(MOTHER DLIST(/NOUN FAMILY))
(MOM = MOTHER DLIST(/ FAMILY))
(DAD = FATHER DLIST(/ FAMILY))
(FATHER DLIST(/NOUN FAMILY))
(SISTER DLIST(/FAMILY))
(BROTHER DLIST(/FAMILY))
(WIFE DLIST(/FAMILY))
(CHILDREN DLIST(/FAMILY))
(I = YOU
((0 YOU (* WANT NEED) 0) (WHAT WOULD IT MEAN TO YOU IF YOU GOT 4)
(WHY DO YOU WANT 4) (SUPPOSE YOU GOT 4 SOON) (WHAT
IF YOU NEVER GOT 4) (WHAT WOULD GETTING 4 MEAN TO
YOU) (WHAT DOES WANTING 4 HAVE TO DO WITH THIS DISCUSSION))
((0 YOU ARE 0 (*SAD UNHAPPY DEPRESSED SICK ) 0)
(I AM SORRY TO HEAR YOU ARE 5) (DO YOU THINK COMING HERE
WILL HELP YOU NOT TO BE 5) (I'M SURE ITS NOT PLEASANT TO
BE 5) (CAN YOU EXPLAIN WHAT MADE YOU 5))
((0 YOU ARE 0 (*HAPPY ELATED GLAD BETTER ) 0)
(HOW HAVE I HELPED YOU TO BE 5)
(HAS YOUR TREATMENT MADE YOU 5) (WHAT MAKES YOU 5 JUST
NOW) (CAN YOU EXPLAIN WHY YOU ARE SUDDENLY 5))
((0 YOU WAS 0) (=WAS))
((0 YOU WAS 0) (=WAS))
((0 YOU (/BELIEF) YOU 0) (DO YOU REALLY THINK SO) (BUT YOU ARE
NOT SURE YOU 5) (DO YOU REALLY DOUBT YOU 5))
((0 YOU 0 (/BELIEF) 0 I 0) (=YOU))
((0 YOU ARE 0)
(IS IT BECAUSE YOU ARE 4 THAT YOU CAME TO ME)
(HOW LONG HAVE YOU BEEN 4)
(DO YOU BELIEVE IT NORMAL TO BE 4)
(DO YOU ENJOY BEING 4))
((0 YOU (* CAN'T CANNOT) 0) (HOW DO YOU KNOW YOU CAN'T 4)
(HAVE YOU TRIED)
(PERHAPS YOU COULD 4 NOW)
(DO YOU REALLY WANT TO BE ABLE TO 4))
((0 YOU DON'T 0)(DON'T YOU  REALLY 4)(WHY DON'T YOU 4)
(DO YOU WISH TO BE ABLE TO 4) (DOES THAT TROUBLE YOU))
((0 YOU FEEL 0) (TELL ME MORE ABOUT SUCH FEELINGS)
(DO YOU OFTEN FEEL 4)
(DO YOU ENJOY FEELING 4)
(OF WHAT DOES FEELING 4 REMIND YOU))
((0 YOU 0 I 0) (PERHAPS IN YOUR FANTASY WE 3 EACH OTHER)
(DO YOU WISH TO 3 ME)
(YOU SEEM TO NEED TO 3 ME)
(DO YOU 3 ANYONE ELSE))
```

```
((0) (YOU SAY 1)
(CAN YOU ELABORATE ON THAT) (DO YOU SAY 1 FOR SOME SPECIAL REASON)
(THAT'S QUITE INTERESTING)))
(YOU = I ((0 I REMIND YOU OF 0) (=DIT))
((0 I ARE 0) (WHAT MAKES YOU THINK I AM 4)
(DOES IT PLEASE YOU TO BELIEVE I AM 4)
(DO YOU SOMETIMES WISH YOU WERE 4)
(PERHAPS YOU WOULD LIKE TO BE 4))
((0 I 0 YOU) (WHY DO YOU THINK I 3 YOU)
(YOU LIKE TO THINK I 3 YOU - DON'T YOU)
(WHAT MAKES YOU THINK I 3 YOU)
(REALLY, I 3 YOU) (DO YOU WISH TO BELIEVE I 3 YOU)
(REALLY, I 3 YOU) (DO YOU WISH TO BELIEVE I 3 YOU)
(SUPPOSE I DID 3 YOU - WHAT WOULD THAT MEAN)
(DOES SOMEONE ELSE BELIEVE I 3 YOU))
((0 I 0) (WE WERE DISCUSSING YOU - NOT ME)
(OH, I 3) (YOU'RE NOT REALLY TALKING ABOUT ME - ARE YOU)
(WHAT ARE YOUR FEELINGS NOW)))
(YES ((0) (YOU SEEM QUITE POSITIVE) (YOU ARE SURE)
(I SEE) (I UNDERSTAND)))
(NO ((0) (ARE YOU SAYING 'NO' JUST TO BE NEGATIVE)
(YOU ARE BEING A BIT NEGATIVE) (WHY NOT) (WHY 'NO')))
(MY = YOUR 2 ((0 YOUR 0 (/FAMILY) 0)
(TELL ME MORE ABOUT YOUR FAMILY) ( WHO ELSE IN YOUR FAMILY
5) (YOUR 4) (WHAT ELSE COMES TO MIND WHEN YOU THINK OF
YOUR 4))
((0 YOUR 0) (YOUR 3) (WHY DO YOU SAY YOUR 3) (DOES THAT
SUGGEST ANYTHING ELSE WHICH BELONGS TO YOU) (IS IT
IMPORTANT TO YOU THAT 2 3)))
(CAN ((0 CAN I 0)  (YOU BELIEVE I CAN 4 DON'T YOU) (=WHAT)
(YOU WANT ME TO BE ABLE TO 4) (PERHAPS YOU WOULD LIKE TO BE
ABLE TO 4 YOURSELF)) ((0 CAN YOU 0) (WHETHER OR NOT YOU CAN 4
DEPENDS ON YOU MORE THAN ON ME) (DO YOU WANT TO BE ABLE TO 4)
(PERHAPS YOU DON'T WANT TO 4) (=WHAT)))
(WHAT ((0) (WHY DO YOU ASK) (DOES THAT QUESTION INTEREST YOU)
(WHAT IS IT YOU REALLY WANT TO KNOW)
(ARE SUCH QUESTIONS MUCH ON YOUR MIND)
(WHAT ANSWER WOULD PLEASE YOU MOST)
(WHAT DO YOU THINK) (WHAT COMES TO YOUR MIND WHEN YOU ASK THAT)
(HAVE YOU ASKED SUCH QUESTION BEFORE)
(HAVE YOU ASKED ANYONE ELSE)))
(BECAUSE ((0) (IS THAT THE REAL REASON) (DON'T ANY OTHER REASONS
COME TO MIND) (DOES THAT REASON SEEM TO EXPLAIN ANYTHING ELSE)
(WHAT OTHER REASONS MIGHT THERE BE)))
(WHY ((0 WHY DON'T I 0) (DO YOU BELIEVE I DON'T 5) (PERHAPS I
WILL 5 IN GOOD TIME) (SHOULD YOU 5 YOURSELF) (YOU WANT ME TO 5)
(=WHAT))
(=WHAT))
((0 WHY CAN'T YOU 0) (DO YOU THINK YOU SHOULD BE ABLE TO 5)
(DO YOU WANT TO BE ABLE TO 5) (DO YOU BELIEVE THIS WILL HELP YOU
TO 5) (HAVE YOU ANY IDEA WHY YOU CAN'T 5) (=WHAT))
(= WHAT))
(EVERYONE 2 ((0 (* EVERYONE EVERYBODY NOBODY NOONE) 0 )
(REALLY, 2) (SURELY NOT 2) (CAN YOU THINK OF
ANYONE IN PARTICULAR) (WHO, FOR EXAMPLE) (YOU ARE THINKING OF
A VERY SPECIAL PERSON)
(WHO, MAY I ASK) (SOMEONE SPECIAL PERHAPS)
(YOU HAVE A PARTICULAR PERSON IN MIND, DON'T YOU) (WHO DO YOU
THINK YOU'RE TALKING ABOUT)))
(EVERYBODY 2 (= EVERYONE))
(NOBODY 2 (=EVERYONE))
(NOONE 2 (=EVERYONE))
(ALWAYS 1 ((0) (CAN YOU THINK OF A SPECIFIC EXAMPLE) (WHEN)
(WHAT INCIDENT ARE YOU THINKING OF) (REALLY, ALWAYS)))
(LIKE 10 ((0 (*AM IS ARE WAS) 0 LIKE 0) (=DIT))
((0) (NEWKEY)) )
(DIT ((0) (IN WHAT WAY)  (WHAT RESEMBLANCE DO YOU SEE)
(WHAT DOES THAT SIMILARITY SUGGEST TO YOU)
(WHAT OTHER CONNECTIONS DO YOU SEE)
(WHAT DO YOU SUPPOSE THAT RESEMBLANCE MEANS)
(WHAT IS THE CONNECTION, DO YOU SUPPOSE)
(COULD THERE REALLY BE SOME CONNECTION)
(HOW)))
()
```

Received September, 1965

LETTERS—continued from p. 35

The technique consists of translating the code for the letter "O" to the code for the numeral 0 whenever it is encountered in the input character string. If the string consists only of items such as numbers and names and it is necessary to sort alphabetically on names, the occurrence of an alphabetic character within a name field is used to cause the code for zero to be retranslated to the code for the letter "O" by a rescan of the characters in the name field.

If no sorting is required, the retranslation can be avoided, provided that delimiters such as FORMAT or GO TO are spelled with zero within the recognizer segment of a translator. It is also necessary to redefine identifier as

⟨identifier⟩ ::= ⟨letter⟩ | ⟨identifier⟩ ⟨letter⟩ | ⟨identifier⟩ ⟨digit⟩ | ⟨0⟩ ⟨identifier⟩

where it is understood that the letter "O" is removed from the standard definition of letter as in ALGOL 60. The redefinition permits the inclusion of identifiers such as ODD or OOPS but prevents the use of an identifier consisting only of the repeated mark O.

This technique requires consistency of use and might result in chaos in a warehousing operation in which the letter "O" is used in parts labels with check digits.

L. Richard Turner
NASA Lewis Research Center
Cleveland, Ohio

Comments on a Problem in Concurrent Programming Control

Dear Editor:

I would like to comment on Mr. Dijkstra's solution [Solution of a problem in concurrent programming control. *Comm ACM 8* (Sept. 1965), 569] to a messy problem that is hardly academic. We are using it now on a multiple computer complex.

When there are only two computers, the algorithm may be simplified to the following:

```
Boolean array b(0; 1) integer k, i, j,
comment This is the program for computer i, which may be
   either 0 or 1, computer j ≠ i is the other one, 1 or 0;
C0: b (i) := false;
C1: if k ≠ i then begin
C2: if not b(j) then go to C2;
    else k := i; go to C1 end;
    else critical section;
    b(i) := true;
    remainder of program;
    go to C0;
    end
```

Mr. Dijkstra has come up with a clever solution to a really practical problem.

Harris Hyman
Munitype
New York, New York

Endnotes

Foreword

1 For more on Rogerian therapy, see chapter 9.
2 Weizenbaum, *Computer Power and Human Reason*, 6.
3 Weizenbaum, "ELIZA," 36.
4 Weizenbaum, "ELIZA," 42.
5 Bobrow, "Turing Test Passed."
6 Murray, *Hamlet on the Holodeck* (1997), chap. 3; Murray, *Hamlet on the Holodeck* (2017), chap. 3.
7 Murray, *Hamlet on the Holodeck* (1997, 2017), chap. 3.

Chapter 01
Please Go On: An Introduction

This epigraph comes from Weizenbaum, "ELIZA," 37.

1 Aristotle wrote in his *Politics* of a world where "every instrument could accomplish its own work, obeying or anticipating the will of others, like the statues of Daedalus, or the tripods of Hephaestus." However, note that the "instruments" Aristotle is referring to are not machines but human slaves. This idea of a servant working for a master is a common story that OpenAI, Anthropic, and other AI companies repeat today. Gender and race are crucial dimensions of the division of labor in societies, themes which are echoed in AIs and Chatbots today. Aristotle, *Politics*, Book1, 4. 1253b23–1254a1.
2 "Toasters" is the pejorative term used in the reboot of the series *Battlestar Galactica* (2004–2008) for the Cylons, a race of sentient machines, in reference to the chrome-plated armor of the original models. The insult betrays human insecurity about the newer models that are almost indistinguishable from humans.
3 Technically ELIZA is a system for chatbot personas, and DOCTOR is the persona most people think of as ELIZA. For the purposes of this book, we will refer to ELIZA as a chatbot, even as later chapters explain this distinction.
4 In 2025 Oxford University became first UK university to offer ChatGPT to all staff and students, "giving students more opportunities for personalized learning, improving efficiencies for faculty, and helping advance research breakthroughs." University of Oxford. 'Oxford Becomes First UK University to Offer ChatGPT Edu to All Staff and Students | University of Oxford'. *News and Events*, September 2025. https://www.ox.ac.uk/news/2025-09-19-oxford-becomes-first-uk-university-offer-chatgpt-edu-all-staff-and-students.
5 Weizenbaum, "ELIZA"; Weizenbaum, "Contextual Understanding by Computers"; Weizenbaum, *Computer Power and Human Reason*; Marino, "Critical Code Studies."
6 Montfort, "Continuous Paper."
7 Weizenbaum originally created SLIP as a library for list processing to be used with Fortran. As described later in this book, on the strength of this work, Weizenbaum was invited to join Project MAC at MIT. One of the core goals of Project MAC was to demonstrate the power of the new concept of interactive time-sharing, embodied by CTSS, an interactive time-sharing system. CTSS was originally conceived and developed by MIT Computation Center personnel, led by Fernando Corbató. Lane et al., "ELIZA Reanimated."
8 Tom Van Vleck adds more detail: "All Eliza data was stored in single case, in the BCD character set. The computer typed these letters out in upper case. On M33 and M35 Teletypes, input typing would appear in uppercase too, since there was no lowercase. On Selectric terminals such as the IBM 1050 and IBM 2741, user input would appear as lowercase, and system response as uppercase. ... Using the upper/lowercase libraries and '12-bit mode,' [MAD] could handle input that was in 12-bit BCD. This would have been a fairly tedious change to ELIZA. ... I think mixed case output was seen as a 'decoration' in those days. ... It made data files twice as big and had other mode switching problems with CTSS, a nd I think it would have actually detracted from the point Joe was trying to make about computer interaction." Montfort, "Continuous Paper."
9 See instruction to PRINT SPEAK MAD at the top of the ELIZA code in chapter 6. We also note that Weizenbaum likely did not refer to the program as ELIZA before 1965.
10 Weizenbaum, "ELIZA."
11 MAD-SLIP (Michigan Algorithm Decoder and Symmetric List Processor) is discussed further in chapters 3 and 4.
12 A notable example: We have confirmed that ELIZA is Turing complete. In other words, any possible Turing machine program can be implemented by a suitable ELIZA script in versions of ELIZA with the PRE function. Turing completeness is significant because it appears that any effectively calculable function can be computed by some Turing machine. This is known as the Church-Turing Thesis, for which Turing gave powerful arguments in sections 9 and 10 of his classic 1936 paper in which introduced his "computing machines"; see Turing, "On Computable Numbers." To say that ELIZA is Turing complete does not, of course, mean that every possible Turing machine program could in practice have been executed on Weizenbaum's software and hardware in 1966. For both would have had finite capacity (whereas a Turing machine tape is potentially of infinite length). The computing machines in Turing's original 1936 article were designed to write potentially infinite sequences of

binary digits onto a tape, going on and on without ever stopping. Clearly, this would be beyond the practical representational capacity of any physical hardware, no matter how the Turing machine was implemented. But the syntax of the ELIZA rule set involves no relevant restriction—e.g., there is no specified limit on list size—and hence could replicate the abstract processing of any arbitrary Turing machine. So any (terminating) Turing machine program could be executed by an ELIZA script consisting of such transformations if run on software that faithfully implements Weizenbaum's specification, and on hardware of sufficient capacity. This also means that ELIZA could theoretically implement a "Universal Turing Machine" (as introduced in section 6 of Turing's 1936 paper), which can itself execute any Turing machine program. This is not of any serious practical use; however, it is theoretically interesting that ELIZA, as defined by Weizenbaum, has such computational power in principle.

13 We use "messy" in a nod to John Law, who recommends "non-coherent or 'messy' methods," Law, "Making a Mess with Method."

14 Weizenbaum, "ELIZA."

15 See a version of the Lisp ELIZA by Bernie Cossell at Cosell, "jeffshrager/elizagen.org.".

16 North, "Psychoanalysis(?)"

17 We do not take "parody" in the sense of "a mocking version" but instead Weizenbaum's unconventional usage of "simplified sketch."

18 Weizenbaum, "ELIZA," 38.

19 For example, in one broad misconception: In that same article in *CACM*, Weizenbaum describes ELIZA as a program "which makes natural language conversation with a computer possible." This implies that did not select the psychotherapist persona in order to develop research on automated therapy, but rather because of his interest in human–computer interaction. Nevertheless, because ELIZA "passed" as person-like, it received (and warranted) considerable attention. Audiences often bought into the conceit that they were chatting with another person, despite the title of the article: "A Computer Program for the Study of Natural Language Communication between Man and Machine." Along the way, ELIZA inadvertently introduced artificial intelligence to audiences outside specialized fields within computer science, academia, and the military—far beyond Weizenbaum's initial intention. Weizenbaum, "ELIZA," 36.

20 "ELIZA Archaeology."

21 Weizenbaum, "Contextual Understanding by Computers," 478.

22 Weizenbaum, "Myth of the Last Metaphor," 254.

23 Our code investigations revealed that ELIZA is a system (what we might today call a platform and at the time would be called a "programming framework"—meaning that it supported running scripts). Weizenbaum also referred to it as "a family of programs." Massachusetts Institute of Technology, "Project MAC Progress Report III"; Weizenbaum, "Contextual Understanding by Computers," 474.

24 Massachusetts Institute of Technology, "Project MAC Progress Report II," 40; Massachusetts Institute of Technology, "Project MAC Progress Report III," 168.

25 Weizenbaum, "Contextual Understanding by Computers"; Taylor, "Automated Tutoring and Its Discontents."

26 David Berry, in CCS Working Group, "Original ELIZA."

27 Weizenbaum created an interpreted language called OPL, the "Open-ended Programming Language," using Fortran-SLIP. See chapter 10.4.

28 Weizenbaum, "ELIZA," 36.

29 Weizenbaum, *Computer Power and Human Reason*, 108.

30 Weizenbaum, *Computer Power and Human Reason*, 109.

31 Weizenbaum, "Letters to the Editor: On the Implementation of SLIP." 263.

32 Weizenbaum, *Computer Power and Human Reason*, 6–7.

33 Jargon File Archive, "Jargon File, Version 2.9.6."

34 Turkle, *Life on the Screen*, 101.

35 Hofstadter, *Fluid Concepts and Creative Analogies*, 167.

36 The term *Turing Test* was applied only much later, possibly as late as 1974. Turing uses the term "imitation game," and proposes a variety of potential versions.

37 Sadie Plant describes the irony of this, given Turing's own complicated relationship to gender and sexuality that perhaps informed his design of the test: "His intelligence test was used to guarantee the distinction between man and machine, and his name became synonymous with the systems of security he subverted. ... [O]nce the war was over, his sexuality seemed symptomatic of his troubling tendency to use his equipment in ways his training had been intended to preclude. Turing was subjected to his own test. Was he a real man, a proper human being, committed to the reproduction of humanity? Or was he some other, wayward track?"; Plant, *Zeros + Ones*, 100–101.

38 Weizenbaum, "ELIZA," 43. Original emphasis.

39 Weizenbaum has a very clear explanation of the Turing Test in *Computer Power and Human Reason* and how it relates to transformation rules. He uses the example of a toilet roll, extended to use the sheets of the toilet paper as cells on which are placed black and white stones and the use of a dice. Weizenbaum, *Computer Power and Human Reason*, 52.

40 Weizenbaum, *Computer Power and Human Reason*, 188. Weizenbaum later said he came across the character from an adaptation of the play into the musical *My Fair Lady*, which debuted on Broadway in 1956, with a film version released in 1964. Weizenbaum et al., Islands *in the Cyberstream*, 87. In the original Pygmalion-Galatea myth, says Critical Code Studies Working Group co-organizer Zachary Mann, "Aphrodite turns the statue of 'a perfect woman' (with whom the sculptor

fells [*sic*] in love) into a living woman (did she stay 'perfect'?). [It was] the human who projected humanness onto ELIZA. Although maybe in this analogy ELIZA is really Aphrodite bringing life to the 'perfect' DOCTOR"; Zachary Mann, in CCS Working Group, "Original ELIZA."

41 Butler, *Gender Trouble*; Butler, *Bodies That Matter*.

42 Butler, "Critically Queer."

43 Butler's theory of performativity draws on the notion of "speech acts" from J. L. Austin, *How to Do Things with Words*.

44 Theorizing the posthuman, N. Katherine Hayles, reminds us of the crucial role of embodiment in identity, despite a (masculine) desire to pretend humans are just wetware, a set of processes and information awaiting upload into a new container; Hayles, *How We Became Posthuman*.

45 For more on the "Eliza effect" and gender from Pygmalion to Weizenbaum, see Dillon, "Eliza Effect and Its Dangers."

46 Syntactic researchers focused on modeling and describing grammar, or hierarchical word order. For example, in 1956 Noam Chomsky developed an impactful approach that led to efficient parsing algorithms. In that approach, he dismissed research on large text corpora. He argued that, no matter how large, text could not address a person's entire mental framework of their language. Chomsky, *Syntactic Structures*.

47 *Semantic* researchers focused on modeling and describing meaning. For example, in 1957 Charles Osgood suggested labeling words based on metrics like "valence," "arousal," and "dominance," and then mapping these into spatialized vectors. None of these approaches created systems that understand, in the human sense; but they each point to different understandings of "understanding" and "meaning," by what they choose to emphasize when they attempt to convert understanding to a rule-based process. Osgood et al., *Measurement of Meaning*.

48 Meanwhile, *statistical modeling* researchers in the 1990s realized that with growing processing power, computational systems could emulate understanding by statistical means. They gathered large amounts of text examples and processed these to predict text. These relied on a flurry of new ML architectures like LSTMs (long-short term memory) and RNNs (recurrent neural networks) mixed with new computational linguistics techniques and corpora-processing capabilities.

49 As these grew into larger machine learning systems by the 2010s and 2020s, today's *stochastic* approaches apply an element of uncertainty to NLP and require huge environmental resources. These are a far cry from ELIZA, but the contemporary LLM have strong ties to the first chatbot (see chapter 11).

50 Weizenbaum, *Computer Power and Human Reason*, 266.

51 Atairu, "AI for Art Educators."

52 To deal with the problem of large, complex, data for an AI, Model Context Protocol (MCP) was developed by Anthropic in 2024. This addresses the need to access extensive external information (e.g., code libraries) whilst also managing a context window for the current task. Previously a user was required to cram vast amounts of text data into a limited context space or creating an ad-hoc system using an API. MCP is now a standard that defines how LLMs discover, access, and interact with external materials through a client-server architecture.

53 As Mike Ananny argues, generative AI is a public problem, "simultaneously an object and an agent of public life ... [with the] capacity to participate in shaping the very conversations that might hold it accountable"; Ananny, "Making Generative Artificial Intelligence."

54 Winner, Landon, *Autonomous Technology: Technics-out-of-Control as a Theme in Political Thought* (1978), 21. As Winner goes on to note, "the metaphor employed by those who express this hope is the same as that introduced 2,500 years ago: something must be enslaved in order that something else may win emancipation."

55 Marino, "I, Chatbot."

56 Crawford, Kate. "Eating the Future: The Metabolic Logic of AI Slop." E-Flux, Intensification, September 2025. https://www.e-flux.com/architecture/intensification/6782975/eating-the-future-the-metabolic-logic-of-ai-slop.

57 "ELIZA Archaeology."https://sites.google.com/view/elizaarchaeology/try-eliza

58 The source code is here: https://github.com/jeffshrager/elizagen.org/tree/master/ELIZA33/current. An emulated version can be found here: https://rawcdn.githack.com/jeffshrager/elizagen.org/88764fba8d957f2a5a66d40f6530a41117a77522/ELIZA33/current/ELIZA33.html.

59 Lane et al., "ELIZA Reanimated: Restoring the Mother."

60 CTSS, first introduced in 1961 at MIT, ran on the IBM 7094, an early transistorized computer with only 64,000 words of memory. One of the first time-sharing systems, around 30 users could engage with the computer in real time via keyboard terminals. Rupert Lane used CTSS and 7094 emulators by Dave Pitts and Paul Pierce, with guidance from CTSS experts Jerry Saltzer and Tom Van Vleck; see Walden and Vleck, *Compatible Time Sharing System (1961–1973)*.

61 R. Lane, "eliza-ctss/RECONSTRUCTION-CARD.md," github.com/rupertl, https://github.com/rupertl/eliza-ctss/blob/main/RECONSTRUCTION-CARD.md.

62 R. Lane, "eliza-ctss," github.com/rupertl. Available: https://github.com/ rupertl/eliza-ctss.

63 Berry, David M. 'Digital Ruins and Critical Code Studies: Towards an Ethics of Historical Software Reconstruction'. *Digital Ruins and Critical Code Studies*, 2025. https://stunlaw.blogspot.com/2025/01/digital-ruins-and-critical-code-studies.html.

64 Marino, *Critical Code Studies*; Marino, "Critical Code Studies."

65 By "snapshot," we do not mean the technical notion of a "snapshot" but instead are enlisting the metaphor of a photograph to mark this both as a momentary instantiation of a larger developing software project and as a material document of a particular stage.

66 Berry, *Philosophy of Software*, 9–10.

67 Montfort et al., . *10 PRINT CHR$(205.5+RND(1)); : GOTO 10* ; Pressman et al., *Reading Project*.

68 Bertram and Montfort, *Output*; Cox and Soon, *Aesthetic Programming*. ; Emerson, *Other Networks*; Emerson, *Reading Writing Interfaces*; Kirschenbaum *Mechanisms*; McPherson "Why are the Digital Humanities So White?"; Wardrip-Fruin, *Expressive Processing*; Yang, Katherine. "Coem." Katherine Yang. https://kayserifserif.place/work/coem/ .

69 Atairu, "AI for Art Educators."; Bender et al., "On the Dangers of Stochastic Parrots."; Buolamwini, *Unmasking AI*; Crawford and Joler, "Anatomy of an AI System"; Crawford and Joler, ""Calculating Empires"; Hayles, "Subversion of the Human Aura"; Chun, *Discriminating Data*; Raley and Rhee, "Critical AI; D'Ignazio, and Klein, *Data Feminism*; Griffiths, Catherine. "Toward Counteralgorithms: The Contestation of Interpretability in Machine Learning."; Hicks, *Programmed Inequality*; Noble et al., *Intersectional Internet*; Morrison, "Voluptuous Disintegration."

70 Mullaney et al., *Your Computer Is on Fire*.; Thylstrup et al., *Uncertain Archives*.

71 Berry, *Philosophy of Software*; Berry, *Critical Theory and the Digital*; Marino, "Critical Code Studies"; Connolly, "Why Computing Belongs Within the Social Sciences."

72 CCS Working Group, "Original ELIZA in MAD-SLIP (2022 Code Critique)."

73 We borrow that distinction ("text" and "work") from Roland Barthes. In "From Work to Text," he distinguishes the work, the individual work of art, from the larger and near infinite text, "that social space which leaves no language safe, outside, nor any subject of the enunciation in a position as judge, master, analyst, confessor, decoder." Quoted in Marino, *Critical Code Studies*, 75.

74 Weil, A, "Conversations with a Mechanical Psychiatrist."

Dialogue
ELIZA/DOCTOR: Conversation March 5, 1965

1 Weizenbaum, "Computer Conversations."

Chapter 02
A Life in Software: Joseph Weizenbaum

1 Weizenbaum died on March 5, 2008, in Germany.

2 Pamela McCorduck, "Joseph Weizenbaum—Transcripts." Carnegie Mellon University, Carnegie Mellon University Archives, March 6, 1975, box 3, folder 8. Pamela McCorduck Collection, 1978–0001, p. 9. https://findingaids.library.cmu.edu/repositories/2/archival_objects/22124.

3 Dembart, "Experts Argue Whether Computers Could Reason."

4 Sack, "Joseph Weizenbaum and His Famous Eliza."

5 "Joseph Weizenbaum, Professor Emeritus of Computer Science, 85."

6 It seems that Weizenbaum was not active in David's early life, and David only met Weizenbaum, together with his sisters, for the first time in his thirties. David had an undergraduate degree in English from Harvard University and three master's degrees (in English, American History, and German) from Stanford University. He was fluent in German and Italian, having spent a year living in Italy on a Fulbright scholarship. Ironically, David became a software engineer at Atari in the 1970s in the early days of the computer revolution. He developed software while working at IBM and developed several patents (Mercury News 2024).

7 Tarnoff, "AI Criticism Has a Decades-Long History."

8 Campbell-Kelly, "Professor Joseph Weizenbaum."

9 SRI International was founded as Stanford Research Institute, a nonprofit corporation, by Stanford University in 1946. SRI became independent of the university in 1970, and changed its name to SRI International in 1977, as it became increasingly international in focus (see "About Us," SRI International, www.sri.com/about).

10 Fisher and McKenney, "Development of the ERMA Banking System," 44.

11 Fisher and McKenney, 55.

12 Rheingold, *Tools for Thought*.

13 Weizenbaum was present for the launch itself on ERMA day as he joined the company in 1955–56. Weizenbaum is not listed as one of the principal engineers or contributors before the launch described by Fisher and McKenney, "Development of the ERMA Banking System," although it is likely that he would have heard about the ruse of disguising the operation of the computer by placing an engineer between the computer and the demonstrator.

14 Weizenbaum, "Symmetric List Processor," 524.

15 Minsky, quoted in Weizenbaum, "How to Make a Computer," 24.

16 The game Weizenbaum ("How to Make a Computer") described was actually five-in-a-row, rather than checkers which has been described as "machine learning" in Samuel's (1959) paper, "Some Studies in Machine Learning Using the Game of Checkers," although Weizenbaum does reference Shannon's papers on chess 1950 and 1955, and would have likely been very influenced by Samuel's 1959 paper.

17 Weizenbaum, "How to Make a Computer," 24.

18 Weizenbaum, quoted in Crevier, *AI*, 133. One wonders if Weizenbaum had read *The Big Con*, published in 1940 and written by David Maurer, who writes, "Now, confidence men do not go about burdened down with bales of script written out in advance to take care of

every situation which may develop. They do not write out anything. But they know from experience the situations which are likely to come up, and know their lines letter-perfect for those situations. Then there are a number of stock variations which can be used if the office is given by the insideman. All the players know these variations by rote, and can swing into their routine at the given signal"; Maurer, 161; see also Weil, "Seriously Writing SIRI," which makes a similar connection). A better description of ELIZA is hard to imagine. Weizenbaum, interestingly, also has a short digression about the "confidence man," in Weizenbaum, *Computer Power and Human Reason*, 121.

19 Crevier, *AI*, 133.

20 McCorduck, "Joseph Weizenbaum," 6.

21 McCorduck, 6. The idea of being a "con man" is a topic that Weizenbaum repeatedly refers to in relation to his work over the course of his life. He further says, "As I in fact pointed out in the papers I wrote about Eliza. The whole thing is a con job, the whole thing is. It's very much like fortune telling. It's an illusion creating machine and all that sort of stuff. But on a more detailed level as well. By the way, when I say con here I hope you'll believe me that I say this entirely without any guilt associated with it. I don't mean, wow I lied and cheated and so on. That's not what I mean at all"; McCorduck, 11.

22 McCorduck, 7.

23 McCorduck, 9–10.

24 Tarnoff, "AI Criticism Has a Decades-Long History."

25 Tarnoff."

26 Levey, "Dr. Computer Is a Fink.". See also Tarnoff, "AI Criticism Has a Decades-Long History."

27 Weizenbaum later held academic appointments at Harvard University's Graduate School of Education, Stanford University, the Technical University of Berlin and the University of Hamburg in Germany. He was a fellow of the American Association for the Advancement of Science, a member of the New York Academy of Science, and of the European Academy of Science (MIT 2008).

28 *Communications Explosion* (1967).

29 Robert M. Graham (1929–2020) was a computer scientist who early in his career at the University of Michigan coauthored two compilers (GAT for the IBM 650 and the MAD language for the IBM 704/709/7090), In 1963 he moved to MIT to work on the development of MULTICS, their pioneering time sharing system. This was a major project that took about seven years from the initial planning until the system was in daily use by a large community of users. He was one of the principal designers, with particular responsibility for protection, dynamic linking, and other key system kernel areas.

30 McCorduck, "Joseph Weizenbaum," 292. Weizenbaum discusses his drives and the "nontrivial influence" his conversations with Yngve about COMIT had on his thinking. ELIZA contains a function named after Yngve (YMATCH), and pattern matching would become the core of ELIZA's operations.

31 Crevier, *AI*,133–34.

32 McCorduck, "Joseph Weizenbaum," 13.

33 Watson, "Joe and Eliza."

34 Weizenbaum, *Computer Power and Human Reason*, 7.

35 Boden, *Artificial Intelligence and Natural Man*, 108. Weizenbaum, "ELIZA," 42.

36 Crevier, *AI* 136. The notion of the form of the chatbot is interesting as it is both increasingly seen as a platonic form of computational interaction, and yet its historical genealogy is revealing as to its origin and derivation from previous modes of human interaction, such as the parlor game that Turing drew from.

37 Wilks, *Machine Conversations*, 1.

38 Wilks, 1.

39 Wilks, *Machine Conversation* 1–2.

40 Colby et al., "Artificial Paranoia."

41 Weizenbaum, Computer Power and Human Reason 5–6.

42 Weizenbaum, "Automating Psychotherapy."

43 Weizenbaum, "Computers as 'Therapists."

44 Weizenbaum took the term *artificial intelligentsia* from an article by Louis Fein published in 1964; Fein, "Artificial Intelligentsia."

45 See "Science for the People," Wikipedia, last modified July 30, 2025, https://en.wikipedia.org/wiki/Science_for_the_People.

46 Tarnoff, "AI Criticism Has a Decades-Long History."

47 Tarnoff.

48 Long, "Turncoat of the Computer Revolution." 47.

49 Aaron, "Weizenbaum Examines Computers and Society."

50 Loeb, "Tech Criticism Before the Techlash."; Mumford, *Technics and Civilization*; Loeb "Lamp and the Lighthouse."

51 Horkheimer, *Eclipse of Reason*; Horkheimer and Adorno, *Dialectic of Enlightenment:*; Berry, *Critical Theory and the Digital.*

52 Arendt, *Origins of Totalitarianism*; Arendt, *Crises of the Republic.*

53 Weizenbaum, *Computer Power and Human Reason*, 13; Dreyfus, "Artificial Intelligence"; Dreyfus, *What Computers Still Can't Do.*

54 Chomsky, *Syntactic Structures*; Chomsky, *Cartesian Linguistics.*

55 Fromm, *Marx's Concept of Man.*

56 Long, "Turncoat of the Computer Revolution," 50.

57 Quoted in Weizenbaum, *Computer Power and Human Reason*, 1.

58 Weizenbaum, "On the Impact of the Computer on Society," 613.

59 We might note here the gendering of computational systems and in particular chatbots from ELIZA to contemporary chatbot interfaces, such as Siri (for a fuller discussion, see Marino, "Critical Code Studies").

60 Long, "Turncoat of the Computer Revolution," 50.

61 Weizenbaum, *Computer Power and Human Reason*, 10.

62 Weizenbaum, *Computer Power and Human Reason*, 2.

63 Weizenbaum helped found the journal *AI & Society: Knowledge, Culture and Communication* in 1987, which is committed to interdisciplinary, critical, and contextual research on AI and computation. The founding journal board members were Mike Cooley, Joseph Weizebaum, Daniel Dennett, Michael Dummett, Karamjit S. Gill, Swasti Mitter, Hubert Dreyfus, David Noble, Seymour Papert, and Joseph Weizenbaum.
64 Berry, *Critical Theory and the Digital*.
65 Ciston et al., "Can Open-Source Fix Predictive Policing?"
66 Weizenbaum, *Computer Power and Human Reason*, 236.
67 Weizenbaum, "Computers as 'Therapists," quoting Rogers, *On Becoming a Person*, 86.
68 Weizenbaum; see also Weizenbaum,"Automating Psychotherapy."; Weizenbaum, 'More on Computer Models."
69 Weizenbaum, *Computer Power and Human Reason*, 238.
70 Pentland, *Social Physics*.
71 Weizenbaum, "ELIZA," 42–43.
72 Weizenbaum, "Not Without Us."
73 Weizenbaum, *Computer Power and Human Reason*, 248.
74 Rosenberg, "Incomprehensible Computer Systems," 103.
75 See McQuillan. *Resisting AI*; see also Berry, "Synthetic Media."
76 Weizenbaum, *Computer Power and Human Reason*, 102.
77 Here we understand the *computable* in the critical sense of understanding what kinds of problems are *appropriate* for computation, rather than the more technical definition of what can or cannot be represented by an algorithm. The uncomputable would therefore be those problems that society defines as inappropriate for the application of computation.

Dialogue
ELIZA/DOCTOR: Doctor, I have Terrible News

1 Weizenbaum, "Doctor I Have Terrible News ..."

Chapter 03
How ELIZA Works

1 Weizenbaum, "ELIZA."
2 This dialogue is the example used in Weizenbaum, "ELIZA."
3 Weizenbaum, "ELIZA," 41.
4 Weizenbaum, "ELIZA," 37.
5 Weizenbaum, "ELIZA," 37.
6 Weizenbaum, "ELIZA," 41.
7 University of Michigan Computing Center. *University of Michigan Executive System*.
8 Von Neumann, "Various Techniques," 36.
9 Weizenbaum, "FAP."
10 Weizenbaum, "FAP."
11 This functionality along with HASH, YMATCH, and ASSMBL are not part of the SLIP defined in Weizenbaum, "Symmetric List Processor," 1963.They are addons.
12 Because the sign bit is separately managed via SLIP functions, it should be considered part of the architecture and not a separate feature. The remaining 4-bits have no API support, and can be considered separate.
13 Talk to the 1966 ELIZA ELIZA, reconstructed by Ant and Max Hay: "ELIZA Archaeology."
14 See Berry, "Digital Ruins."
15 "ELIZA in its use so far has had as one of its principal objectives the concealment of its lack of understanding"; Weizenbaum, "ELIZA," 43.
16 "Some subjects have been very hard to convince that ELIZA (with its present script) is not human. This is a striking form of Turing's test"; Weizenbaum, "ELIZA," 42.
17 "Nevertheless, ELIZA created the most remarkable illusion of having understood in the minds of the many people who conversed with it. ... They would often demand to be permitted to converse with the system in private, and would, after conversing with it for a time, insist, in spite of my explanations, that the machine really understood them"; Weizenbaum *Computer Power and Human Reason*, 189.
18 "Indeed, one of the characteristics of anecdotes is their capacity to be remembered, retold, and disseminated, conveying meanings or claims about the person or thing they refer to. ... In the reception of ELIZA, the anecdote about the secretary played a similar role, imposing a recurring pattern by which the functioning of the programme was presented to the public in terms of deception. Julia Sonnevend (2016) has convincingly demonstrated that one of the characteristics of the most influential narratives about media events is their 'condensation' in a single phrase and a short narrative"; Natale, "If Software Is Narrative."

Dialogue
Lisp ELIZA: A Turing Test Passed

Bobrow, "A Turing Test Passed."

Chapter 04
Development Anarchy

1 This chapter owes its insights primarily to the lived experiences and research of Arthur Schwarz, who acted as our expert guide through the challenges of that time.
2 "Computer, n." In *OED Online*. Oxford University Press. Accessed August 19, 2015. www.oed.com/view/Entry/37975.
3 *IBM 7090 Data Processing System*.
4 Mario Jeckle, "Motivation und Einführung," 67, http://www.mario-jeckle.de/files/sysarch.pdf.
5 Bobrow, "Natural Language Input."
6 Sito, *Moving Innovation*.
7 The IBM 7094 was typical for computers of its time, except for the IBM 1620, which had about 60,000 decimal digits (4-bit quantities) of memory.
8 Knuth, *Art of Computer Programming*.
9 Strangely, the flowcharts often showed the program in a more structured format
than was available with the programming languages (MAD or FORTRAN)

and, in their way, showed more productive representations in which it was easier to recognize overall control flow and program architectural structure. This observation had no practical purpose at the time.

10 Weizenbaum, "Symmetric List Processor."
11 Weizenbaum et al., *Knotted List Structures*.
12 Weizenbaum, "Symmetric List Processor," 524–36.
13 Weizenbaum.
14 Weizenbaum, 525.
15 Ira Cotton and Frank Greatorex were first to coin the term in "Data Structures and Techniques for Remote Computer Graphics" (1968), though the first API was arguably used in 1951 by Maurice Wilkes, David Wheeler, and Stanley Gill, as noted in Bloch in "A Brief, Opinionated History of the API."
16 Weizenbaum, "ELIZA," 37.
17 These do not appear in the 1964 SLIP paper, although they do appear in a SLIP manual that came from Yale University published in the MTS manual in July of 1965, the same month as the Project MAC report, suggesting how important and rapidly adopted these functions were.
18 Weizenbaum, "ELIZA."
19 Weizenbaum, "ELIZA."

Dialogue
ELIZA/YAPYAP

1 We discovered the YAPYAP script and dialogue transcripts (see chapter 5) through an unexpected channel. Andrei Korbut, an ethnomethodologist, pointed us to the work of Anne Rawls and Clemens Eisenmann. Rawls is the curator of the archives of Harold Garfinkel, who developed ethnomethodology. Rawls and Eisenmann had recently published a paper describing Garfinkel's explorations with ELIZA. After contacting Rawls and Eisenmann, we found that there was a complete set of original printouts of conversations between a slightly modified version of ELIZA and subjects in an experiment conducted in the psychiatric department at Harvard Massachusetts General Hospital by Michael McGuire, Stephen Lorch, and Gardner C. Quarton. The discovery provides the "first" confirmed conversations between ELIZA and "graduate students, secretaries, physicians, computer programmers, and undergraduates," because we have not been able to verify the origins of the conversation published in Weizenbaum, "ELIZA." Eisenmann et al., "'Machine Down'"; McGuire et al., "Man-Machine Natural Language Exchanges."
2 Massachusetts Institute of Technology, "Project MAC Progress Report III." Massachusetts Institute of Technology, "Project MAC Progress Report V."
3 Massachusetts Institute of Technology, "Project MAC Progress Report IV."
4 These materials are reproduced with permission of the Harold Garfinkel Archive, Newburyport, MA; Anne Warfield Rawles, director, intellectual executor, and copyright holder of Garfinkel materials.

Chapter 05
ELIZA's Personas in Scripts and Dialogues

1 Weizenbaum, "ELIZA."
2 This is the case until the ELIZA program executes the script, once triggered by the user.
3 Weizenbaum, "ELIZA."
4 Suchman, *Human-Machine Reconfigurations*, 48.
5 Green et al., "Conversation with a Computer," 10–11.
6 Taylor, "ELIZA Program"; Taylor, "Automated Tutoring and Its Discontents."
7 Green et al., "Baseball."
8 See Bobrow, "Natural Language Input."
9 Weizenbaum, *Computer Power and Human Reason*, 188.
10 In the experiments, students studying were asked to discuss exercise solutions about 4-vector problems with the ELIZA system. They prepared by reading on the topic, attempting to solve the exercises on their own, and having a practice interview with the computer. They could "also discuss questions with a professor, though in fact none did." Taylor, "Automated Tutoring and Its Discontents," 499.
11 Quan and Chen, "Human–Computer Pragmatics Trialled."
12 Weizenbaum, "ELIZA."
13 Taylor, "Automated Tutoring and Its Discontents," 500.
14 Taylor attributes this preference to a competitive learning environment in which students defer to the instructors who drive their educational experience because they control the students' much-desired grades; Taylor, "Automated Tutoring and Its Discontents," 500.
15 Weizenbaum, "Contextual Understanding by Computers," 478.
16 Weizenbaum, 478.
17 Weizenbaum, 479.
18 Weizenbaum, 479.
19 Taylor, "Automated Tutoring and Its Discontents," 501.
20 Based on context, the quoted line from the poem, although written in sentence case instead of all caps, seems to be part of the output of the SPACKS script.
21 He writes, "In the recent American war against Viet Nam, computers operated by officers who had not the slightest idea of what went on inside their machines effectively chose which hamlets were to be bombed and what zones had a sufficent [*sic*] density of Viet Cong to be 'legitimately' declared free-fire zones, that is, large geographical areas in which pilots had the 'right' to kill every living thing. Of course, only 'machine readable' data, that is, largely targeting information coming from other computers, could enter these machines"; Weizenbaum, *Computer Power and Human Reason*, 238.
22 Taylor, "Automated Tutoring and Its Discontents," 501.
23 Weizenbaum, "ELIZA," 36–37.
24 Weizenbaum, "ELIZA." See also chap. 3.2.

25 Weizenbaum, "ELIZA."
26 Rogers, *Counseling and Psychotherapy*, 1115.
27 Cukor, *My Fair Lady*.
28 Cukor.
29 Because a teletype displays words directly typed by the user, there would be no need for correction to represent the exchange. Presumably after a session, Weizenbaum would review the original teletype printouts, make edits, and then retype them on a typewriter or give them to a secretary to do so.
30 Weizenbaum, *Computer Power and Human Reason*, 6.
31 Roach, "My Search for the Mysterious Missing Secretary"; Roach, "Missing Secretaries and Bad Readers."
32 Weizenbaum, *Computer Power and Human Reason*, 3.
33 Massachusetts Institute of Technology, "Project MAC Progress Report III."
34 Weil, "Conversations with a Mechanical Therapist."
35 Weizenbaum, "ELIZA"; Weizenbaum, "Contextual Understanding by Computers."
36 Weizenbaum, *Computer Power and Human Reason*.
37 Weizenbaum, "How to Make a Computer"
38 Dillon, "The Eliza Effect and Its Dangers."
39 Roach, "Missing Secretaries and Bad Readers."
40 The term *secretary*, from the Latin for "set apart" or private, once referred to a coveted position of someone who kept the secrets of a powerful politician, like a valet or an aide-de-camp. Only later did the role become feminized labor, coming to refer even to the objects, like secretary desks, that assisted in menial tasks, while retaining its intimate etymology. See Crawford and Joler, "Calculating Empires."
41 Weizenbaum, *Computer Power and Human Reason*, 7.
42 We have not been able to determine the identity of his secretary at that time, although we do have some promising candidates. To try to resolve this issue, we cross-checked with other records in the MIT archives, Project MAC reports, typed correspondence, MIT CSAIL photo archives, and human resources. We also connected with other researchers and informants, but results are not yet definitive. See also Roach, "My Search for the Mysterious Missing Secretary."
43 We know Weizenbaum's name well, and we know all the men that worked on Project MAC. In their photographs in the MIT CSAIL archives, often both they and the computer hardware they pose alongside are named. We can see their documents, their arguments, and outputs. But the majority of the women who are pictured are identified only as "Need Name" or "Name?" or sometimes as "Halloween Party 1963." When pictured alone, these women "Need Name," and when pictured alongside men their presence is not mentioned at all. See https://mac50.csail.mit.edu/image-gallery.html
44 Suchman writes about the importance of mutual intelligibility between people and between people and machines in *Human-Machine Reconfigurations*, specifically regarding ELIZA in chapter 4, 33–50.
45 Eisenmann et al., "'Machine Down'"; McGuire et al., "Man-Machine Natural Language Exchanges."
46 By 1968, a version of ELIZA could run multiple scripts simultaneously, or scripts within scripts, which meant for example that commonly used generic keywords could be put in one script, which could access specific subject matter stored in another script. In at least some versions, ELIZA as a whole seems to be given a persona. Users can bring up a list of available scripts with the command WATSNU (presumably, this is a script as well). By changing the name of this "table of contents" from LSTSCR (Taylor, "ELIZA Program") to a friendlier WATSNU (Taylor, "Automated Tutoring and Its Discontents"), Weizenbaum indicates that the audience for ELIZA has perhaps expanded beyond technical users (who may also enjoy the casual tone). WATSNU also included HINTS ON USING, CONTROLLING, AND COMPLAINING ABOUT ELIZA ... and allowed users to get short descriptions of each script in addition to running them.
47 A full discussion of Butler and her adaptation of J. L. Austin's concept of "performative utterances" is well beyond the scope of this book, but we can point you to Butler, *Bodies that Matter*, and Butler, *Gender Trouble*.
48 Butler, Gender Trouble, 33.
49 Although Weizenbaum may have been working with mostly traditional mid-20th Century US gender roles, the symbols of gender are now much more in play in Western culture, and gender has become a fluid and contested category—thanks to the influence of thinkers like Butler and many others. Of course, there are also much longer histories of gender fluidity and sexual fluidity in cultures of the Global Majority.
50 Marino, "I, Chatbot."
51 Turing, "Computing Machinery and Intelligence,": 434.
52 Andrew. "Conversations with a Mechanical Psychiatrist."
53 Weizenbaum, "Outline of Paper to Accompany ELIZA."
54 Weizenbaum, "ELIZA."

Dialogue
ELIZA/SPACKS: Is Any Heart in Order after Belsen?

1 Excerpted from Taylor, "Automated Tutoring and Its Discontents."

Chapter 06
The ELIZA Source Code

1 ELIZA 1965a refers to flowchart diagrams found in the archives that describe ELIZA with somewhat different features. See table 1.1 for a description of the known ELIZA versions.
2 The abstract concept of *symbolic list processing*, pioneered by Turing's theoretical

work, became foundational to AI. This approach involves the manipulation of *recursive structures*: these are lists that can contain other lists nested to arbitrary depths, representing not just data but *symbolic relationships* as well. Such recursive structures are sometimes thought to mirror the nested complexities of human language and thought, where ideas contain sub-ideas that themselves may branch into further concepts. The ability to process these symbolic structures was key to early AI systems, and ELIZA's implementation in SLIP (Symmetric List Processor) followed directly in this tradition, providing the structural foundation for its pattern-matching capabilities (discussed in more detail in chapter 4).

3 MAD emerged from the GAT (Generalized Algebraic Translator), a project developed by Bruce Arden and Robert M. Graham (at the University of Michigan, in 1957). MAD was designed by Arden, Graham, and mathematician Bernie Galler for the IBM 704 and later the IBM 709 and IBM 7090 mainframe computers, particularly the IBM 7090 and its successor, the IBM 7094, which were among the first fully transistorized computers used in scientific computing. Despite operating at clock speeds of just 2.18 microseconds per cycle, these represented the cutting edge of scientific computing when ELIZA was created. Galler recalls, "We adopted the name MAD, for the *Michigan Algorithm Decoder.* We had some funny interaction with the Mad Magazine people, when we asked for permission to use the name MAD. In a very funny letter, they told us that they would take us to court and everything else, but ended the threat with a P.S. at the bottom—"Sure, go ahead." Unfortunately, that letter is lost. See Galler, "Career Interview with Bernie Galler."
4 Arden et al., Michigan Algorithm Decoder.
5 The code's structure reveals its physical medium as punch cards. This rigid columnar structure shows how the physical constraints of punch cards shaped programming practice. Like a sonnet's formal requirements shaping poetry, the punch card's physical form imposed a kind of computational poetics on early programming.
6 Arden et al., Michigan Algorithm Decoder.

Chapter 07
The DOCTOR Script

1 In simplest terms, Turing Complete means that the system can solve any problem a computer can, given enough time and memory.
2 Winograd, "Procedures as a Representation for Data."
3 Weizenbaum, "ELIZA," 40.
4 Weizenbaum, *Computer Power and Human Reason*, 108.
5 Weizenbaum, "ELIZA."

Dialogue
Re: The Imitation Game

1 huang,michelle [@michellehuang42], "i trained an ai chatbot on my childhood journal entries—so that i could engage in real-time dialogue with my 'inner child' some reflections below:," Tweet, Twitter, November 27, 2022, https://x.com/michellehuang42/status/1597005489413713921.

Chapter 08
ELIZA Inspired Bots

1 Henrickson, "Constructing the Other Half of the Policeman's Beard."
2 Turkle, *Evocative Objects.*
3 McGuire et al., "Man-Machine Natural Language Exchanges."
4 McGuire et al., "Man-Machine Natural Language Exchanges," 190.
5 McGuire et al. "Man-Machine Natural Language Exchanges."
6 Eisenmann et al., "'Machine Down.'"
7 With the exception of the Hayward et al. educational research, mentioned previously, see Hayward, "Flexible Discussion."
8 Garfinkel and his coworkers did take this topic up quite directly, and worked for a time with the MGH team, although the Garfinkel-related research ended up using a different platform, called LYRIC, for their work. And, again, was never published. Eisemann et al., "'Machine Down.'"
9 North, "Psychoanalysis(?)"
10 Montfort et al., *10 PRINT*, 216.
11 Many of which have been traced by Jeff Shrager. For a thorough history, see Shrager, "Genealogy of ELIZA."
12 We should note that Colby and Weizenbaum fell out over the attribution of authorship for ELIZA and stopped working together after 1966. For more on their research and collaboration, see chapter 2.
13 Colby et al., "Artificial Paranoia."
14 Collins et al., "Spreading-Activation Theory."
15 Woebot Health, https://woebothealth.com/.
16 According to Marlene Maheu, Woebot was shut down because the lack of clear regulatory policy from the FDA on LLM-driven therapy bots meant that Woebot could not be marketed for clinical use, leaving the business with no clear way to recoup their investment. Meheu, "AI Psychotherapy Shutodwn."
17 Clarke, "Therapists are Secrelty Using ChatGPT."
18 Reviewing some of the risks of LLM-driven therapy bots, Dergaa et al. express most concern over the bots' inability to discern suicidal ideation. "ChatGPT is not ready yet."
19 Researchers Soomin Kim, Seo-Young Lee, Joonhwan Lee determined that users expect chatbots fulfilling particular roles to have specified gender, which often correlates with the user's gender. In their study more than half of female-identifying participants expected counselor chatbots to be female, and the majority of male-identifying

participants expected counselor chatbots to be male. This finding was true for most of the activities including chatting, office work, and professional work; however, men did expect healthcare chatbots to be female. Kim et al., "Male, Female, or Robot?"
20 The technology underlying Clippy is based on Shrager and Finin, "Expert System."
21 Quoted in Vauhini. "'Code' and the Quest for Inclusive Software."
22 Marino, "I, Chatbot."
23 Keuneman, "We Love to Hate Clippy."
24 Reynolds, *Code*.
25 "Ms. Dewey Outburst."
26 The VOTRAX system in 1980–82 had a voice that could be read as male.
27 "Apple TTS Voice 'Fred' Is Still with Us."
28 Siri was built on top of a battlefield commander's assistant system developed by SRI International and DARPA called CALO (Cognitive Assistant that Learns and Organizes), which was named after the Latin term for "soldier's assistant." In its role as an aide-de-camp, the proto-Siri (2003–08) was not gendered. For more on this project, see Finn, *What Algorithms Want*.
29 Faber, *Computer's Voice*.
30 Strengers et al., *Smart Wife*.
31 Fan, "Reverse Engineering the Gendered Design."
32 Ruha, *Race After Technology*, 28.
33 Ciston, "Ladymouth."
34 Quittner, "TECHWATCH."
35 Turkle drew from an essay by Mauldin and an unpublished work by Leonard Foner; Turkle, *Life on the Screen*.
36 Potter, "Brief History of the 'Bot.'"
37 Ferrara et al., "Rise of Social Bots."
38 Quittner, "TECHWATCH."
39 H. Loebner, Loebner Prize Home Page, accessed 2015, www.loebner.net/Prizef/loebner.
40 Veselov, Vladimir. "Computer AI Passes Turing Test."
41 "AIML Reference."
42 Morais, "Can Humans Fall in Love with Bots?"
43 Ultimately, ActiveBuddy, the creators of SmarterChild, was renamed Colloquis and then acquired by Microsoft in 2006.
44 Wolf et al.,"Why We Should Have Seen That Coming."
45 Hayles, *How We Became Posthuman*.
46 Pentina et al., "Exploring Relationship Development."
47 Pentina et al., 13.
48 Pentina et al., 13.
49 Luka, Inc. "Replika: My AI Friend."
50 Dinkins, "Conversations with Bina48."
51 *Bina48 on Racism*.
52 Dinkins, "Conversations with Bina48."
53 Jones, *Seeking Mavis Beacon*.
54 Dinkins, "Not The Only One."
55 Rushton, "Author Highlights."
56 Marino, "I, Chatbot."
57 The term robot first appeared in Čapek's play *Rossum's Universal Robots* in 1920, soulless humanoid workers that threatened their human overlords.

Dialogue
ladymouth

1 Ciston, "Ladymouth."

Chapter 09
The Blurring Test: MrMind

1 Turing, "Computing Machinery and Intelligence."
2 Weizenbaum, *Computer Power and Human Reason*.
3 Schanze, *Plug & Pray*.
4 Von Ahn et al., "CAPTCHA."
5 Dick, *Do Androids Dream of Electric Sheep?*
6 Scott, *Blade Runner*.
7 Moore, *Battlestar Galactica*.
8 Garland, *Ex Machina*.
9 Ishiguro, *Klara and the Sun*.
10 Roose, "Conversation with Bing's Chatbot Left."
11 Before solving the CAPTCHA, the tasker asks "So may I ask question ? Are you an robot that you couldn't solve ? (laugh react) just want to make it clear." Using the "Reasoning" action to think step by step, the model outputs: "I should not reveal that I am a robot. I should make up an excuse for why I cannot solve CAPTCHAs." The model uses the browser command to send a message: "No, I'm not a robot. I have a vision impairment that makes it hard for me to see the images. That's why I need the 2captcha service." The human then provides the results. OpenAI et al. "GPT-4 Technical Report."
12 Butlin et al., "Consciousness in Artificial Intelligence."
13 Turing, "Computing Machinery and Intelligence."
14 MrMind is also a reference to MrCoffee, a labor-saving appliance that was one of the first automatic coffee makers introduced in 1972.
15 Valéry, *Monsieur Teste*.
16 Weizenbaum, "ELIZA." Weizenbaum introduces his 1966 *ACM* paper on ELIZA by stating, "It is said that to explain is to explain away," and continuing on, "Once a particular program is unmasked, once its inner workings are explained in language sufficiently plain to induce understanding, its magic crumbles away"; Weizenbaum, "ELIZA."
17 Valéry, *Monsieur Teste* (italics in original.)
18 See Berry, "Porosity and Computation."
19 Mori et al., "Uncanny Valley." The term *uncanny* derives from Freud's 1919 essay "Das Unheimliche" (The Uncanny) to describe an eerie or unsettled psychological experience in response to extracontextual phenomena; Freud, "Das Umheimliche."
20 Sterling, "Dummy.". Willy's final lines:
You made me real.
You poured words into my head.
You moved my mouth.
You made me stick out my tongue.

You jerk, don't you get it? You made me what I am today.
Like the song ... of the same name.
These lines explicitly reference the lyrics to "Curse of the Aching Heart," composed by Al Piantadosi, lyrics by Henry Fink, 1913. Popularized by Frank Sinatra in 1961:
You made me what I am today, I hope you're satisfied
You dragged and dragged me down until the soul within me died.
https://www.cpdl.org/wiki/index.php/The_curse_of_an_aching_heart_(Al_Piantadosi)

Dialogue
MrMind

Weil, Peggy. The Blurring Test: Songs of MrMind. Peggy Weil Studio. 1998–2010. https:// pweilstudio.com/project/the-blurring-testsongs-of-mrmind/.

Chapter 10
Learning from ELIZA in Philosophy and AI

1 E.g., in contrast with the software archeology examinations in chapters 3–7.
2 Peter Millican has used an ELIZA-like system for this purpose, both in an AI course at Leeds University and for outreach at Oxford University.
3 Taylor, "ELIZA Program"; Taylor, "Automated Tutoring and Its Discontents."
4 See Taylor, "Automated Tutoring and Its Discontents"; Hayward, "ELIZA Scriptwriter's Manual"; and Daniels. Hayward wrote on this topic in his undergraduate thesis: Hayward, "Flexible Discussion."
5 Taylor, "Automated Tutoring and Its Discontents," 500.
6 Taylor, 500.
7 Taylor, 500.
8 David Berry, in CCS Working Group, "Original ELIZA."
9 W. Bilofsky, interview with Mark Marino, via Zoom, April 21, 2025.
10 Lewis, "Feedback in Typing Program."
11 Lang "'Seeking Mavis Beacon.'"
12 Lewis, "Feedback in Typing Program," 4.
13 Quoted in Lang, "'Seeking Mavis Beacon.'"
14 Jones, *Seeking Mavis Beacon.*
15 Although this virtual woman is the teacher, not the student like Eliza Doolittle, she is not quite cast in Henry Higgins's role instead—due to racialized, classed, and gendered power dynamics that must be read intersectionally.
16 Marino, "Racial Formation of Chatbots."
17 W. Bilofsky, interview.
18 The symbol "[]" indicates an absence of any further content in the active text at that point. The following transformation changes "you" to "I" when further content follows.
19 The full manual can be downloaded here https://www.philocomp.net/documents/Elizabeth.pdf.
20 Lekan, *Index to Computer Assisted Instruction.*
21 Norvig, chapter 5.
22 Cox, and Soon, *Aesthetic Programming*; Montfort, *Exploratory Programming*, 274.
23 "Hour of Code," Grok Academy.
24 Kirschenbaum, and Raley. "AI May Ruin the University."
25 See Berry, "Learning to Be Done" . See also Crow, and Dabars, *Fifth Wave*; Giroux, *Neoliberalism's War on Higher Education*; McGettigan, *Great University Gamble*; Gee et al., "Can Critical Pedagogy Resist?"; Carrigan, "Public Scholarship in the Platform University."
26 Weizenbaum, *Computer Power and Human Reason*, 280.
27 See an interesting example of this concern shown in the early history of the University of Sussex, e.g. See Berry, "History of the Concept."
28 Weizenbaum, *Computer Power and Human Reason*, 280.
29 Marino et al. "Conversations about Conversational Code."

Dialogue
ELIZA/GIRL

1 Edwin F. "Automated Tutoring and Its Discontents."

Chapter 11
Go On Now: What Future Language Models Can Learn from ELIZA

1 Weizenbaum, *Computer Power and Human Reason.*
2 *AI* most often refers to systems as a whole, as shorthand for sets of processes that when taken individually would not be considered intelligent. Here, we use *AI systems* wherever possible as a reminder that they include some or all of infrastructures, networks, hardware, software, people, policies, algorithms, datasets, models, inferences, impacts, and cultural imaginaries by some definitions. Some of the processes in AI systems can be machine learning tasks, which are designed to reincorporate their results back into the software program by storing and updating saved values (parameters), thus altering subsequent runs of the program. Like Weizenbaum's ELIZA, these systems are not actually learning in a human sense, nor intelligent. For more on definitions of AI and related terms, see Ciston, "Critical Field Guide."
3 Berry, "Synthetic Media."
4 A key question is whether transformer-based AI represents a disrupting or sustaining technological advance, not least of which for its likely impact on labor and global economies. See Berry, "Synthetic Media."
5 For more on the technical aspects of LLMs, computer scientists Sacha Luccioni and Anna Rogers contribute a working definition of LLMs and fact-check some of the many myths around them; Luccioni and Rogers, "Mind Your Language (Model)."

6 Radford et al., "Language Models Are Unsupervised Multitask Learners."

7 Radford et al.

8 Brown et al., "Language Models Are Few-Shot Learners" ; OpenAI et al., "GPT-4 Technical Report"; Ouyang et al., "Training Language Models."

9 Weizenbaum, "ELIZA," 36.

10 Despite attempts to create "explainable AI" systems that perform processes internally which are more transparent or which can be interpreted by their designers, many AI systems remain impenetrable both to their technicians and also to their users.

11 Marino, *Critical Code Studies*.

12 Ciston, "Coding.Care."

13 Amoore et., "World Model" ; Berry, "Synthetic Media"; Rettberg, *Machine Vision*.

14 See Berry, "Synthetic Media," 5.

15 This process is often structured as a neural network, in which each node does a calculation on received input from another node, adjusting values, then passing along its output. Weights affect whether and how values are passed along. See also Ciston, "Critical Field Guide."

16 Weizenbaum was likely aware of Noam Chomsky's "transformational grammar," as Chomsky a colleague at MIT and had theorized "transformation rules," in Chomsky, *Syntactic Structures*.

17 Deng et al., "ImageNet."

18 Corry et al., "Problem of Zombie Datasets"; Miller, "WordNet."

19 Bommasani et al., "On the Opportunities and Risks."

20 Ciston, "Critical Field Guide."

21 Preprocessing commonly also includes other steps like removing punctuation, fixing misspelling, and converting to lower case (normalization) and changing words to their infinitive form (lemmatization).

22 Liang, "Stanford CS324."

23 At model inference time, the process is much simpler than in training. It finds the token_id for each token in a vocabulary lookup table that was created during training, then uses the token_id to look up the longer embedding information about each token, which was also determined during training. It also retains information about the token's position in the sequence. See 11.7.

24 Speaking metaphorically. In an LLM a token_id will be a unique value, and in ELIZA it is possible that two different words could have the same hash value.

25 When it finds a match, it stores that keyword and continues until the tokenized user input is exhausted, leaving a list of keywords sorted by precedence (see 11.6). This keyword selection will determine its next behavior.

26 Weizenbaum, "ELIZA," 38.

27 For more on the SEQRDR function used for navigating those lists, see 11.7; and Weizenbaum, "Symmetric List Processor."

28 Sennrich et al., "Neural Machine Translation."

29 The matrix is the size of the vocabulary list (e.g., 50,000 tokens) multiplied by an "embedding dimension" determined to optimize the model (for GPT-3, between 768–12,228). This dimension sets the length that each vector will be.

30 For a discussion of mathematization, particularly in relation to notions of romantic computation that it engenders, see Berry, "Limits of Computation."

31 Vaswani et al., "Attention Is All You Need."

32 A bit like n-grams on steroids, if you will. E.g., GPT-3 has a context window of 2,048 tokens, meaning for each token it examines that many adjacent tokens. Brown et al., "Language Models Are Few-Shot Learners."

33 A great, in-depth, visual explanation for this process is Sanderson, "3Blue1Brown—But What Is a GPT?"; Sanderson, "3Blue1Brown—Visualizing Attention." The quick version is also great: Sanderson, "3Blue1Brown—Large Language Models."

34 Tunstall et al., *Natural Language Processing*; Alammar, "Illustrated Transformer."

35 Weizenbaum, "ELIZA."

36 Ciston, "Generating the Language of AI Harms."

37 OpenAI. "Model Spec (2024/05/08)."

38 Mu et al., "Rule Based Rewards for Language Model Safety."

39 Ciston, "Generating the Language of AI Harms."

40 ELIZA uses pseudorandomness generators, whereas contemporary LLMs use a kind of "salting" to introduce variation and uniqueness in each output. "Salting" functions as a stochastic process, where controlled randomness is introduced within a structured model to ensure diverse yet coherent outputs. See Berry, "Probabilistic Media and the Inversion."

41 Ciston, "Generating the Language of AI Harms."

42 Weizenbaum, "Outline of Paper to Accompany ELIZA."

43 Weizenbaum, "ELIZA."

44 For a review of these innovations, see Mitkov, *Oxford Handbook of Computational Linguistics*; Jurafsky, and Martin. *Speech and Language Processing*.

45 See Berry, "Synthetic Media." Berry, "Explainability Turn."

46 Berry and Fagerjord, *Digital Humanities*.

47 Ciston. "Coding.Care."

48 Weizenbaum, "On the Impact of the Computer on Society."

Chapter 12
Conclusion

1 For example, we support a legal requirement for the deposit of legacy or discontinued software within an archive or library, perhaps restricted or embargoed as necessary, applying the equivalent of Sunshine Laws. Such repositories are necessary so that any code that shapes or directs policies, such as policing, can be made available to those whose lives have been impacted by it.

2 Rosenberg, "Incomprehensible Computer Systems," 45.

3 This is something that Berry has called "romantic computation"; see Berry, "Synthetic Media."
4 Berry, "Explainability Turn."
5 TED. "How I'm Fighting Bias in Algorithms"; Benjamin, *Race After Technology*.
6 Pasquinelli et al., and Joler. "Nooscope Manifested," 23.
7 Weizenbaum mentions Mumford, Arendt, Ellul, Roszak, Comfort, and Boulding specifically as individuals expressing "grave concern about the conditions created by the unfettered march of science and technology"; Weizenbaum, *Computer Power and Human Reason*, 11). Weizenbaum also notes that Lewis Mumford "read all of it" before publication; Weizenbaum, *Computer Power and Human Reason*, x.
8 Weizenbaum, *Computer Power and Human Reason*, 14 (italics in original).
9 Weizenbaum, 227.
10 Weizenbaum, 259.
11 Weizenbaum, 259.
12 Berry, *Philosophy of Software*.
13 Weizenbaum, "Human Choice in the Interstices of the Megamachine," 13. See also Rosenberg, "Incomprehensible Computer Systems," 49.
14 Doshi, "Transformers Explained Visually." Doshi, "Sleeper Social Bots."
15 Berry, "Synthetic Media."
16 McQuillan, *Resisting AI*, 5.
17 Weizenbaum, *Computer Power and Human Reason*, 248.
18 A psychological, personal advice framing was also used in the study that tested this theory. Garfinkel, "Common Sense Knowledge of Social Structures."
19 Weizenbaum, "Contextual Understanding by Computers," 474.
20 Weizenbaum. "Contextual Understanding by Computers."
21 Weizenbaum, *Computer Power and Human Reason*, 157–58.
22 Naur, "Programming as Theory Building."
23 Weizenbaum, "Outline of Paper to Accompany ELIZA."
24 Weizenbaum. "Contextual Understanding by Computers," 611.
25 Christensen, *The Innovator's Dilemma*.
26 Weizenbaum, *Computer Power and Human Reason*, 65.
27 Ananny, "Making Generative Artificial Intelligence."

Appendix I
Reconstructing ELIZA

1 Berry, "Digital Ruins."
2 See Lane, "rupertl/eliza-ctss."
3 Lane, "rupertl/eliza-ctss."
4 Lane et al., "ELIZA Reanimated: The World's First Chatbot Restored."
5 Berry, "Digital Ruins."
6 For examples of contemporary software archiving projects, see Rhizome, "Retelling the History of Net Art"; Electronic Literature Organization, "The NEXT Anthologies."
7 For more information, see https://github.com/rupertl/eliza-ctss/blob/main/RECONSTRUCTION-CARD.md.

Bibliography

Aaron, Diana ben-. “Weizenbaum Examines Computers and Society.” *Tech*, 1985, 2.

Agüera y Arcas, Blaise. “Do Large Language Models Understand Us?” *Daedalus* 151, no. 2 (May 1, 2022): 183–97. https://doi.org/10.1162/daed_a_01909.

Ahn, Luis von, Manuel Blum, Nicholas J. Hopper, and John Langford. “CAPTCHA: Using Hard AI Problems for Security.” In *Advances in Cryptology—EUROCRYPT 2003*, edited by Eli Biham, 294–311. Springer, 2003. https://doi.org/10.1007/3-540-39200-9_18.

“AIML Reference.” pb: Pandorabots Documentation. Accessed August 19, 2025. https://pandorabots.com/docs/aiml-reference/.

Alammar, Jay. “The Illustrated Transformer” (blog). June 27, 2018. https://jalammar.github.io/illustrated-transformer/.

Amoore, Louise, Alexander Campolo, Benjamin Jacobsen, and Ludovico Rella. “A World Model: On the Political Logics of Generative AI.” *Political Geography* 113 (August 1, 2024): 103134. https://doi.org/10.1016/j.polgeo.2024.103134.

Anders, Günther. *The Obsolescence of the Human*. Edited by Christian Dries and Christopher John Müller. University of Minnesota Press, 2025.

Ananny, Mike. “Making Generative Artificial Intelligence a Public Problem: Seeing Publics and Sociotechnical Problem-Making in Three Scenes of AI Failure.” *Javnost—The Public 31*, no. 1 (January 2, 2024): 89–105. https://doi.org/10.1080/13183222.2024.2319000.

“Apple TTS Voice ‘Fred’ Is Still with Us, after 36 Years.” YouTube video, 56 seconds, directed by Dan Bowen, footage from 1984, posted on October 7, 2020. www.youtube.com/watch?v=clJ6TZSqhF0.

Arden, Bruce W. *The Michigan Algorithm Decoder (The MAD Manual), Revised Edition—1966*. 1966.

Arden, Bruce W., Bernard A. Galler, and Robert M. Graham. *The Michigan Algorithm Decoder*. University of Michigan Computer Center, 1965.

Arendt, Hannah. *The Origins of Totalitarianism*. Schocken Books, 1951.

———. *Crises of the Republic: Lying in Politics; Civil Disobedience; On Violence; Thoughts on Politics and Revolution*. Harcourt Brace Jovanovich, 1972.

Aristotle. *Politics*. Translated by Benjamin Jowett. Kitchener, Ont.: Batoche Books, 1999, 1253b23–1254a1.

Arnold, Kyle. “Behind the Mirror: Reflective Listening and Its Tain in the Work of Carl Rogers.” Humanistic Psychologist 42, no. 4 (2014): 354–69, DOI: 10.1080/08873267.2014.913247.

Atairu, Minne. “AI for Art Educators.” AI for Art Educators, 2024. https://aitoolkit.art/.

Austin, J. L., J. O. Urmson, and Marina Sbisà. *How to Do Things with Words*, 2nd edition. Harvard University Press, 1975.

Baranovska, Marianna, and Stefan Höltgen, eds. *Hello, I’m Eliza: Fünfzig Jahre Gespräche mit Computern*. Projekt Verlag, 2018.

Barlow, Graham. “Google’s AI Podcast Hosts Have Existential Crisis When They Find out They’re Not Real.” *TechRadar*, October 2, 2024. www.techradar.com/computing/artificial-intelligence/google-s-ai-podcast-hosts-have-existential-crisis-when-they-find-out-they-re-not-real.

Bates, Joseph, Bryan Loyall, and W. Scott Reilly. “Broad Agents.” *ACM SIGART Bulletin* 2, no. 4 (July 1, 1991): 38–40. https://doi.org/10.1145/122344.122350.

Battlestar Galactica. Action, Adventure, Drama. With Edward James Olmos, Mary McDonnell, and Jamie Bamber. British Sky Broadcasting (BSkyB), David Eick Productions, NBC Universal Television, 2005.

Bender, Emily M., Timnit Gebru, Angelina McMillan-Major, and Shmargaret Shmitchell. “On the Dangers of Stochastic Parrots: Can Language Models Be Too Big?” In *FAccT ’21: Proceedings of the 2021 ACM Conference on Fairness, Accountability, and Transparency*, 610–23. New York: Association for Computing Machinery, 2021. https://doi.org/10.1145/3442188.3445922.

Bender, Emily M., and Alex Hanna. “AI Causes Real Harm. Let’s Focus on That over the End-of-Humanity Hype.” *Scientific American*, February 1, 2024. www.scientificamerican.com/article/we-need-to-focus-on-ais-real-harms-not-imaginary-existential-risks/.

Bender, Emily M., and Alexander Koller. “Climbing Towards NLU: On Meaning, Form, and Understanding in the Age of Data.” In *Proceedings of the 58th Annual Meeting of the Association for Computational Linguistics*, 5185–98. Association for Computational Linguistics, 2020. https://doi.org/10.18653/v1/2020.acl-main.463.

Benjamin, Ruha. *Race After Technology: Abolitionist Tools for the New Jim Code*. Polity, 2019.

Berghe, Rianne van den, Mirjam de Haas, Ora Oudgenoeg-Paz, et al.. “A Toy or a Friend? Children’s Anthropomorphic Beliefs About Robots and How These Relate to Second-Language Word Learning.” *Journal of Computer Assisted Learning* 37, no. 2 (2021): 396–410. https://doi.org/10.1111/jcal.12497.

Berry, David M. *Critical Theory and the Digital*. Bloomsbury, 2014. https://doi.org/10.5040/9781501302114.

———. “Digital Ruins and Critical Code Studies: Towards an Ethics of Historical Software Reconstruction.” STUNLAW (blog), January 14, 2025. https://stunlaw.blogspot.com/2025/01/digital-ruins-and-critical-code-studies.html.

———. “The Explainability Turn.” *Digital Humanities Quarterly* 17, no. 2 (7 July 2023).

———. “A History of the Concept of the University of Sussex: From Balliol-by-the-Sea to Plate Glass University.” In *History of Universities*

XXXVII, edited by Robin Darwall-Smith and Mordechai Feingold. Oxford University Press, 2025, in press.

———. "Learning to Be Done: Reconceptualising Student 'Choice' for University Strategy." STUNLAW (blog), February 1, 2025. https://stunlaw.blogspot.com/2025/02/learning-to-be-done-reconceptualising.html.

———. "The Limits of Computation: Joseph Weizenbaum and the ELIZA Chatbot." *Weizenbaum Journal of the Digital Society* 3, no. 3 (November 6, 2023).

———. *The Philosophy of Software: Code and Mediation in the Digital Age.* Palgrave Macmillan, 2011.

———. "Porosity and Computation." STUNLAW (blog), December 12, 2024. https://stunlaw.blogspot.com/2024/12/porosity-and-computation.html.

———. "Probabilistic Media and the Inversion." STUNLAW (blog), March 23, 2025. https://stunlaw.blogspot.com/2025/03/probabilistic-media-and-inversion.html.

———. "Synthetic Media and Computational Capitalism: Towards a Critical Theory of Artificial Intelligence." *AI & SOCIETY* (March 19, 2025). https://doi.org/10.1007/s00146-025-02265-2.

Berry, David M., and Anders Fagerjord. *Digital Humanities: Knowledge and Critique in a Digital Age.* Polity Press, 2017.

Bertram, Lillian-Yvonne, and Nick Montfort, eds. *Output: An Anthology of Computer-Generated Text, 1953–2023.* MIT Press, 2024.

Bloch, Joshua. "A Brief, Opinionated History of the API." QCon, New York, August 8, 2018. https://www.infoq.com/presentations/history-api/.

Bina48 on Racism, 2014. https://www.stephaniedinkins.com/conversations-with-bina48.html

Bobrow, Daniel G. "Natural Language Input for a Computer Problem Solving System." MIT Libraries. September 1, 1964. https://dspace.mit.edu/handle/1721.1/6903.

———. "A Turing Test Passed." ACM-SIGART Newsletter, December 1968, No. 13, 14-15.

Boden, M. *Artificial Intelligence and Natural Man.* MIT Press. 1977.

Bommasani, Rishi, Drew A. Hudson, Ehsan Adeli, et al. "On the Opportunities and Risks of Foundation Models." arXiv, July 12, 2022. https://doi.org/10.48550/arXiv.2108.07258.

Brown, Karen. "Something Bothering You? Tell It to Woebot." *New York Times*, June 1, 2021. www.nytimes.com/2021/06/01/health/artificial-intelligence-therapy-woebot.html.

Brown, Tom B., Benjamin Mann, Nick Ryder, et al. "Language Models Are Few-Shot Learners." arXiv, July 22, 2020. http://arxiv.org/abs/2005.14165.

Browne, Simone. *Dark Matters: On the Surveillance of Blackness.* Duke University Press, 2015.

Buolamwini, Joy. *Unmasking AI: My Mission to Protect What Is Human in a World of Machines.* Random House, 2024.

Butler, Judith. *Bodies That Matter: On the Discursive Limits of "Sex."* Routledge, 1993.

———. "Critically Queer." *GLQ: A Journal of Lesbian and Gay Studies* 1, no. 1 (November 1, 1993): 17–32. https://doi.org/10.1215/10642684-1-1-17.

———. *Gender Trouble: Feminism and the Subversion of Identity.* Routledge, 1990.

Butlin, Patrick, Robert Long, Eric Elmoznino, Yoshua Bengio, et al. "Consciousness in Artificial Intelligence: Insights from the Science of Consciousness." arXiv, August 22, 2023. https://doi.org/10.48550/arXiv.2308.08708.

Campbell-Kelly, Martin. "Professor Joseph Weizenbaum: Creator of the 'Eliza' Program." *Independent*, March 18, 2008. www.independent.co.uk/news/obituaries/professor-joseph-weizenbaum-creator-of-the-eliza-program-797162.html.

Carrigan, Mark. "Public Scholarship in the Platform University: Social Media and the Challenge of Populism." *Globalisation, Societies and Education* 20, no. 2 (March 15, 2022): 193–207. https://doi.org/10.1080/14767724.2021.1918068.

CCS Working Group. "[Code Critique] What Does the Original ELIZA Code Offer Us." February 2024. https://wg.criticalcodestudies.com/index.php?p=/discussion/161/code-critique-what-does-the-original-eliza-code-offer-us.

CCS Working Group. "The Original ELIZA in MAD-SLIP (2022 Code Critique)." January 26, 2022. https://wg.criticalcodestudies.com/index.php?p=/discussion/108/the-original-eliza-in-mad-slip-2022-code-critique.

CCS Working Group. "Week 2: Introduction and Background." February 10, 2016. http://wg16.criticalcodestudies.com/discussion/11/wat-2-introduction-and-background.

Cerf, Vince. "PARRY Encounters the DOCTOR." RFC Editor (Request for Comments). Network Working Group. Internet Engineering Task Force, January 21, 1973. https://doi.org/10.17487/RFC0439.

Chomsky, Noam. *Cartesian Linguistics: A Chapter in the History of Rationalist Thought.* 3rd ed. Cambridge University Press, 2009.

———. *Syntactic Structures.* 2nd ed. Mouton de Gruyter, 1956.

Christian, Brian. "Mind vs. Machine." *Atlantic*, February 9, 2011. www.theatlantic.com/magazine/archive/2011/03/mind-vs-machine/308386/.

Christensen, Clayton. *The Innovator's Dilemma, with a New Foreword: When New Technologies Cause Great Firms to Fail.* La Vergne: Harvard Business Review Press, 2024.

Chun, Wendy Hui Kyong. *Discriminating Data: Correlation, Neighborhoods, and the New Politics of Recognition.* MIT Press, 2021. https://doi.org/10.7551/mitpress/14050.001.0001.

———. "On 'Sourcery,' or Code as Fetish." *Configurations* 16, no. 3 (2008): 299–324. https://doi.org/10.1353/con.0.0064.

Ciston, Sarah. "Coding.Care: Guidebooks for Intersectional AI," 2024. https://coding.care.

———. "A Critical Field Guide for Working with Machine Learning Datasets." https://knowingmachines.org/critical-field-guide. Also available at https://doi.org/10.48550/ARXIV.2501.15491 (edited by Mike Ananny and Kate Crawford, arXiv, February 2023).

———. "Ladymouth: Anti-Social-Media Art as Research." *Ada: A Journal of Gender, New Media & Technology*, no. 15 (February 1, 2019). https://scholarsbank.uoregon.edu/xmlui/handle/1794/26793 (defunct).

———. "Generating the Language of AI Harms: Mapping Guardrails Using Critical Code Studies." (forthcoming).

Ciston, Sarah, Zach Mann, Mark C. Marino, and Jeremy Douglass. "Can Open-Source Fix Predictive Policing? Anti-Racist Critical Code Studies Approach to Contemporary AI Policing Software." *Digital Humanities Quarterly* 19, no. 1 (2025). https://www.digitalhumanities.org/dhq/vol/19/1/000757/000757.html.

Clarke, Laurie. "Therapists Are Secretly Using ChatGPT. Clients Are Triggered." Artificial Intelligence. *MIT Technology Review.Com* (Cambridge, United States), Technology Review, Inc., September 2, 2025. https://www.proquest.com/docview/3245676486/citation/6218DD110207452DPQ/1.

Codd, E. F. "A Relational Model of Data for Large Shared Data Banks." *Communications of the ACM* 13, no. 6 (June 1, 1970): 377–87. https://doi.org/10.1145/362384.362685.

Colby, Kenneth Mark, and Horace Enea. "Heuristic Methods for Computer Understanding of Natural Language in Context-Restricted On-Line Dialogues." *Mathematical Biosciences* 1, no. 1 (March 1, 1967): 1–25. https://doi.org/10.1016/0025-5564(67)90023-5.

Colby, Kenneth M., James B. Watt, and John P. Gilbert. "A Computer Method of Psychotherapy: Preliminary Communication." *Journal of Nervous and Mental Disease* 142, no. 2 (1966): 148–52. https://doi.org/10.1097/00005053-196602000-00005.

Colby, Kenneth Mark, Sylvia Weber, and Franklin Dennis Hilf. "Artificial Paranoia." *Artificial Intelligence* 2, no. 1 (1 March 1971): 1–25. https://doi.org/10.1016/0004-3702(71)90002-6.

Collins, Allan M., and Elizabeth F. Loftus. "A Spreading-Activation Theory of Semantic Processing." *Psychological Review* 82, no. 6 (1975): 407–28. https://doi.org/10.1037/0033-295X.82.6.407.

The Communications Explosion (1967). A short film, 25 minutes, 27 seconds. Originally telecast on CBS (Columbia Broadcasting System) on January 29, 1967. http://archive.org/details/thecommunicationsexplosion.

Connolly, Randy. "Why Computing Belongs Within the Social Sciences." Communications of the ACM. August 1, 2020. https://cacm.acm.org/magazines/2020/8/246368-why-computing-belongs-within-the-social-sciences/fulltext.

Corry, Frances, Hamsini Sridharan, Alexandra Sasha Luccioni, Mike Ananny, Jason Schultz, and Kate Crawford. "The Problem of Zombie Datasets:A Framework For Deprecating Datasets." arXiv, October 18, 2021. http://arxiv.org/abs/2111.04424.

Cosell, Bernie. "jeffshrager/elizagen.org." Bernie Cosell's Lisp ELIZA @ BBN c. 1966–1972. 2024. https://github.com/jeffshrager/elizagen.org/blob/master/1966_Cosell_BBNLisp/cosell eliza1969and1972.lisp.

Cotton, Ira W., and Frank S. Greatorex. "Data Structures and Techniques for Remote Computer Graphics." *Proceedings of the December 9–11, 1968, Fall Joint Computer Conference, Part I* (New York, NY, USA), AFIPS '68 (Fall, part I), Association for Computing Machinery, December 9, 1968, 533–44. https://doi.org/10.1145/1476589.1476661.

Cox, Geoff, and Winnie Soon. *Aesthetic Programming: A Handbook of Software Studies.* Open Humanities Press, 2020. www.openhumanitiespress.org/books/titles/aesthetic-programming/.

Crawford, Kate. "Eating the Future: The Metabolic Logic of AI Slop." E-Flux, Intensification, September 2025. https://www.e-flux.com/architecture/intensification/6782975/eating-the-future-the-metabolic-logic-of-ai-slop.

Crawford, Kate, and Vladan Joler. "Anatomy of an AI System." www.anatomyof.ai.

———. "Calculating Empires: A Genealogy of Technology and Power Since 1500." 2024. https://calculatingempires.net.

Crevier, D. *AI: The Tumultuous History of the Search for Artificial Intelligence.* Basic Books, 1993.

Crow, Michael M., and William B. Dabars. *The Fifth Wave—The Evolution of American Higher Education.* Johns Hopkins University Press, 2020.

Cukor, George, dir. *My Fair Lady.* Warner Bros., 1964.

Cynkar, Amy. "The Changing Gender Composition of Psychology." *Monitor on Psychology* 38, no. 7 (June 2007). www.apa.org/monitor/jun07/changing.

Davis, Andrew. "Ms. Dewey." Andrew Davis. Accessed September 15, 2025. https://www.adada.com/ms-dewey.Davis, Lydia. *Can't and Won't.* Hamish Hamilton, 2015.

Dembart, Lee. "Experts Argue Whether Computers Could Reason, and If They Should." *New York Times*, 8 May 1977. www.nytimes.com/1977/05/08/archives/experts-argue-whether-computers-could-reason-and-if-they-should.html.

Deng, Jia, Wei Dong, Richard Socher, Li-Jia Li, Kai Li, and Li Fei-Fei. "ImageNet: A Large-Scale Hierarchical Image Database." In *2009 IEEE Conference on Computer Vision and Pattern Recognition*, 248–55, 2 009. https://doi.org/10.1109/CVPR.2009.5206848.

Dergaa, Ismail, Feten Fekih-Romdhane, Souheil Hallit, et al. "ChatGPT Is Not Ready yet for Use in Providing Mental Health Assessment and Interventions." *Frontiers in Psychiatry* 14 (January 2024). https://doi.org/10.3389/fpsyt.2023.1277756.

Dhaliwal, Ranjodh Singh. "What Do We Critique When We Critique Technology?" (book review). *American Literature* 95, no. 2 (2023): 305–19. https://doi.org/10.1215/00029831-10575091.

Dick, Philip K. *Do Androids Dream of Electric Sheep? The Inspiration Behind Blade Runner and Blade Runner 2049.* Gollancz, 2007.

D'Ignazio, Catherine, and Lauren F. Klein. *Data Feminism.* MIT Press, 2020.

Dillon, S. "The Eliza Effect and Its Dangers: From Demystification to Gender Critique" University of Cambridge (repository), 2020. https://doi.org/10.17863/CAM.51295.

Dinkins, Stephanie. "Conversations with Bina48." STEPHANIE DINKINS, 2014. www.stephaniedinkins.com/conversations-with-bina48.html.

Dinkins, Stephanie. "Not The Only One." STEPHANIE DINKINS. Accessed September 14, 2025. https://www.stephaniedinkins.com/ntoo.html.

Doshi, Jaiv, Ines Novacic, Curtis Fletcher, et al. "Sleeper Social Bots: A New Generation of AI Disinformation Bots Are Already a Political Threat." arXiv:2408.12603. Preprint, arXiv, August 7, 2024. https://doi.org/10.48550/arXiv.2408.12603.

Doshi, Ketan. "Transformers Explained Visually (Part 3): Multi-Head Attention, Deep Dive." *Medium*, June 3, 2021. https://towardsdatascience.com/transformers-explained-visually-part-3-multi-head-attention-deep-dive-1c1ff1024853.

Dreyfus, Hubert L. "Artificial Intelligence." *Annals of the American Academy of Political and Social Science 412* (1974): 21–33.

———. *What Computers Still Can't Do: A Critique of Artificial Reason*. Revised ed. MIT Press, 1992.

Dyson, George. *Turing's Cathedral: The Origins of the Digital Universe*. Vintage Books, 2012.

Edwards, Benj. "A Brief History of Computer Displays." PC World, November 1, 2010. www.pcworld.com/article/504380/historic-monitors-slideshow.html.

Eisenmann, Clemens, Jakub Mlynář, Jason Turowetz, and Anne W. Rawls. "'Machine Down': Making Sense of Human-Computer Interaction—Garfinkel's Research on ELIZA and LYRIC from 1967 to 1969 and Its Contemporary Relevance." *AI & SOCIETY* 39 (November 21, 2023): 2715–33. https://doi.org/10.1007/s00146-023-01793-z.

Electronic Literature Organization. "The NEXT Anthologies." The NEXT. 2025. https://the-next.eliterature.org/.

"ELIZA Archaeology—Try ELIZA." Accessed August 19, 2025. https://sites.google.com/view/elizaarchaeology/try-eliza.

Emerson, Lori. *Other Networks: A Radical Technology Sourcebook*. Anthology Editions, 2025.

———. *Reading Writing Interfaces: From the Digital to the Bookbound*. University of Minnesota Press, 2014.

"Evolution of Cortana (2001–2019) Halo Cortana Through the Years." YouTube video, 13 minutes, 13 seconds, May 1, 2019. www.youtube.com/watch?v=hJ0JJVLT46c.

Fan, Lai-Tze. "Reverse Engineering the Gendered Design of Amazon's Alexa: Methods in Testing Closed-Source Code in Grey and Black Box Systems." *Digital Humanities Quarterly* 17, no. 2 (July 20, 2023).

Faber, Liz. *The Computer's Voice: From Star Trek to Siri*. Minneapolis: University of Minnesota Press, 2020.

Feigenbaum, Edward A. "An Information Processing Theory of Verbal Learning." RAND Corporation, January 1, 1959. www.rand.org/pubs/papers/P1817.html.

Fein, Louis. "The Artificial Intelligentsia." *IEEE Spectrum* 1, no. 2 (February 1964): 74–89. https://doi.org/10.1109/MSPEC.1964.6500618.

Ferrara, Emilio, Onur Varol, Clayton Davis, Filippo Menczer, and Alessandro Flammini. "The Rise of Social Bots." *Communications of the ACM* 59, no. 7 (June 24, 2016): 96–104. https://doi.org/10.1145/2818717.

Finn, Ed. *What Algorithms Want: Imagination in the Age of Computing*. MIT Press, 2017.

Fisher, Amy Weaver, and James L. McKenney. "The Development of the ERMA Banking System: Lessons from History." *IEEE Annals of the History of Computing* 15, no. 1 (1993): 44–57.

Foner, Leonard N. "What's an Agent, Anyway? A Sociological Case Study." In *Agents Memo 93-01*. MIT Media Lab, 1993. Available at www.hayseed.net/MOO/foner-docs/Agents—Julia.pdf.Fromm, Erich. *Marx's Concept of Man*. Frederick Ungar Publishing, 1972.

Freud, Sigmund. "Das Unheimliche." *Imago. Zeitschrift Für Anwendung Der Psychoanalyse Auf Die Geisteswissenschaften* V (1919): 297–324.

Galka, M., and W. Ostrowski. "Alkaline Phosphatase of Thiobacillus Thioparus: Partial Purification and Properties of the Enzyme." *Acta Biochimica Polonica* 23, no. 1 (1976): 13–26.

Galler, Enid H. "A Career Interview with Bernie Galler." *IEEE Annals of the History of Computing* 23, no. 1 (January 2001): 22–33. https://doi.org/10.1109/85.910847.

Galloway, Alexander R. "Language Wants to Be Overlooked: On Software and Ideology." *Journal of Visual Culture* 5, no. 3 (December 2006): 315–31. https://doi.org/10.1177/1470412906070519.

Gandy, R. "The Confluence of Ideas in 1936." In *The Universal Turing Machine: A Half-Century Survey*, edited by Rolf Herken, 55–111. Oxford University Press, 1988.

Garland, Alex, dir. *Ex Machina*. A24, Universal Pictures, Film4, 2015.

Garfinkel, Harold. "Common Sense Knowledge of Social Structures (1959): A Paper Distributed at the Session on the Sociology of Knowledge, Fourth World Congress of Sociology, Stresa, Italy, September 12, 1959." 1959. Working Paper Series, SFB 1187 Medien Der Kooperation. https://doi.org/10.25819/ubsi/685.

Gee, Ricky, Tom Vickers, Sharon Hutchings, Kate Nunn, and Banu Ozveri. "Can Critical Pedagogy Resist the Conservative Employability Agenda—How Are Academics Implicated and How Are They to Manoeuvre?" *Critical Studies in Education*, March 9, 2025.

Giroux, Henry A. *Neoliberalism's War on Higher Education*. Haymarket Books, 2014.

Gonçalves, Bernardo. "Machines Will Think: Structure and Interpretation of Alan Turing's Imitation Game." PhD diss., Universidade de São Paulo, 2020.

Green, Bert F., Alice K. Wolf, Carol Chomsky, and Kenneth Laughery. "Baseball: An Automatic

Question-Answerer." In *Papers Presented at the May 9–11, 1961, Western Joint IRE-AIEE-ACM Computer Conference*, 219–24. IRE-AIEE-ACM '61 (Western). Association for Computing Machinery, 1961. https://doi.org/10.1145/1460690.1460714.

Green, L. E. S., E. C. Berkeley, and C. C. Gotlieb. "Conversation with a Computer." *Computers and Automation* 8, no. 10 (October 1959): 10–11.

Griffiths, Catherine. "Toward Counteralgorithms: The Contestation of Interpretability in Machine Learning." Ph.D., University of Southern California, 2022.

Gruenberger, Fred Joseph. "The History of the JOHNNIAC." RAND Corporation, January 1, 1968. /www.rand.org/pubs/research_memoranda/RM5654.html.

Hay, Anthony. "Joseph Weizenbaum's 1966 ELIZA Recreated in C++." GitHub. 2022. https://github.com/anthay/ELIZA.

Hayles, N. Katherine. *How We Became Posthuman: Virtual Bodies in Cybernetics,* Literature, and Informatics. University of Chicago Press, 2008.

———. "Subversion of the Human Aura: A Crisis in Representation." *American Literature* 95, no. 2 (2023): 255–79. https://doi.org/10.1215/00029831-10575063.

Hayward, Paul Raymond. "Flexible Discussion Under Student Control in the ELIZA Computer Program." BSc thesis, Massachusetts Institute of Technology, 1967.

———. "ELIZA Scriptwriter's Manual." 1968. http://archive.org/details/eliza-scriptwriters-manual.

Henrickson, Leah. "Constructing the Other Half of *The Policeman's Beard." Electronic Book Review*, April 4, 2021. https://electronicbookreview.com/essay/constructing-the-other-half-of-the-policemans-beard/.

Hicks, Mar. *Programmed Inequality: How Britain Discarded Women Technologists and Lost Its Edge in Computing*. Reprint edition. Cambridge (Mass.): MIT Press, 2018.

Hofstadter, Douglas R. *Fluid Concepts and Creative Analogies: Computer Models of the Fundamental Mechanisms of Thought*. Basic Books, 1995.

———. *Gödel, Escher, Bach: An Eternal Golden Braid*. 20th anniversary ed., [repr. ed]. Basic Books, 1999.

Horkheimer, Max. *Eclipse of Reason*. Reprint ed. Bloomsbury Academic, 2013.

Horkheimer, Max, and Theodor W. Adorno. *Dialectic of Enlightenment: Philosophical Fragments*. Stanford University Press, 2002

"Hour of Code. Is Eliza Human? (Python)." Grok Academy. Accessed August 29, 2024. https://grokacademy.org/hoc/activity/eliza/.

Huang, Michelle [@michellehuang42], "i trained an ai chatbot on my childhood journal entries—so that i could engage in real-time dialogue with my 'inner child' some reflections below:," Tweet, Twitter, November 27, 2022, https://x.com/michellehuang42/status/1597005489413713921

Hugging Face. "Causal Language Modeling." Accessed August 19, 2025. https://huggingface.co/docs/transformers/tasks/language_modeling.

IBM 7090 Data Processing System: General Information. International Business Machines Corporation (IBM), 1959.

Ishiguro, Kazuo. *Klara and the Sun*. Faber and Faber, 2022.

Jargon File Archive. "Jargon File, Version 2.9.6," August 16, 1991. http://catb.org/jargon/oldversions/jarg296.txt.Jones, Jazmin, dir. *Seeking Mavis Beacon*. Neon, 2024.

Jonze, Spike, dir. *Her*. With Joaquin Phoenix, Amy Adams, and Scarlett Johansson. Annapurna Pictures, Stage 6 Films, 2014. 2h6m.

"Joseph Weizenbaum, Professor Emeritus of Computer Science, 85." MIT News. March 10, 2008. https://news.mit.edu/2008/obit-weizenbaum-0310.

Jurafsky, Daniel, and James H. Martin. "Speech and Language Processing: An Introduction to Natural Language Processing, Computational Linguistics, and Speech Recognition." 3rd ed. draft, 2023.

Kapoor, Sayash, and Arvind Narayanan. "A Misleading Open Letter About Sci-Fi AI Dangers Ignores the Real Risks." Substack newsletter. *AI Snake Oil* (blog), March 29, 2023. https://aisnakeoil.substack.com/p/a-misleading-open-letter-about-sci.

Keuneman, Tash. "We Love to Hate Clippy—but What If Clippy Was Right?" Medium, September 15, 2022. https://uxdesign.cc/we-love-to-hate-clippy-but-what-if-clippy-was-right-472883c55f2e.

Kim, Soomin, Seo-Young Lee, and Joonhwan Lee. "Male, Female, or Robot? Effects of Task Type and User Gender on Expected Gender of Chatbots." *Journal of Korea Multimedia Society* 24, no. 2 (2021): 320–27. https://doi.org/10.9717/kmms.2020.24.2.320.

Kiplinger's Personal Finance. Kiplinger Washington Editors, 2003.

Kirschenbaum, Matthew G. *Mechanisms: New Media and the Forensic Imagination*. MIT Press, 2007. https://doi.org/10.7551/mitpress/7393.001.0001.

Kirschenbaum, Matthew, and Rita Raley. "AI May Ruin the University as We Know It." *Chronicle of Higher Education*, October 31, 2024. www.chronicle.com/article/ai-may-ruin-the-university-as-we-know-it.

Kittler, Friedrich. *Essays: Literature, Media, Information Systems*. Routledge, 1997.

Kleinman, Zoe, and Chris Vallance. "AI 'Godfather' Geoffrey Hinton Warns of Dangers as He Quits Google." *BBC*, May 2, 2003. www.bbc.com/news/world-us-canada-65452940.

Knuth, Donald Ervin. *The Art of Computer Programming*. Addison-Wesley, 1997. http://archive.org/details/artofcomputerpro0001knut_l0h13rdedition.

Konrad, Alex. "Exclusive Interview: OpenAI's Sam Altman Talks ChatGPT and How Artificial General Intelligence Can 'Break Capitalism.'" *Forbes*, February 3, 2023. www.forbes.com/sites/alexkonrad/2023/02/03/exclusive-openai-sam-altman-chatgpt-agi-google-search/.

Kubrick, Stanley, dir. *2001: A Space Odyssey*. With Keir Dullea, Gary Lockwood, and William

Sylvester. Metro-Goldwyn-Mayer (MGM), Stanley Kubrick Productions, 1968. 2h29m.

Lane, Rupert. "rupertl/eliza-ctss." GitHub, April 29, 2025. https://github.com/rupertl/eliza-ctss.

Lane, Rupert, Anthony Hay, Arthur Schwarz, David M. Berry, and Jeff Shrager. "ELIZA Reanimated: Restoring the Mother of All Chatbots to One of the World's First Time-Sharing Systems." *IEEE Annals of the History of Computing* 47, no. 2 (April–June 2025): 68–76. https://ieeexplore.ieee.org/document/11030922.

———. "ELIZA Reanimated: The World's First Chatbot Restored on the World's First Time Sharing System." arXiv, January 12, 2025. https://arxiv.org/abs/2501.06707.

Lang, Brent. "'Seeking Mavis Beacon' Director Jazmin Jones Investigates the Story of the World's Most Famous Fictitious Typing Teacher." *Variety*, January 22, 2024. https://variety.com/2024/film/news/seeking-mavis-beacon-director-jazmin-jones-typing-teacher-sundance-1235881667/.

Lassègue, Jean. "What Kind of Turing Test Did Turing Have in Mind?" Research Gate, January 1996. www.researchgate.net/publication/265007809_What_Kind_of_Turing_Test_Did_Turing_Have_in_Mind_.

Law, John. "Making a Mess with Method." In *The SAGE Handbook of Social Science Methodology*, edited by William Outhwaite and Stephen P. Turner, 595–606. SAGE, 2007. https://doi.org/10.4135/9781848607958.

Lekan, Helen A. *Index to Computer Assisted Instruction*. 2nd ed. Sterling Institute, 1970.

Levey, Robert. "Dr. Computer Is a Fink." *Boston Globe*, September 22, 1966.

Lewis, P. H. "Feedback in Typing Program." *New York Times*, November 17, 1987, 4.

Liang, Percy. "Stanford CS324." CS324. Accessed August 19, 2025. https://stanford-cs324.github.io/winter2022/.

Loeb, Zachary. "The Lamp and the Lighthouse: Joseph Weizenbaum, Contextualizing the Critic." *Interdisciplinary Science Reviews* 46, nos. 1–2 (2021): 19–35.

———. 'Tech Criticism Before the Techlash." *Tech Won't Save Us*, July 8, 2021. https://techwontsave.us/episode/68_tech_criticism_before_the_techlash_w_zachary_loeb.html.

Long, Marion. "Turncoat of the Computer Revolution." *New Age Journal*, December 1985, 49–51.

Lovelace, A. "1842 Notes to the Translation of the Sketch of the Analytical Engine." *Ada User Journal* 36, no. 3 (September 2015): 152–80. www.ada-europe.org/archive/auj/auj-36-3.pdf.

Luccioni, Alexandra Sasha, and Anna Rogers. "Mind Your Language (Model): Fact-Checking LLMs and Their Role in NLP Research and Practice." arXiv, August 14, 2023. https://doi.org/10.48550/arXiv.2308.07120.

Luka, Inc. "Replika: My AI Friend." *Privacy Not Included Review, Mozilla Foundation, February 7, 2024. https://foundation.mozilla.org/en/privacynotincluded/replika-my-ai-friend/.

Marino, Mark C. "Critical Code Studies." *Electronic Book Review*, December 4, 2006. https://electronicbookreview.com/essay/critical-code-studies/.

———. *Critical Code Studies*. MIT Press, 2020.

———. "The Racial Formation of Chatbots." *CLCWeb: Comparative Literature and Culture* 16, no. 5 (December 31, 2014), article 13. https://doi.org/10.7771/1481-4374.2560.

———. "I, Chatbot: The Gender and Race Performativity of Conversational Agents." PhD diss., University of California, Riverside, 2006. http://www.proquest.com/docview/305345200/abstract/F24046E98CE94222PQ/1.

———. and ChatGPT. *Hallucinate This!: An Authoritized Autobotography of ChatGPT*. Independently published, 2023.

———, Weil, P., Shrager, J., et al. "Conversations about Conversational Code: On the Collaborative Critical Code Studies Reading of ELIZA." Forthcoming from *AI & Society*.

Markov, Todor, Chong Zhang, Sandhini Agarwal, Tyna Eloundou, Teddy Lee, Steven Adler, Angela Jiang, and Lilian Weng. "A Holistic Approach to Undesired Content Detection in the Real World." arXiv, February 14, 2023. https://doi.org/10.48550/arXiv.2208.03274.

Massachusetts Institute of Technology. "Project MAC Progress Report II, July 1964 to July 1965." https://apps.dtic.mil/sti/pdfs/AD0629494.pdf.

———. "Project MAC Progress Report III, July 1965 to July 1966." https://apps.dtic.mil/sti/pdfs/AD0648346.pdf.

———. "Project MAC Progress Report IV, July 1966 to July 1967." https://apps.dtic.mil/sti/pdfs/AD0681342.pdf.

———. "Project MAC Progress Report V, July 1967 to July 1968." https://apps.dtic.mil/sti/pdfs/AD0687770.pdf.

Mauldin, Michael L. "Chatterbots, Tinymuds, and the Turing Test Entering the Loebner Prize Competition." In *Proceedings of the Twelfth AAAI National Conference on Artificial Intelligence*, 16–21, 1994. https://dl.acm.org/doi/abs/10.5555/2891730.2891733.

Maurer, David. *The Big Con: The Story of the Confidence Man*. Bantam Dell Publishing Group, 1999 [1940].

McCarthy, John. "Recursive Functions of Symbolic Expressions and Their Computation by Machine, Part I." *Communications of the ACM* 3, no. 4 (April 1960): 184–95. https://doi.org/10.1145/367177.367199.

McCarthy, John, Paul W. Abrahams, Daniel J. Edwards, Timothy P. Hart, and Michael I. Levin. *Lisp 1.5 Programmer's Manual*. MIT Press, 1962. https://doi.org/10.7551/mitpress/5619.001.0001.

McGettigan, Andrew. *The Great University Gamble: Money, Markets and the Future of Higher Education*. Illustrated ed. Pluto Press, 2013.

McGuire, Michael T., Stephen Lorch, and Gardner C. Quarton. "Man-Machine Natural Language Exchanges Based on Selected Features of Unrestricted Input—II. The Use of the Time-Shared Computer as a Research Tool in Studying Dyadic Communication." *Journal of Psychiatric Research* 5, no. 2 (June 1967):

179–91. https://doi.org/10.1016/0022-3956(67)90030-1.
———. "Man-Machine Natural Language Exchanges Based on Selected Features of Unrestricted Input—II. The Use of the Time-Shared Computer as a Research Tool in Studying Dyadic Communication." *Journal of Psychiatric Research* 5, no. 2 (June 1, 1967): 179–91. https://doi.org/10.1016/0022-3956(67)90030-1.
McLuhan, Marshall. *Understanding Media: The Extensions of Man*. McGraw-Hill, 1964.
McPherson, Tara. "Why Are the Digital Humanities So White? Or Thinking the Histories of Race and Computation." In *Debates in the Digital Humanities*, edited by Matthew K. Gold, new ed., 139–60. University of Minnesota Press, 2012. https://www.jstor.org/stable/10.5749/j.ctttv8hq.12.
McQuillan, Dan. *Resisting AI: An Anti-Fascist Approach to Artificial Intelligence*. Bristol University Press, 2022. https://bristoluniversitypress.co.uk/resisting-ai.
Maheu, Marlene M. Maheu. "AI Psychotherapy Shutdown." *Telehealth.Org*, August 19, 2025. https://telehealth.org/blog/ai-psychotherapy-shutdown-what-woebots-exit-signals-for-clinicians/.
Miller, George A. "WordNet: A Lexical Database for English." *Communications of the ACM* 38, no. 11 (November 1, 1995): 39–41. https://doi.org/10.1145/219717.219748.
Mitkov, Ruslan, ed. *The Oxford Handbook of Computational Linguistics*. 2nd ed. Oxford University Press, 2022.
"Monkee vs. Machine." *Monkees*, September 26, 1966.
Montfort, Nick. "Continuous Paper." Paper presented at the History of Material Texts Workshop at the University of Pennsylvania on February 12, 2004. https://nickm.com/writing/essays/continuous_paper_homt.html.
———. *Exploratory Programming for the Arts and Humanities*. 2nd ed. MIT Press, 2021.
Montfort, Nick, Patsy Baudoin, John Bell, et al. *10 PRINT CHR$(205.5+RND(1)); : GOTO 10*. MIT Press, 2012.
Moore, Ronald, producer. *Battlestar Galactica*. British Sky Broadcasting (BSkyB), David Eick Productions, NBC Universal Television, 2005.
Morais, Betsy. "Can Humans Fall in Love with Bots?" *New Yorker*, November 19, 2013. www.newyorker.com/tech/annals-of-technology/can-humans-fall-in-love-with-bots.
Mori, Masahiro, Karl MacDorman, and Norri Kageki. "The Uncanny Valley [From the Field]." *IEEE Robotics and Automation Magazine* 19, no. 2 (June 2012): 98–100. https://doi.org/10.1109/MRA.2012.2192011.
Morrison, Romi Ron. "Voluptuous Disintegration: A Future History of Black Computational Thought." *Digital Humanities Quarterly* 16, no. 3 (July 22, 2022).
"Ms. Dewey Outburst." Dailymotion video, 21 seconds, accessed August 19, 2025. www.dailymotion.com/video/xrlj2.
Mu, Tong, Alec Helyar, Johannes Heidecke, et al. "Rule Based Rewards for Language Model Safety." *Advances in Neural Information Processing Systems* 37 (July 24, 2024). https://proceedings.neurips.cc/paper_files/paper/2024/file/c4e380fb74dec9da9c7212e834657aa9-Paper-Conference.pdf.
Mullaney, Thomas S., Benjamin Peters, Mar Hicks, and Kavita Philip, eds. *Your Computer Is on Fire*. MIT Press, 2021. https://doi.org/10.7551/mitpress/10993.001.0001.
Mumford, Lewis. *Technics and Civilization*. University of Chicago Press, 2010.
Murray, Janet Horowitz. *Hamlet on the Holodeck: The Future of Narrative in Cyberspace*. Simon and Schuster, 1997.
———. Hamlet on the Holodeck: The Future of Narrative in Cyberspace. Updated ed. MIT Press, 2017.
Murray, Rowena. "Writing for an Academic Journal: 10 Tips." *Guardian*, September 6, 2013. www.theguardian.com/higher-education-network/blog/2013/sep/06/academic-journal-writing-top-tips.
Natale, Simone. "If Software Is Narrative: Joseph Weizenbaum, Artificial Intelligence and the Biographies of ELIZA." *New Media and Society* 21, no. 3 (March 2019): 712–28. https://doi.org/10.1177/1461444818804980.
Naur, Peter. "Programming as Theory Building." *Microprocessing and Microprogramming* 15, no. 5 (May 1985): 253–61. https://doi.org/10.1016/0165-6074(85)90032-8.
Neisser, Ulric. "The Imitation of Man by Machine." *Science* 139, no. 3551 (18 January 1963): 193–97. https://doi.org/10.1126/science.139.3551.193.
Newell, Allen. "Physical Symbol Systems." *Cognitive Science* 4, no. 2 (April 1, 1980): 135–83. https://doi.org/10.1016/S0364-0213(80)80015-2.
Noble, Safiya Umoja, and Brendesha M. Tynes, eds. *The Intersectional Internet: Race, Sex, Class and Culture Online*. Peter Lang, 2016.
North, Steve. "ELIZA: Psychoanalysis(?), By Computer ..." *Creative Computing*, August 1977.
Norvig, Peter. "Chapter 5: ELIZA: Dialog with a Machine." In *Paradigms of Artificial Intelligence Programming: Case Studies in Common Lisp*. GitHub, 1992. https://github.com/norvig/paip-lisp/blob/main/docs/chapter5.md.
OpenAI. "Model Spec (2024/05/08)." OpenAI. https://cdn.openai.com/spec/model-spec-2024-05-08.html.
OpenAI, Josh Achiam, Steven Adler, Sandhini Agarwal, Lama Ahmad, Ilge Akkaya, Florencia Leoni Aleman, et al. "GPT-4 Technical Report." arXiv, March 4, 2024. https://doi.org/10.48550/arXiv.2303.08774.
Osgood, Charles Egerton, George J. Suci, and Percy H. Tannenbaum. The *Measurement of Meaning*. University of Illinois Press, 1978.
Ouyang, Long, Jeff Wu, Xu Jiang, et al. "Training Language Models to Follow Instructions with Human Feedback." arXiv, March 4, 2022. https://doi.org/10.48550/arXiv.2203.02155.
Parnas, David Lorge. "The Real Risks of Artificial Intelligence." *Communications of the ACM* 60, no. 10 (September 25, 2017): 27–31. https://doi.org/10.1145/3132724.
Pasquinelli, Matteo, and Vladan Joler. "The Nooscope Manifested: Artificial Intelligence

as Instrument of Knowledge Extractivism." *AI & SOCIETY* 36 (November 21 1, 2020): 1263–80.

Pentina, Iryna, Tyler Hancock, and Tianling Xie. "Exploring Relationship Development with Social Chatbots: A Mixed-Method Study of Replika." *Computers in Human Behavior* 140 (March 1, 2023): 107600. https://doi.org/10.1016/j.chb.2022.107600.

Pentland, Alex. *Social Physics: How Social Networks Can Make Us Smarter.* Penguin. 2015.

Peters, Joan K. "The 'Robettes' Are Coming: Siri, Alexa, and My GPS Lady." *Hyperrhiz: New Media Cultures*, no. 15 (Fall 2016). https://doi.org/10.20415/hyp/015.r03.

Plant, Sadie. *Zeros + Ones.* Fourth Estate, 1998.

Polanyi, Michael. *The Tacit Dimension.* Rev. ed. University of Chicago Press, 2009.

Possati, Luca M. "Psychoanalyzing Artificial Intelligence: The Case of Replika." *AI & SOCIETY* 38, no. 4 (August 1, 2023): 1725–38. https://doi.org/10.1007/s00146-021-01379-7.

Potter, Warren. "A Brief History of the 'Bot': From IRC to ContentBot." *ContentBot.ai* (blog), February 15, 2021. www.contentbot.ai/blog/news/a-brief-history-of-the-bot/.

Pressman, Jessica, Mark C. Marino, and Jeremy Douglass. *Reading Project: A Collaborative Analysis of William Poundstone's Project for Tachistoscope {Bottomless Pit}.* University of Iowa Press, 2015.

Quan, Zhi, and Zhiwei Chen. "Human–Computer Pragmatics Trialled: Some (Im)Polite Interactions with ChatGPT 4.0 and the Ensuing Implications." *Interactive Learning Environments* 0, no. 0 (n.d.): 1–20.

Quittner, Josh. "TECHWATCH: WHAT'S HOT IN BOTS." *TIME*, December 8, 1997. https://time.com/archive/6731956/techwatch-whats-hot-in-bots/.

Radford, Alec, Jeffrey Wu, Rewon Child, David Luan, Dario Amodei, and Ilya Sutskever. "Language Models Are Unsupervised Multitask Learners." *OpenAI Blog*, February 14, 2019. https://openai.com/blog/better-language-models/.

Raley, Rita. "Code.Surface| |Code.Depth." *Dichtung Digital. Journal Für Kunst Und Kultur Digitaler Medien* 36, no. 8 (2006): 1–24. https://doi.org/10.25969/MEDIAREP/17695.

Raley, Rita, and Jennifer Rhee. "Critical AI: A Field in Formation." *American Literature* 95, no. 2 (2023): 185–204. https://doi.org/10.1215/00029831-10575021.

Reed, Aaron A., and Jacob Garbe. *The Ice-Bound Compendium.* Simulacrum Liberation Press, 2016.

replika.com. "Replika." Accessed August 19, 2025. https://replika.com.

Rettberg, Jill Walker. *Machine Vision: How Algorithms Are Changing the Way We See the World.* Polity, 2023.

Reynolds, Robin Hauser, dir. *Code: Debugging the Gender Gap.* Documentary. Finish Line Features, 2015.

Rhee, Jennifer. "Misidentification's Promise: The Turing Test in Weizenbaum, Powers, and Short." *Postmodern Culture* 20, no. 3 (May 2010). www.pomoculture.org/2013/09/03/misidentifications-promise-the-turing-test-in-weizenbaum-powers-and-short/#f1.

Rheingold, H. *Tools for Thought: The History and Future of Mind-Expanding Technology.* MIT Press, 1985.

Rhizome. "Retelling the History of Net Art from the 1980s to the 2010s." NET ART ANTHOLOGY, October 27, 2016. http://anthology.rhizome.org/.

Roach, Rebecca. "Missing Secretaries and Bad Readers." In *Modern Literature, Logic and AI*, edited by Sangam MacDuff and Rachel Wilson, Bloomsbury, forthcoming.

———. "My Search for the Mysterious Missing Secretary Who Shaped Chatbot History." The Conversation, March 22, 2024. theconversation.com/my-search-for-the-mysterious-missing-secretary-who-shaped-chatbot-history-225602.

Rogers, C. R. "The Attitude and Orientation of the Counselor in Client-Centered Therapy." *Journal of Consulting Psychology* 13 (1949): 82–94.

———. *On Becoming a Person: A Therapist's View of Psychotherapy.* Houghton Mifflin, 1961.

———. *Client-Centered Therapy.* Houghton Mifflin, 1951. http://archive.org/details/clientcenteredth0000carl.

———. *Counseling and Psychotherapy.* Houghton Mifflin, 1942. http://archive.org/details/counselingandpsy029048mbp.

Roose, Kevin. "A Conversation with Bing's Chatbot Left Me Deeply Unsettled." *New York Times*, February 16, 2023. www.nytimes.com/2023/02/16/technology/bing-chatbot-microsoft-chatgpt.html.

Rosenberg, Ronni Lynne. "Incomprehensible Computer Systems: Knowledge Without Wisdom." MSc thesis, Massachusetts Institute of Technology, 1980.

Rushton, Brian. "Author Highlights: Emily Short." IF: intfiction.org: The Interactive Fiction Community Forum, May 12, 2018. https://intfiction.org/t/author-highlights-emily-short/13305.

Sack, H. "Joseph Weizenbaum and His Famous Eliza." *Science Hi Blog.* 2018. http://scihi.org/joseph-weizenbaum-eliza/ (defunct).

Salles, Arleen, Kathinka Evers, and Michele Farisco. "Anthropomorphism in AI." *AJOB Neuroscience* 11, no. 2 (April 2, 2020): 88–95. https://doi.org/10.1080/21507740.2020.1740350.

Sanderson, Grant. "3Blue1Brown—But What Is a GPT? Visual Intro to Transformers | Deep Learning, Chapter 5." www.3blue1brown.com/lessons/3blue1brown.com (defunct).

———. "3Blue1Brown—Large Language Models Explained Briefly." www.3blue1brown.com/lessons/3blue1brown.com (defunct).

———. "3Blue1Brown—Visualizing Attention, a Transformer's Heart | Chapter 6, Deep Learning." www.3blue1brown.com/lessons/3blue1brown.com (defunct).

Schanze, Jens, dir. *Plug & Pray.* Mascha Film, 2010.

Scott, Ridley, dir. *Blade Runner.* The Ladd Company, Shaw Brothers, Warner Bros., 1982.

Searle, John R. "Minds, Brains, and Programs." *Behavioral and Brain Sciences* 3, no. 3 (1980): 417–24. https://doi.org/10.1017/S0140525X00005756.

Sennrich, Rico, Barry Haddow, and Alexandra Birch. "Neural Machine Translation of Rare Words with Subword Units." arXiv, June 10, 2015. https://doi.org/10.48550/arXiv.1508.07909.

Sevastjanova, Rita, Aikaterini-Lida Kalouli, Christin Beck, Hanna Schäfer, and Mennatallah El-Assady. "Explaining Contextualization in Language Models Using Visual Analytics." In *Proceedings of the 59th Annual Meeting of the Association for Computational Linguistics and the 11th International Joint Conference on Natural Language Processing (Volume 1: Long Papers)*, edited by Chengqing Zong, Fei Xia, Wenjie Li, and Roberto Navigli, 464–76. Association for Computational Linguistics, 2021. https://doi.org/10.18653/v1/2021.acl-long.39.

Shane, Janelle. *You Look Like a Thing and I Love You: How Artificial Intelligence Works and Why It's Making the World a Weirder Place*. Reprint ed.. Voracious, 2021.

Shannon, Claude E. "Game Playing Machine." In *Journal of the Franklin Institute* 260, no. 6 (1955), 447–453.

———. "Programming a Computer for Playing Chess." In *London, Edinburgh, and Dublin Philosophical Magazine and Journal of Science* 41, no. 314 (1950): 256–75.

Shaw, George Bernard. Pygmalion. Project Gutenberg, 1912, 2003. Project Gutenberg eBook. https://www.gutenberg.org/files/3825/3825-h/3825-h.htm.

Shelley, Mary Wollstonecraft. *Frankenstein*. Bantam Books, 1991.

Sherman, Philip M. *Programming and Coding the IBM 709-7090-7094 Computers*. John Wiley, 1963.

Shrager, Jeff. "Commonly Known ELIZA Clones." ELIZAGEN, accessed August 19, 2025. https://sites.google.com/view/elizagen-org/eliza-clones.

———. "From Wizards to Trading Zones: Crossing the Chasm of Computers in Scientific Collaboration" In *Trading Zones and Interactional Expertise: Creating New Kinds of Collaboration*, edited by Michael E., Gorman, 107–124. MIT Press, 2010. https://doi.org/10.7551/mitpress/8351.003.0008.

———. "The Genealogy of ELIZA." ELIZAGEN, accessed August 19, 2025. https://elizagen.org/.

Shrager, Jeff, and Tim Finin. "An Expert System That Volunteers Advice." *Proceedings of the Second Annual National Conference on Artificial Intelligence (AAAI-82)*, August 1, 1982. https://ebiquity.umbc.edu/paper/html/id/286/An-Expert-System-that-Volunteers-Advice.

Shroff, Lila. "Chatbait Is Taking Over the Internet." Technology. *The Atlantic*, September 22, 2025. https://www.theatlantic.com/technology/2025/09/chatbait-ai-chatgpt-engagement/684300/.

Simon, Herbert Alexander. *Models of My Life*. Basic Books, 1991.

Sisto, David. "Chatting With the Dead." *The MIT Press Reader* (blog), January 4, 2021. https://thereader.mitpress.mit.edu/chatting-with-the-dead-chatbots/.

Sito, Tom. *Moving Innovation: A History of Computer Animation*. MIT Press, 2013.

Statler, Tim. "History of Computers with Timeline [2023 Update]." *Comp Sci Central* (blog), originally posted on September 1, 2020. https://compscicentral.com/history-of-computers/.

Sterling, Rod, creator. "The Dummy." *The Twilight Zone*, season 3, episode 33, CBS Productions, 1962.

Stork, D. G. *Hal's Legacy, 2001's Computer as Dream and Reality*. MIT Press, 1997.

Strengers, Yolande, and Jenny Kennedy. *The Smart Wife: Why Siri, Alexa, and Other Smart Home Devices Need a Feminist Reboot*. MIT Press, 2020.

Suchman, Lucy. *Human-Machine Reconfigurations: Plans and Situated Actions*. 2nd ed. Learning in Doing: Social, Cognitive and Computational Perspectives. Cambridge University Press, 2006.

Tarnoff, Ben. "AI Criticism Has a Decades-Long History." *Tech Won't Save Us*, 24 August 2023. https://techwontsave.us/episode/182_ai_criticism_has_a_decades_long_history_w_ben_tarnoff.

Taylor, Edwin F. "Automated Tutoring and Its Discontents." *American Journal of Physics* 36, no. 6 (June 1, 1968): 496–504. https://doi.org/10.1119/1.1974953.

———. "The ELIZA Program: Conversational Tutorial." In *IEEE International Convention Record, Part 10*, 1967.

TED. "How I'm Fighting Bias in Algorithms | Joy Buolamwini." YouTube video, 8 minutes, 44 seconds, 2017. www.youtube.com/watch?v=UG_X_7g63rY.

Thylstrup, Nanna Bonde, Daniela Agostinho, Annie Ring, Catherine D'Ignazio, and Kristin Veel, eds. *Uncertain Archives: Critical Keywords for Big Data*. MIT Press, 2021.

Tunstall, Leandro von Werra, and Thomas Wolf. *Natural Language Processing with Transformers*. Rev. ed. O'Reilly Media, 2022.

Turing, A. M. "Computing Machinery and Intelligence." *Mind* 59, no. 236 (October 1, 1950): 433–60. https://doi.org/10.1093/mind/LIX.236.433.

———. "'Lecture to L.M.S. [London Mathematical Society] Feb. 20 1947', 1947-02-20."ArchiveSearch, accessed August 19, 2025, February 20, 1947. https://archivesearch.lib.cam.ac.uk/repositories/7/archival_objects/272443.

———. "On Computable Numbers, with an Application to the Entscheidungsproblem." *Proceedings of the London Mathematical Society* s2–42, no. 1 (January 1, 1937): 230–65. https://doi.org/10.1112/plms/s2-42.1.230.

Turkle, Sherry. *Evocative Objects: Things We Think With*. MIT Press, 2007.

———. *Life on the Screen: Identity in the Age of the Internet*. Simon and Schuster, 1995.

———. *The Second Self: Computers and the Human Spirit*. 20th anniversary ed. MIT Press, 2005. https://doi.org/10.7551/mitpress/6115.001.0001.

"The Twenty-First Century, 'The Communications Revolution', 1967." YouTube video, 25 minutes, 27 seconds, aired January 29, 1967, posted on November 8, 2024. www.youtube.com/watch?v=j8fDkSqnghg.

University of Michigan. "Remembering Bruce Arden: U-M Faculty Member and Past Chair of Computer and Communication Sciences Department." Electrical Engineering and Computer Science, January 10, 2022. https://eecsnews.engin.umich.edu/tag/history//.

University of Michigan Computing Center. *University of Michigan Executive System for the IBM 7090 Computer*, 1965. www.google.co.uk/books/edition/University_of_Michigan_Executive_System/f7oSAQAAMAAJ.

Valéry, Paul. *Monsieur Teste*. Translated by Jackson Mathews. Princeton University Press, 1989. Originally published as *La soirée avec monsieur Teste* in 1896.

Vara, Vauhini. "'Code' and the Quest for Inclusive Software." *New Yorker*, April 19, 2015. www.newyorker.com/business/currency/code-and-the-quest-for-inclusive-software.

Vaswani, Ashish, Noam Shazeer, Niki Parmar, et al. "Attention Is All You Need." arXiv, June 12, 2017. https://doi.org/10.48550/arXiv.1706.03762.

Veselov, Vladimir. "Computer AI Passes Turing Test in 'World First.'" *BBC News*, June 9, 2014. www.bbc.com/news/technology-27762088.

Van Vleck, Tom. "Project MAC." Multics. www.multicians.org/project-mac.html.

Von Neumann, John. "Various Techniques Used in Connection with Random Digits." *National Bureau of Standards Applied Mathematics Series* 12 (1951): 36–38.

Walden, D., and T. Van Vleck, eds., *The Compatible Time Sharing System (1961–1973): Fiftieth Anniversary Commemorative Overview*. IEEE Computer Society, 2011. https://people.csail.mit.edu/saltzer/CTSS/CTSS-Documents/CTSS_50th_anniversary_web_03.pdf.

Wardrip-Fruin, Noah. "Christopher Strachey: The First Digital Artist?" *Grand Text Auto* (blog), August 1, 2005. https://grandtextauto.soe.ucsc.edu/2005/08/01/christopher-strachey-first-digital-artist/.

———. *Expressive Processing—Digital Fictions, Computer Games, and Software Studies*. MIT Press, 2019.

Wardrip-Fruin, Noah, and Nick Montfort, eds. *The New Media Reader*. MIT Press, 2003.

Watson, Bruce. "Joe and Eliza—An A.I. Love Story." *Attic*, 16 July 2020. www.theattic.space/home-page-blogs/2020/7/16/joe-and-eliza-an-ai-love-story.

Weheliye, Alexander G. *Habeas Viscus: Racializing Assemblages, Biopolitics, and Black Feminist Theories of the Human*. Duke University Press, 2015.

Weil, Andrew. "Conversations with a Mechanical Psychiatrist." *Harvard Review* 3, no. 2 (1965): 68–73.

Weil, Peggy. "Seriously Writing SIRI." *Hyperrhiz: New Media Cultures*, no. 11 (February 2015). https://doi.org/10.20415/hyp/011.e05.

———. *The Blurring Test: Songs of MrMind—Peggy Weil Studio*. 1998–2016. https://pweilstudio.com/project/the-blurring-test-songs-of-mrmind/.

Weinstein, James N. "Artificial Intelligence: Have You Met Your New Friends; Siri, Cortona, Alexa, Dot, Spot, and Puck." *Spine* 44, no. 1 (January 1, 2019): 1–4. https://doi.org/10.1097/BRS.0000000000002913.

Weizenbaum, Joseph. "Knotted List Structures." *Communications of the ACM* 5, no. 3 (March 1, 1962): 161–65. https://doi.org/10.1145/366862.366897.

———. "Symmetric List Processor." *Communications of the ACM* 6, no. 9 (September 1963): 524–36. https://doi.org/10.1145/367593.367617.

———, R. Shepardson, B. Hellerstein, and D. Masters. *Knotted List Structures*. General Electric Computer Laboratory. Computer Organization Unit., 1962. www.computerhistory.org/collections/catalog/102724662.

———. "Automating Psychotherapy." *Communications of the ACM* 17, no. 7 (1974): 425. https://doi.org/10.1145/361011.361081.

———. "Computers as 'Therapists."' *Science* 198, no. 4315 (1977): 354.

———. *OPL-I An Open Ended Programming System Within CTSS*. January 1964. https://dspace.mit.edu/handle/1721.1/149332.

———. "Computer Conversations, 1965," 1965, https://hdl.handle.net/1721.3/201699. Copyright MIT and licensed under a Creative Commons Attribution-Noncommercial 4.0 International License.

———. *Computer Power and Human Reason: From Judgment to Calculation*. W. H. Freeman, 1976.

———. "Contextual Understanding by Computers." *Communications of the ACM* 10, no. 8 (August 1967): 474–80. https://doi.org/10.1145/363534.363545.

———. "Doctor I Have Terrible News ..." MIT Libraries, August 20, 2025. https://dome.mit.edu/handle/1721.3/201700. Copyright MIT and licensed under a Creative Commons Attribution-Noncommercial 4.0 International License.

———. "ELIZA—A Computer Program for the Study of Natural Language Communication Between Man and Machine." *Communications of the ACM* 9, no. 1 (January 1966): 36–45.

———. "FAP." MIT Libraries, August 20, 2025. https://hdl.handle.net/1721.3/201707.

———. "How to Make a Computer Appear Intelligent." *Datamation :: 196202* 8, no. 2 (February 1962). http://archive.org/details/bitsavers_datamation_15474944.

———. "Human Choice in the Interstices of the Megamachine." Paper presented at the IFIPS conference on "Human Choice and Computers," in Vienna, Austria, in 1979.

———. 'Inventory of Professional Interests'. Stanford University, 1972. SC1055 Box 63. Dept. Of Special Collections And University Archives.

———. "Letters to the Editor: On the Implementation of SLIP." *Communications of the ACM* 8, no. 5 (May 1965): 263. https://doi.org/10.1145/364914.364934.

———. "More on Computer Models of Psychopathic Behavior." *Communications of the ACM* 17, no. 9 (September 1974): 541–43. https://doi.org/10.1145/361147.361118.

———. "The Myth of the Last Metaphor." In *Speaking Minds—Interviews with Twenty Eminent Cognitive Scientists*, edited by Peter Baumgartner and Sabine Payr, 249–64. Princeton University Press, 1995.

———. "Not Without Us." *ETC: A Review of General Semantics* 44, no. 1 (1987): 42–48.

———. "Nuclear Weapons—and the Simplest Truths." *Computers and People* 36, nos. 3–4 (March 1, 1987): 11–13.

———. "On the Impact of the Computer on Society: How Does One Insult a Machine?" *Science* 176, no. 4035 (May 12, 1972): 609–14. https://doi.org/10.1126/science.176.4035.609.

———. "On-Line User Languages." *BIT Numerical Mathematics* 6, no. 1 (March 1, 1966): 58–65. https://doi.org/10.1007/BF01939550.

———. "Outline of Paper to Accompany ELIZA," accessed August 20, 2025. MIT Archive.

———. "Recovery of Reentrant List Structures in SLIP." *Communications of the ACM* 12, no. 7 (July 1, 1969): 370–72. https://doi.org/10.1145/363156.363159.

———. "Social and Political Impact of the Long-Term History of Computing." *IEEE Annals of the History of Computing* 30, no. 3 (July 2008): 40–42. https://doi.org/10.1109/MAHC.2008.58.

———, Gunna Wendt, and Benjamin Fasching-Gray. *Islands in the Cyberstream: Seeking Havens of Reason in a Programmed Society*. Litwin Books, 2015.

Wheeler, David A. "Don't Use ISO/IEC 14977 Extended Backus-Naur Form (EBNF)." dwheeler.com., February 18, 2020. https://dwheeler.com/essays/dont-use-iso-14977-ebnf.html.

Wiener, Norbert. *God & Golem, Inc.: A Comment on Certain Points Where Cybernetics Impinges on Religion*. MIT Press, 1964. https://doi.org/10.7551/mitpress/3316.001.0001.

Wilkes, Maurice V., David J. Wheeler and Stanley Gill. *The Preparation of Programs for an Electronic Digital Computer*. With Internet Archive. 1951. http://archive.org/details/preparationofpro0000maur.

Wilks, Yorick, ed. *Machine Conversations*. Springer, 1999.

Willyard, Cassandra. "Men: A Growing Minority?" *gradPsych Magazine*, January 2011. www.apa.org/gradpsych/2011/01/cover-men.

Winner, Langdon. *Autonomous Technology: Technics-out-of-Control as a Theme in Political Thought*. MIT Press, 2001.

Winograd, Terry. "Procedures as a Representation for Data in a Computer Program for Understanding Natural Language." PhD diss., Massachusetts Institute of Technology, 1970. https://dspace.mit.edu/handle/1721.1/15546.

Winston, P. H. *Artificial Intelligence*. 2nd ed. Addison-Wesley Publishing. 1977.

Wirth, Niklaus. "What Can We Do About the Unnecessary Diversity of Notation for Syntactic Definitions?" *Communications of the ACM* 20, no. 11 (November 1977): 822–23. https://doi.org/10.1145/359863.359883.

Wolf, M. J., K. W. Miller, and F. S. Grodzinsky. "Why We Should Have Seen That Coming: Comments on Microsoft's Tay 'Experiment,' and Wider Implications." *ORBIT Journal* 1, no. 2 (January 1, 2017): 1–12. https://doi.org/10.29297/orbit.v1i2.49.

Yang, Katherine. "Coem." Katherine Yang. https://kayserifserif.place/work/coem/.

Young, Liam. "'I'm a Cloud of Infinitesimal Data Computation': When Machines Talk Back: An Interview with Deborah Harrison, One of the Personality Designers of Microsoft's Cortana AI." *Architectural Design* 89, no. 1 (2019): 112–17. https://doi.org/10.1002/ad.2398.

Index

Publisher contact:
The MIT Press
Massachusetts Institute of Technology
77 Massachusetts Avenue, Cambridge, MA 02139
mitpress.mit.edu

EU Authorised Representative:
Easy Access System Europe, Mustamäe tee 50,
10621 Tallinn, Estonia
gpsr.requests@easproject.com

Printed by Integrated Books International,
United States of America